# Bioelectrochemistry III

## Charge Separation Across Biomembranes

**ETTORE MAJORANA**
**INTERNATIONAL SCIENCE SERIES**
Series Editor:
**Antonino Zichichi**
European Physical Society
Geneva, Switzerland

---

**(PHYSICAL SCIENCES)**

*Recent volumes in the series:*

**Volume 41** **PROGRESS IN MICROEMULSIONS**
Edited by S. Martellucci and A. N. Chester

**Volume 42** **DIGITAL SEISMOLOGY AND FINE MODELING OF THE LITHOSPHERE**
Edited by R. Cassinis, G. Nolet, and G. F. Panza

**Volume 43** **NONSMOOTH OPTIMIZATION AND RELATED TOPICS**
Edited by F. H. Clarke, V. F. Dem'yanov, and F. Giannessi

**Volume 44** **HEAVY FLAVOURS AND HIGH-ENERGY COLLISIONS IN THE 1–100 TeV RANGE**
Edited by A. Ali and L. Cifarelli

**Volume 45** **FRACTALS' PHYSICAL ORIGIN AND PROPERTIES**
Edited by Luciano Pietronero

**Volume 46** **DISORDERED SOLIDS: Structures and Processes**
Edited by Baldassare Di Bartolo

**Volume 47** **ANTIPROTON–NUCLEON AND ANTIPROTON–NUCLEUS INTERACTIONS**
Edited by F. Bradamante, J.-M. Richard, and R. Klapisch

**Volume 48** **SAFETY, ENVIRONMENTAL IMPACT, AND ECONOMIC PROSPECTS OF NUCLEAR FUSION**
Edited by Bruno Brunelli and Heinz Knoepfel

**Volume 49** **NONLINEAR OPTICS AND OPTICAL COMPUTING**
Edited by S. Martellucci and A. N. Chester

**Volume 50** **HIGGS PARTICLE(S): Physics Issues and Experimental Searches in High-Energy Collisions**
Edited by A. Ali

**Volume 51** **BIOELECTROCHEMISTRY III: Charge Separation Across Biomembranes**
Edited by G. Milazzo and M. Blank

---

# Bioelectrochemistry III

## Charge Separation Across Biomembranes

Edited by

**G. Milazzo**

Bioelectrochemical Society
Rome, Italy

and

**M. Blank**

College of Physicians and Surgeons
Columbia University
New York, New York

Plenum Press • New York and London

Library of Congress Cataloging in Publication Data

International School of Biophysics (19th: 1988: Erice, Italy)
Bioelectrochemistry III: charge separation across biomembranes / edited by G. Milazzo and M. Blank.
p. cm.—(Ettore Majorana international science series. Physical sciences; v. 51)
"Proceedings of the Nineteenth Course of the International School of Biophysics, Bioelectrochemistry III—Charge separation across biomembranes, held November 3–13, 1988, in Erice, Italy"—T.p. verso.
Includes bibliographical references.
Includes index.
ISBN 0-306-43606-X
1. Membranes (Biology)—Congresses. 2. Ion channels—Congresses. 3. Biological transport, Active—Congresses. 4. Bioelectrochemistry—Congresses. I. Milazzo, Giulio. II. Blank, Martin, date. III. Title. IV. Series.
[DNLM: 1. Biochemistry—congresses. 2. Biological Transport—congresses. 3. Cell Membrane—metabolism—congresses. 4. Electrochemistry—congresses. QH 634.5 I61b 1988]
QH601.I52 1988
574.87′5—dc20
DNLM/DLC 90-7463
for Library of Congress CIP

*Bioelectrochemistry I: Biological Redox Reactions* edited by G. Milazzo and Martin Blank (Plenum Press: 1983) was published as Volume 11 in Ettore Majorana International Science Series—Life Sciences, which has since been discontinued.

Proceedings of the Nineteenth Course of the International School of Biophysics, Bioelectrochemistry III—Charge Separation Across Biomembranes, held November 3–13, 1988, in Erice, Italy


A Division of Plenum Publishing Corporation
233 Spring Street, New York, N.Y. 10013



Printed in the United States of America

# PREFACE

This book contains a series of review papers related to the lectures given at the Third Course on Bioelectrochemistry held at Erice in November 1988, in the framework of the International School of Biophysics.

The topics covered by this course, "Charge Separation Across Biomembranes," deal with the electrochemical aspects of some basic phenomena in biological systems, such as transport of ions, ATP synthesis, formation and maintenance of ionic and protonic gradients. In the first part of the course some preliminary lectures introduce the students to the most basic phenomena and technical aspects of membrane bioelectrochemistry. The remaining part of the course is devoted to the description of a selected group of membrane–enzyme systems, capable of promoting, or exploiting, the processes of separation of electrically charged entities (electrons or ions) across the membrane barrier. These systems are systematically discussed both from a structural and functional point of view.

The effort of the many distinguished lecturers who contributed to the course is aimed at offering a unifying treatement of the electrogenic systems operating in biological membranes, underlying the fundamental differences in the molecular mechanisms of charge translocation.

As is usual in multiauthored books, a certain lack of homogeneity in length and depth of each subtheme is also present here, but the abundance of quoted papers and reviews at the end of every chapter should be helpful to readers wishing to deepen their understanding of the topics treated.

CONTENTS

## SYMBOLS AND ACRONYMS

For the sake of consistency and to ensure immediate understanding, the symbols of the most frequently occurring quantities and the acronyms of the organic chemicals are collected here. Consulting this list, attention must be given to the following points:

1. Only the most common symbols are included. Some, only seldom used, are not included to avoid confusion. Their meaning is given in the text.
2. Since the number of all quantities symbolized in chemistry, physics, biology, *etc.*, and officially accepted by the corresponding International Unions, even using different characters (roman, boldface, italic, *etc.*) is remarkably larger than the number of available letters, it occurs that the use of the same symbols for different quantities becomes sometimes unavoidable, and was accepted by the International Unions (for example the symbol *G* for the free enthalpy and for the electric conductance, or the symbol *A* for the area and for the optical absorbance). But the quantity to be correctly considered unambiguously results from the text.

### Latin alphabet

| | |
|---|---|
| [...] | concentration of the species (mol / $dm^3$) |
| *A* | absorbance, area, preexponential factor |
| A | ampere |
| *a* | activity |
| *a.c.* | alternating current |
| AcCh | acetyl choline |
| AcChR | acetyl choline receptor |
| ADP | adenosine diphosphate |
| ANS | anilino–naphthalene sulphonate |
| a.p. | action potential |
| ATP | adenosine triphosphate |
| Bgt (Btx) | bungarotoxin |
| BLM | bilayer membrane, black lipid membrane |
| BR | bacteriorhodopsin |
| *C* | capacitance |
| C | coulomb |
| °C | CELSIUS degree |
| *c* | concentration |
| | centi |
| cyt | cytochrome |
| F | farad |
| F | FARADAY's constant |
| *f*(...) | function of... |

| | |
|---|---|
| f | femto |
| FCCP | carbonylcyanide – *p* – trifluoro- – methoxy – phenyl hydrazone |
| FPLC | fast protein liquid chromatography |
| FTIR | FOURIER transformed infrared spectroscopy |
| *G* | free enthalpy (GIBBS free energy); conductance |
| GTP | general insertion protein |
| *H* | enthalpy |
| h.p.l.c. | high performance liquid chromatography |
| HR | halorhodopsin |
| Hz | hertz |
| *I* | current intensity; light intensity |
| i.c.p.a.e.s | inductively coupled plasma atomic emission spectroscopy |
| I.R. | infrared |
| J | joule |
| *j* | current density |
| $K_m$ | MICHAELIS MENTEN constant |
| K | KELVIN degree |
| *k* | BOLTZMANN constant, rate constant |
| lg | logarithm, decadic |
| ln | logarithm, natural |
| LDAO | dodecyldimethyl amine |
| *M* | concentration (mol / $dm^3$, molar) |
| m | meter, milli |
| m.c.d. | magnetic circular dichroism |
| MES | morpholino – ethane sulfonic acid |
| min | minute |
| MIT | monoiodotyrosine |
| MOPS | N – morpholino – propane sulfonic acid |
| MPP | mitochondrial processing peptidase |
| *m.w.* | molecular weight |
| *N,n* | number of... |
| N | newton |
| NEM | N – methyl maleinimide |
| n.m.r. | nuclear magnetic resonance |
| *P* | permeability |
| p | pico |
| p.a.g.e. | polyacryl amide gel electrophoresis |
| PEP | processing enhancing protein |
| PERS | protein electric response signal |
| p.i.x.e. | proton induced X ray emission |
| pm | purple membrane |
| *Q* | electric charge |
| *R* | electric resistance, gas constant |
| *r* | radius |
| RNA | ribonucleic acid |
| RR | resonance Raman |
| S | siemens |
| s | second |
| *sat* | (subscript) saturated |
| SDS | sodium dodecyl sulfate |
| SKL | serine – lysine – leucine |
| SR | sarcoplasmic reticulum; sensory rhodopsin |
| SRP | signal recognition particle |
| SV | slow vacuolar |
| *T* | KELVIN temperature |
| *t* | time |
| TEMED | tetramethylethylene diamine |
| TMPD | tetramethyl – *p* – phenylene diamine |
| $TPMP^+$ | triphenylmethyl phosphonium |
| $TPP^+$ | tetraphenyl phosphonium |
| *U* | electric potential difference |
| UV | ultraviolet |
| V | volt |
| *v* | velocity, reaction rate |
| VDAC | voltage dependent anion channel |

| | |
|---|---|
| W | watt |
| $w$ | weight |
| $z$ | ionic charge |

**Greek alphabet**

| | |
|---|---|
| α | polarizability |
| Δ | difference |
| ε | dielectric constant |
| ζ | zeta potential |
| θ | angle |
| λ | wavelength |
| μ | dipole moment |
| $\tilde{\mu}$ | electrochemical potential |
| ϱ | resistivity |
| Σ | sum |
| τ | time (as special quantity) |
| φ | internal electric potential |

CODES FOR AMINO ACIDS

| Amino acid | Three – letter abbreviation | One – letter symbol |
|---|---|---|
| Alanine | Ala | A |
| Arginine | Arg | R |
| Asparagine | Asn | N |
| Aspartic acid | Asp | D |
| Cysteine | Cys | C |
| Glutamine | Gln | Q |
| Glutamic acid | Glu | E |
| Glycine | Gly | G |
| Histidine | His | H |
| Isoleucine | Ile | I |
| Leucine | Leu | L |
| Lysine | Lys | K |
| Methionine | Met | M |
| Phenylalanine | Phe | F |
| Proline | Pro | P |
| Serine | Ser | S |
| Threonine | Thr | T |
| Tryptophan | Trp | W |
| Tyrosine | Tyr | Y |
| Valine | Val | V |

# MOLECULAR MECHANISMS OF ION TRANSPORT: NEW INSIGHTS BY PATCH–CLAMP STUDIES

R. HEDRICH*, W. STÜHMER• and B.U. KELLER•

* *Pflanzenphysiologisches Institut, Universität Göttingen*
• *Max Planck Institut für biophysikalische Chemie*
*3400 Göttingen, West Germany*

Contents



*Bioelectrochemistry III*
Edited by G. Milazzo and M. Blank
Plenum Press, New York, 1990

## 1. Introduction

New insights into the molecular processes involved in ion and nutrient transport across membranes of animal and plant cells were obtained since the application of the patch–clamp technique to isolated cells, protoplasts (wall–free plant cells) and organelles. While excitable electrical behaviour was first observed in plant cells about a century ago, the underlying mechanisms are only now being directly studied at the molecular level. Ion channels are integral transmembrane proteins which, when open, allow the movement of ions and some non–electrolytes down their electrochemical potential gradients. Although ion currents in plant cells were among the first to be studied in detail, the electrophysiological characterization of plant ion channels has been somewhat slower compared to their animal counterparts. This has been due to problems specific to plants, such as the presence of the cell wall, having the plasma membrane and vacuolar membrane in series separated by only a relatively small cytoplasmic compartment.

These difficulties are rapidly being overcome with the application of the patch–clamp technique. The patch–clamp technique is a revolutionary electrophysiological method allowing high resolution recording of ion currents from biological membranes, both at the single channel level and from whole cells or organelles. The various applicabilities of the patch–clamp technique were used to investigate active and passive mechanisms of solute fluxes across cell membranes and their regulation. In this chapter the role of ion channels in important cellular processes such as osmoregulation, turgor control and the transduction of external and internal signals will be discussed. Electrophysiological methods, including the patch–clamp technique, and their application to record ion fluxes from whole tissues, cells and membrane patches will be described.

## 2. Electrophysiological techniques

### *2.1. Vibrating probe*

Extracellular recorded potential differences, or currents are a reflection of transmembrane ion transport being resolved as current flowing across an extracellular series resistor. Although these types of measurements are relatively easy to perform, they suffer from lack of resolution. A large body of earlier work on whole tissues using extracellular recording electrodes concerned the stimulation and propagation of action potentials.

More recent examples of this approach were the studies from L.F. JAFFE's group on sea urchin eggs and plant pollen using the vibrating probe. The vibrating probe technique is based on the rapid, extracellular measurement of potential difference at two different locations close to one another. Any spatial difference in potential is attributed to the presence of net ion current flow, representing the sum of currents arising from membrane conductances. These currents result in a characteristic topology of the electric field along single cells or whole tissues. The results obtained using this approach led to the suggestion that ionic currents are required as triggers for the establishment of cellular polarity or developmental responses. It has been shown that changes in the intensity of external electrical fields precede changes in cell polarity, and that certain types of tissue growth are associated with transcellular current flow. Nevertheless, the vibrating probe technique is limited because the contribution of individual ion species (*e.g.* $K^+$) to the overall current is not easily distinguishable, and direct access to the source of ion currents, the membrane, is constrained because of the physical size of the probe.

### 2.2. *Standard microelectrodes*

If the composition on either side of a membrane is different, and if the membrane is semipermeable, an electrical potential difference will exist across the membrane, the size of which is a function of the membrane conductance for the permeant ion species. This is the *transmembrane potential*, the difference between the intracellular and extracellular potentials. It can be measured directly using fine intracellular glass microelectrodes or more indirectly using potential–sensitive dyes. Subsequently, with the application of voltage–clamp techniques (clamping the voltage to the zero–potential level by measuring the injected current needed), the measurement of transmembrane ionic current flow became possible. The best example of such an application is the characterization of the ionic basis for the action potential in the squid giant axon by COLE, HODGKIN, HUXLEY and KATZ roughly 40 years ago.

In animal cells, following the macroscopic description of cellular ion currents and their activating mechanisms, much effort was expended in trying to study the elementary characteristics of individual ion channels. At that time, the 1970's, noise analysis was the method of choice and was used to study the fluctuations in current arising from a varying number of open channels in the population present in a single cell. Ion channels are now accepted as the main transporters for ion movement down electrochemical potential gradients. Questions of current interest concern notably the regulation and modulation of ion channel activity by external and internal signals (such as neurotransmitters and second messengers). Thus the physiologist wanted access to living cells to effectively control these signals.

To study small cells of 10–20 micrometer in diameter (a normal cell size), blunt, low resistance microelectrodes were used to penetrate the cell wall (in plant cells only) and/or the plasma membrane, thus leading unavoidably to high leak conductances. These may give rise to uncontrolled shifts in intracellular ion concentrations, and therefore poorly defined equilibrium potential differences, due to excessive leakage into the cell of the electrolyte used to fill the microelectrodes. Thus the inevitable leakage conductances associated with intracellular microelectrode impalements are too large with respect to the inherent membrane conductances to allow reliable interpretation of potential clamp data from small cells. However, many of these limitations have been circumvented in recent years, following the development of the patch–clamp technique.

### 2.3. *Patch clamp*

The patch–clamp technique was first used to obtain direct measurements of the elementary current passing through a single ion channel. Recent advances allow the patch–clamp technique to be used to record ionic currents from membrane patches and from entire small cells.In a typical patch–clamp experiment, measurements are performed in physiological buffer solutions. Cells or organelles are usually allowed to settle on the surface of a petri dish for patch–clamp measurements. A heat–polished glass pipette with a tip diameter on the order of 1 micrometer is pressed against the membrane surface. When suction is applied to the interior of the pipette, a seal forms between the pipette tip and the membrane (cell–attached configuration, Fig. 1). The high seal resistance and small membrane area reduce background noise and ensure that currents passing the channels in the membrane patch will flow into the pipette.

This cell–attached measuring configuration, as well as excised–patch configurations (inside–out and outside–out patch, see Fig. 1), allows the resolution of single–channel currents of less than one picoampere. Figure 2 shows current recordings from the vacuolar membrane of suspension–cultured plant cells. Single–channel currents consist of rectangular pulses of random duration reflecting conformational changes of a macro-molecule. Each downward current step represents the opening, and each upward current step the closing, of a single ion channel. As long as the channel is open, ions pass through it driven by their gradients of electrochemical potential. The current amplitude indicates the number of ions passing through the channel within a given time. The duration of the mean open and closed–times can depend on the applied potential difference and on a

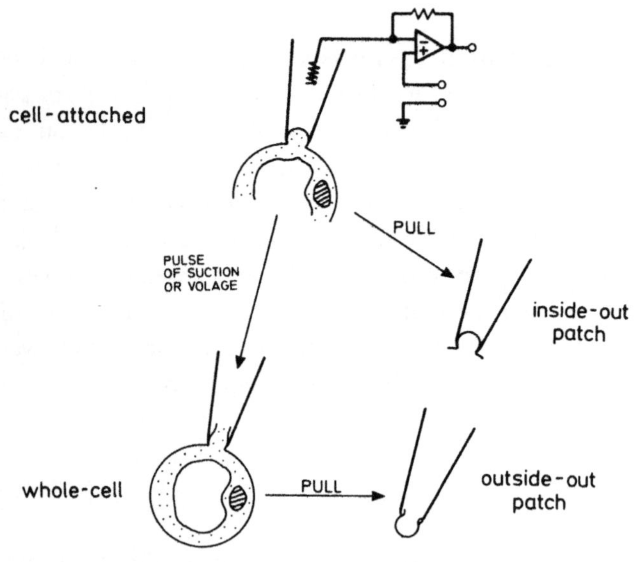


Fig. 1. – Patch–clamp configurations for animal and plant cells.

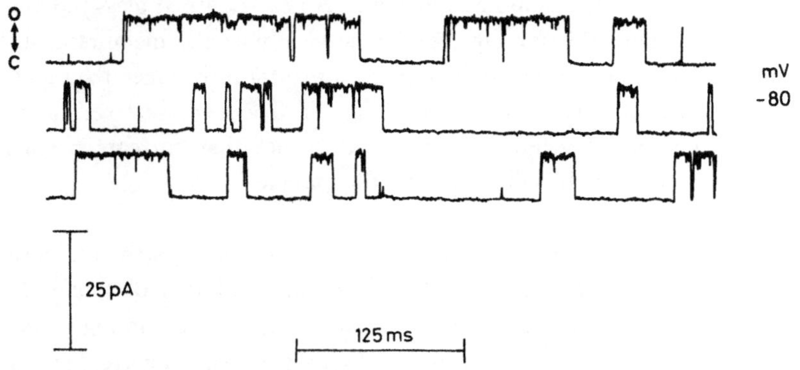


Fig. 2 – Single "SV–type" channels in the vacuolar membrane of *Chenopodium rubrum* recorded at a transmembrane potential of –80 mV inside the vacuole.

variety of chemical interactions with the channel protein (*e.g.* $Ca^{2+}$, pH, second messengers, phospcrylation). A statistical analysis of open and closed time intervals gives insights into the molecular dynamics of the channel protein.

The mechanical stability of the *pipette–to–membrane seal* allows a membrane patch encircled by the pipette tip to be excised (inside–out or outside–out orientation), or to be ruptured without destroying the seal to provide access to the interior of the cell or organelle (*whole–cell* or *organelle* configuration, Fig. 1). With these patch–clamp configurations, the transmembrane potential and the composition of the media on either side of the membrane are well–defined and easy to control. Nevertheless, *wash–out* of cellular factors (into the pipette) essential for ion transport or for membrane–associated processes has to be taken into account. On the other hand, *wash–in* (from the pipette interior) of essential compounds may prevent loss of activity (*rundown*) or even reactivate processes under investigation, thus leading to an understanding of ion transport regulation. The whole–cell configuration enables experiments concerning overall current flow arising from a population of channels or other transporters distributed over the entire membrane surface. Using the whole–cell configuration, it is possible to study the properties of pumps or carriers, even though the current arising from a single pump protein is too small to be detected. Specific stimuli (such as pulses of substrate, light or hormones) can elicit activation of these *low–turnover* transporters in the entire cell, resulting in a summed whole–cell current that can be readily measured and studied. Whereas patch–clamp recordings of pump currents from several animal and plant systems have already been obtained, direct measurements of ion currents produced by carriers are still restricted to a few cases in animal systems. Ionic currents through single channels in animal and plant cells measured with the patch–clamp technique ([1], Figs. 1 and 2) will be discussed in the following sections.

## 3. Channels in the plasma membrane

Patch–clamp techniques have traditionally been used to measure ionic currents through the plasma membrane. Observations of ion channels in the plasma membrane of animal and plant cells is currently increasing. Particular types of ion channels can be classified by the ion species able to permeate the open channel (*e.g.* $K^+$, $Cl^-$ and $Ca^{2+}$).

### 3.1. *K–channels from rat brain: Molecular structure and biological functions*

Injection of mRNA corresponding to a $K^+$ channel from rat brain (BAUMANN *et. al.,* [2]) into *Xenopus* oocytes leads to the functional expression of a $K^+$ current with delayed rectifier properties (Stühmer *et al.,* [3]). The pharmacological characterisation of this channel reveals that it has all the properties attributed to voltage–dependent delayed rectifier channels. The single channel conductance obtained from cell–attached patches of oocytes gives a conductance of 9.3 pS in the main open state. The outward $K^+$–current does not inactivate in the ms scale. Since this channel is derived from cDNA, it is possibile to introduce site–directed mutations in order to study structure–function relations. This example shows how molecular genetics, in combination with patch–clamp techniques, reveals insights into the molecular processes involved in ion channel gating.

### *3.2. $K^+$ channels in plants*

Following the initial characterization of $K^+$ channels in guard cells (Schroeder *et al.,* [4]), the presence of a variety of $K^+$ channels has been reported in the plasma membrane of other higher plant tissues. $K^+$ channels in guard cells and other motor tissues (Schroeder *et al.,* [5]; Schroeder [6]) are strongly regulated by the potential difference across membrane, allowing $K^+$ influx or efflux upon activation [4–8]. Outward $K^+$ current is carried by approximately two hundred $K^+$ channels in the plasma membrane of a guard cell (Schroeder *et al.*, [5]). The $K^+$ channels found in various higher plant protoplasts have properties very similar to the outward $K^+$ conductance in algal cells (Tazawa *et al.,* [9]; Sokolik and Yurin , [10]; Hedrich and Schroeder, [11]). The magnitude of $K^+$ fluxes through $K^+$ channels in guard cells can account for physiological $K^+$ fluxes of 0.7 fmol $s^{-1}$ per guard cell during stomatal closing ([5], Outlaw [12]). Properties of $K^+$ channels in guard cell protoplasts [6] agree with $K^+$ fluxes observed in guard cells embedded in their original environment of the epidermis [12]. It may be concluded that $K^+$ channels represent a major pathway for $K^+$ uptake and release in guard cells, and possibly in plant cells in general.

### *3.3. Chloride channels*

Volume decrease in plant cells is mediated by a reduction of turgor through the release of potassium salts. The detection of $K^+$ selective channels in guard cell plasma membrane has provided evidence that turgor regulation is achieved by potassium efflux through potential–difference–dependent $K^+$ channels. However, the question of how accompanying anion fluxes are mediated across the plasma membrane remained largely unresolved until now. To address this question, the anionic permeability of the plasma membrane was studied with the patch–clamp technique. Experiments from Keller, Hedrich and Raschke (personal communication) demonstrate the existence of strongly potential–difference–dependent anion channels in guard cells. These channels are preferentially activated for membrane depolarisations positive to –60mV. Their single channel conductance is 40 pS in 100 m*M* KCl. The ensemble activity of ion channel in the whole cell generates currents of about 15 to 20 pA. Additionally, J.I. SCHROEDER observed a $Ca^{2+}$–activated chloride conductance (personal communication). This chloride conductance did not show a pronounced potential–difference–dependence. In addition to passive efflux of $K^{+}$, potential–difference–dependent anion channels, a $Ca^{2+}$–modulated chloride conductance may provide the guard cell, or plant cells in general, with a versatile mechanism for anion efflux during volume decrease.

### 3.4. *Stretch–activated channels*

Channels which are activated by stretching of the plasma membrane have been found in various animal cells, plant protoplasts ([11]; EDWARDS and PICKARD, [13]), in yeast and the outer membrane of *E. coli* [14]. In all cases these channels were found to be rather non–selective. It has been suggested that stretch–activated channels may function as both turgor sensors and mechano–sensor.

## 4. **Channels in organelles**

### 4.1. *Vacuoles*

Mature plant cells and yeast cells are characterized by the presence of a large central vacuole. The storage of solutes in vacuoles and their subsequent release is important in cell metabolism and plays a fundamental role in the balance of osmotic pressure and the control of the electrical potential difference across the vacuolar membrane. The vacuolar membrane is able to generate and maintain large concentration gradients of ions and metabolites [15–17]. A patch–clamp survey of the electrical properties of the vacuolar membrane from a large variety of plant material has demonstrated the presence of potential–difference–dependent ion channels in higher plant vacuoles [18, 19].

At high cytoplasmic $Ca^{2+}$ concentrations (> 0.3 μ*M*), the ionic conductance of the vacuolar membrane was found to be accounted for by currents directed into the vacuole. These currents are activated at negative potential differences (negative inside the vacuole, Fig. 2), as well as at slightly positive potential differences [20–24]. The kinetics of activation of these currents are slow ($\tau$ = 100–200 ms) and were therefore termed *SV* (slow–vacuolar)–type currents [25]. Using excised patches, SV–type currents could be resolved at the single–channel level [18, 19]. The reported single–channel conductance obtained under different patch–clamp configurations averaged 60–80 pS in 50–100 m*M* salt solutions [18–23]. SV–channels are rather non–selective with respect to monovalent cations such as $K^+$, $Na^+$, $Li^+$ [2, 23, 26]. The permeability for anions was 2–10 times smaller. Anion–transport blockers strongly affected SV channels in sugar beet. Among them, $ZnCl_2$ in micromolar concentrations effectively and reversibly blocked SV–channels up to 90 %. The stilbene derivatives, known to block anion channels in various cell types [27], caused an irreversible inhibition of ion movement through SV channels [19].

### 4.2. *Endoplasmic reticulum*

Ion transport across subcellular membranes is a central event associated with a variety of physiologically important processes such as secretion or hormone action. In particular, the Endoplasmic Reticulum (ER) seems to be involved in these processes by regulating the free calcium concentration inside the cell. Until recently, it has been a considerable problem to investigate signal transduction across most intracellular membranes of animal cells with the patch clamp technique, mainly because the organelles are too small. The problem of size can be overcome by a recently developed strategy [28] to fuse isolated organelle membranes to form large vesicles of 20–50 μm in diameter. This strategy has been shown to be applicable to a wide variety of membrane systems [28]. As an example, its application will be demonstrated for the endoplasmic reticulum from rat liver cells.

To investigate the ion transport on the molecular level, ER membranes were isolated according to standard purification procedures and carefully dehydrated in a dessicator. Dehydration of isolated membranes takes place in the presence of excess lipids or 5 % polyethyleneglycol (PEG) for protection of the membrane proteins. To induce the fusion process, the slightly dehydrated membrane suspension was covered with buffer solutions conventionally used for patch–clamp experiments [1]. After 1–2 hours, this leads to the formation of giant liposomes at the edge of the dehydrated lipid film. Giant liposomes were transferred to the recording chamber to perform electrophysiological studies.

Several types of ion channels could be observed in giant ER vesicles. The predominant current was mediated by anion selective channels with a single channel conductance of 64 pS in 50 m*M* KCl. These channels displayed a linear current *versus* potential–difference relationship. Their reversal potential in asymmetrical solutions of KCl was close to the reversal potential for chloride, indicating their preferential selectivity for anions. The gating of these anion channels was potential–difference–dependent with large positive or negative transmembrane potentials favouring channel closing. The second type of channel in ER membranes was found in liver cells mediated cation fluxes. This cation–selective channel revealed a conductance of 80–95 pS in 50 m*M* KCl. The selectivity of $Na^+/K^+$ was in the range of 0.8. Together, anion and cation selective channels found in the ER may provide the cell with suitable pathways for the controlled exchange of ions between intracellular stores and the cytosol.

### 4.3. *Mitochondria*

Ion transport across the inner and outer mitochondrial membrane is widely believed to be involved in several regulatory processes of the mitochondria. The exchange of small metabolites, and the import of entire proteins into mitochondria is thought to occur through water-filled, channel–like pathways in the inner and outer membrane [30]. In order to investigate these transport mechanisms with the patch–clamp–technique, two technical problems need to be solved. First, mitochondria have to be enlarged to reach a suitable size for direct patch–clamp measurements, and second, the outer and inner membrane have to be separated. Both problems were recently solved by osmotic swelling of mitochondria [29–31].

Patch-clamp measurements on osmotically swollen mitoplasts (mitochondria without the outer membrane) revealed the existence of strongly potential–difference–dependent ion channels in the inner mitochondrial membrane (IMM). These channels were anion selective and displayed a single channel conductance of 110 pS in solutions containing 150 m*M* KCl. Channel openings were preferentially induced by polarizing the IMM (positive inside), allowing ion exchange between cytosol and mitochondria. At the resting potential–difference (–180 mV, negative inside the mitochondria), these channels were completly closed [30]. Thus, the decrease in membrane conductance allows the mitochondria to maintain the electrical potential difference across the inner mitochondrial membrane necessary for ATP synthesis.

### 4.4 *Photosynthetic membrane (thylakoid)*

Photosynthetic electron transport in the thylakoid generates a proton gradient which in turn drives the ATP–synthase [32]. Thus it was proposed that counterion fluxes across the thylakoid such as chloride and magnesium may short–circuit photosynthetic proton transport [33].

By osmotic swelling of giant *Peperomia metallica* chloroplasts ([34], Fig. 3) the thylakoid can be made accessible to patch pipettes. Potential–difference–dependent

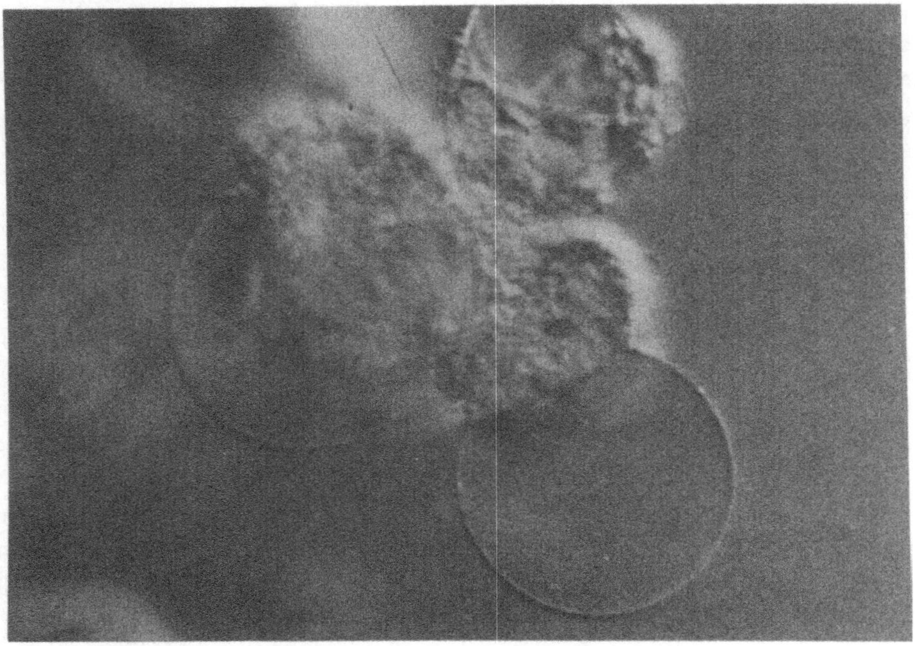

Fig. 3 – Osmotically inflated photosynthetic membrane stack (thylakoid) from giant chloroplasts of *Peperomia metallica*.

chloride–selective channels with a conductance of 80–100 pS (in 100 m*M* KCl) were identified. These anion channels may be essential to balance the transthylakoid potential difference and to establish a pH gradient.

## 5. Concluding remarks

Recent patch–clamp studies have led to the discovery that ion channels occur not only in animal cells but also in membranes of higher plant cells and microbes. As demonstrated for $K^+$ channels in the brain, genetic information in the form of cDNA or mRNA encoding ion–channel proteins from various origins can be isolated, injected into frog oocytes, expressed and incorporated in the oocyte plasma membrane. Patch-clamp measurements on genetically modified channels will further increase our knowledge about the structure and function of channel proteins.

## References


[1] O.P. Hamill, A. Marty, E. Neher, B. Sakmann and F.J. Sigworth, *Plfügers Arch. Ges. Physiol.,* **391,** 85 (1981).

[2] A. Baumann, A. Grupe, A. Ackermann and O. Pongs, *EMBO J.,* **7,** 2457 (1988).

[3] W. Stühmer, M. Stocker, B. Sakmann, P. Seeburg, A. Baumann, A. Grupe and O. Pongs, *FEBS Lett.,* **242,** 199 (1988).

[4] J.I. Schroeder, R. Hedrich and J.M. Fernandez, *Nature (London),* **312,** 361 (1984).

[5] J.I. Schroeder, K. Raschke and E. Neher, *Proc. Natl. Acad. Sci.* USA, **84,** 4108 (1987).

[6] J.I. Schroeder, *J. Gen. Physiol.,* **92,** 667 (1988).

[7] T. Ijima and S. Hagiwara, *J. Membr. Biol.,* **100,** 73 (1987).

[8] R.L. Satter and N. Moran, *Physiol. Plant.,* **72,** 816 (1988).

[9] M. Tazawa, T. Shimmen and T. Mimura, *Annu. Rev. Plant Physiol.,* **38,** 95 (1987).

[10] A.I. Sokolik and V.M. Yurin, *J. Membr. Biol.,* **89,** 9 (1986).

[11] R. Hedrich and J.I. Schroeder, *Annu. Rev. Plant Physiol.,* **40,** 539 (1989).

[12] W.H. Outlaw, *Physiol. Plant.,* **59,** 407 (1983).

[13] K.L. Edwards and B.G. Pickard in: *The Cell Surface in Signal Transduction,* E. Wagner, H. Grepping and B. Biller (Editors), *NATO ASI Series H 12,* Springer–Verlag, Heidelberg (1987), p. 41.

[14] Y. Saimi, B. Martinac, M.C. Gustin, M.R. Culbertson, J. Adler and C. Kung, *Trends Biochem. Sci.,* **13,** 304 (1988).

[15] P. Matile, *Annu. Rev. Plant Physiol.,* **29,** 193 (1978).

[16] R. Gerhardt and H.W. Heldt, *Plant Physiol.,* **83,** 399 (1987).

[17] P.K. Hepler and R. Wayne, *Annu. Rev. Plant. Physiol.,* **36,** 397 (1985).

[18] R. Hedrich, H. Barbier–Brygoo, H. Felle, U.I. Flügge, U. Lüttge, F.J.M. Maathuis, S. Marx, H.B.A. Prins, K. Raschke, H. Schnabl. J.I. Schroeder, I. Struve, L. Taiz and P. Ziegler, *Botan. Acta,* **1,** 7 (1988).

[19] R. Hedrich and A. Kurkdjian, *EMBO J.,* **12,** 3061 (1988).

[20] R. Hedrich, U.L. Flügge and J.M. Fernandez, *FEBS Lett.,* **204,** 228 (1986).

[21] R. Hedrich, J.I. Schroeder and J.M. Fernandez, *Trends Biochem. Sci.,* **12,** 49 (1986).

[22] H.A. Kolb, K. Köhler and E. Martinoia, *J. Membr. Biol.,* **95,** 163 (1987).

[23] L. Coyaud, A. Kurkdjian, R. Kado and R. Hedrich, *Biochim. Biophys. Acta,* **902,** 263 (1987).

[24] R. Colombo, R. Cerana, P. Lado and A. Peres, *J. Membr. Biol.,* **103,** 227 (1988).

[25] R. Hedrich and E. Neher, *Nature (London),* **329,** 833 (1987).

[26] Y. Wada, Y. Ohsumi, M. Tanifuji, M. Kasai and Y. Anraku, *J. Biol. Chem.,* **262,** 17260 (1987).

[27] B. HILLE, *Ionic Channels of Excitable Membranes,* Sinauer Associates, Sunderland MA. (1984).

[28] M. CRIADO and B.U. KELLER, *FEBS Lett.,* **224,** 172 (1987).

[29] M.C. SORGATO, B.U. KELLER and W. STÜHMER, *Nature (London),* **330.** 498 (1987).

[30] B.U. KELLER, R. HEDRICH, W.L.C. VAZ and M. CRIADO, *Pflügers Arch. Ges. Physiol.,* **411,** 94 (1988).

[31] B.U. KELLER, M.C. SORGATO and W. STÜHMER, *Biophys. J.,* **53,** 31 (1988).

[32] W. JUNGE, *Curr. Top. Membrane Biol.,* **98,** 69 (1982).

[33] G. HIND, H.Y. NAKATANI and S. IZAWA, *Proc. Natl. Acad. Sci. USA,* **71,** 1484 (1974).

[34] G. SCHÖNKNECHT, R. HEDRICH, W. JUNGE and K. RASCHKE, *Nature (London),* **336,** 589 (1988).

# PROTEIN TRANSLOCATION ACROSS BIOLOGICAL MEMBRANES

FRANZ-ULRICH HARTL

*Institut für Physiologische Chemie der*
*Universität München*
*Goethestrasse 33, 8000 München 2, FRG*

Contents



*Bioelectrochemistry III*
Edited by G. Milazzo and M. Blank
Plenum Press, New York, 1990

## 1. Introduction

Several thousand different polypeptides are synthesized within an eukaryotic cell. In general, protein synthesis takes place in the cytosol, but many proteins have their functional location in subcellular compartments that are separated from the cytosol by at least one membrane. Certain proteins, for example, those residing in the lumen of chloroplast thylakoids, have to be translocated across as many as three bilayers to reach their destination. Others, for example, proteins exerting transport functions, become integrated into membranes, often in arrangements spanning the bilayer many times. How are proteins targeted to the correct membrane compartment? What are the signals involved, and how can polypeptides, *i.e.* macromolecules containing many hydrophilic groups, penetrate lipid bilayers at all? By describing the main principles of protein translocation across biological membranes I shall try to answer some of these basic questions in molecular cell biology.

A number of different membrane systems competent in protein translocation are known (for reviews see Refs. 1-9), including the bacterial plasma membrane, and the membranes of the endoplasmic reticulum (ER), of mitochondria, chloroplasts and peroxisomes in eukaryotes (Table 1). In recent years results by many groups have demonstrated that with all of these membrane systems protein transport follows certain common principles. Of special importance was the finding that in all cases transport could occur post-translationally, *i.e.* after complete synthesis of the polypeptide chain, at least if suitable experimental conditions were selected. Therefore, in contrast to what had been originally widely assumed for protein transport across the ER membrane, a mechanistic coupling between translation and translocation does not exist. The basic requirements of protein translocation can be summarized as follows [10-12]:

I ) Proteins to be transported must contain targeting signals specific for a certain membrane;

II) the membranes of the target compartment have receptors or receptor-like structures exposed at their cytosolic surface which interact with these targeting signals;

III) translocation requires metabolic energy either in the form of ATP, or of ATP and a membrane potential;

IV) proteins have to assume a loosely-folded conformation in order to be competent for membrane translocation.

Table 1 – Membranes competent in protein translocation*

| Transport system | Targeting signal | Energy requirement | Cytosolic components | Organelle |
|---|---|---|---|---|
| ER | N-terminal; approx. 15 to 35 residues; basic N-terminus, hydrophobic region, small residue at cleavage site | ATP | Signal recognition particle (SRP); 70 kDa heat shock proteins plus additional *"unfoldase"* (not identified) | Docking protein (SRP receptor), signal sequence receptor and signal peptidase in the ER membrane. Immunoglobulin heavy chain binding protein (BiP) at luminal surface of ER membrane |
| Bacterial | Similar to and functionally exchangable with ER signals | ATP plus total protonmotive force | SecA protein, SecB protein, trigger factor, groEL protein | SecY protein, leader peptidase, lipoprotein lipase in the membrane |
| Mitochondria | N-terminal; approx. 12 to 70 residues; rich in basic and hydroxylated residues; additional sorting signals for transport from matrix to intermembrane space similar to bacterial export signals | ATP plus membrane potential | 70 kDa heat shock protein; NEM-sensitive factor | Surface receptors and "general insertion protein" (GIP) in outer membrane (not identified); mitochondrial processing peptidase (MPP), processing enhancing protein (PEP) and groEL homologue hsp60 in matrix; second processing peptidase(s) at outer surface of inner membrane |
| Chloroplasts | Similar to mitochondrial targeting sequences; additional sorting sequences for thylakoids similar to bacterial export signals | ATP | Not known | 31 kD receptor at contact sites? Stroma protease (not identified) and groEL homologue Rubisco binding protein in stroma; thylakoid peptidase in thylakoid membrane |
| Peroxisomes | C-terminal; SKL-motif, additional internal signals possible | ATP | Not known | Surface receptors suggested based on functional data |

* for references see text

The molecular mechanism of the translocation process itself is still unclear. It is assumed that translocation occurs through a hydrophilic, probably proteinaceous, membrane environment. However, penetration of proteins directly into the lipid bilayer seems possible in special cases. For the description of translocation pathways we will focus on the three systems studied in most detail, namely protein export in *E. coli*, and transport of proteins into the ER and into mitochondria.

## 2. Targeting signals

The signals required to direct a protein to its target membrane are contained in contiguous sequences. These targeting sequences are mostly at the amino-terminus as so-called presequences, but can actually be located anywhere in the protein. The secretory protein ovalbumin, colicin E1 of *E. coli* and the ADP/ATP carrier of the inner mitochondrial membrane are examples of proteins having internal targeting sequences [13-15]. With a number of peroxisomal proteins carboxy-terminal targeting signals have been found [16, 17]. Consequently, their mode of translocation has to be essentially post-translational. Most N-terminal targeting signals are cleaved by specific processing peptidases or signal peptidases at the trans-side of the membrane. However, many proteins carry non-cleavable amino-terminal signals, and proteolytic removal of cleavable signals appears not to be mechanistically connected to translocation.

The presence or absence of an amino-terminal targeting sequence alone can determine whether a protein is localized in the cytosol or in a specific membrane compartment. For example, cyclophilin, a cyclosporin. A binding-protein, of *Neurospora crassa* is located in the cytosol and in mitochondria [18]. Two mRNAs are transcribed from the same gene. Only the longer one codes for a precursor form having a cleavable amino-terminal extension which directs the protein to mitochondria. Targeting signals have been defined most clearly by gene fusion experiments demonstrating the ability of the signal sequence to direct the translocation of a non-related (*e.g.* cytosolic) *passenger protein* across the target membrane [19-21].

Targeting signals specific for a certain membrane can vary considerably in length and show no significant sequence similarity. This suggests that the targeting information must reside in secondary- or tertiary-structure elements. Indeed, targeting sequences of precursors destined to the same compartment have common structural features. It is useful to distinguish the following three types of targeting signals:

I) targeting sequences containing extended stretches of hydrophobic amino acid residues (type I signals);

II) targeting signals without such hydrophobic segments having many hydrophilic and basic residues (type II signals);

III) targeting signals consisting of short stretches of residues with positive charges (typeIII signals).

Type I signals are found in the targeting sequences of proteins exported across the plasma membrane in bacteria and in proteins transported into the ER [22-27]. Signals of these two membrane systems (often referred to as *leader peptides* or *signal peptides*) have been shown to be interchangeable with respect to specificity of targeting and proteolytic processing [21, 22]. Three structural motives can be recognized. One to three positively-charged residues are present at the amino-terminus (amino-terminal basic region), followed by a hydrophobic region of 7 to 20 residues length (hydrophobic domain) and a sequence at the carboxy-terminus including a small amino acid, alanine or glycine, at the cleavage site [25-27]. The signal sequences of eukaryotic secretory proteins typically have small or neutral amino acids (alanine, glycine, serine, threonine) at position minus 3 of the cleavage site [27].

Type II signals are found in proteins destined to mitochondria and chloroplasts. Mitochondrial targeting sequences are rich in basic and hydroxylated residues and normally lack acidic residues [4, 5, 28, 29]. The signals of chloroplast proteins appear to be very similar and are functionally exchangeable with mitochondrial targeting sequences to a limited degree [30, 31]. The precursor of the small subunit of the chloroplast protein ribulose-bisphosphate carboxylase (Rubisco) as well as fusion protein between the amino-terminal two-thirds of the Rubisco presequence and dihydrofolate reductase (a cytosolic protein) could be imported into mitochondria *in vitro*, albeit with very low efficiency [31]. Under *in vivo* conditions, however, missorting between mitochondria and chloroplasts presequences tend to form different secondary structure arrangements. Many mitochondrial presequences are assumed to exist as amphiphilic α-helices having the basic residues exposed to one side of the helix and the uncharged and more hydrophobic ones to the opposite side of the helix [29, 30]. So far, this has not been demonstrated for chloroplast presequences [30]. It should be noted that amphiphilicity is an essential structural feature at least for all type I targeting sequences due to their linear arrangement of polar and apolar regions. Chemically-synthesized signal sequences of both type I and II show *detergent*-like properties in experiments with artificial membrane systems [5, 23]. This might reflect a function of these sequences during the initial insertion into the target membrane.

Type III signals are present in polypeptides that are targeted to the nucleus and to peroxisomes [16, 17, 33]. They are characteristically not found at the amino-terminus and they appear to be rather short (as few as 4 residues being sufficient). Both their structural organization as well as their mode of interaction with the target membrane are poorly understood.

## 3. Translocation pathways

### *3.1. Transport across the ER membrane and bacterial protein export*

The mechanisms of transport of proteins into and across the ER membrane in eukaryotes and of protein export in *E. coli* appear to be rather similar (for reviews see Refs. 1-3). Eukaryotic signal peptides function in the procaryotic system and *vice versa*. For example, upon expression in *E. coli*, rat proinsulin is secreted into the periplasmic space of the bacterium directed either by a prokaryotic or by an eukaryotic signal peptide [34]. On the other hand, phage M13 procoat protein (normally assembled into the *E. coli* plasma membrane) can be inserted into dog pancreas microsomes and correctly processed [22, 35]. Obviously, the mechanism for recognition of the signal peptide has been conserved in evolution.

#### 3.1.1. The SRP-cycle

Molecular details of the initial steps of protein secretion have first been analyzed with the mammalian ER and the *signal hypothesis* has been able to explain a number of basic steps [36-42]. Synthesis of secretory proteins starts on free cytosolic ribosomes. Once the signal peptide emerges from the ribosome (*i.e.* after synthesis of approx. 20-40 amino acids) it interacts with a ribonucleoprotein complex, the signal recognition particle (SRP), which consists of six polypeptides and a 7S RNA. The signal recognition activity of SRP results in a block or slowing down of translation activity (a point of controversial debate) and the ternary complex consisting of ribosome, nascent polypetide chain and SRP is delivered to the ER membrane. There SRP binds to a receptor-like structure, docking protein or SRP receptor, which is a 72 kDa integral membrane protein. Upon interaction with docking protein, SRP is released from the nascent chain-ribosome complex and the elongation arrest is relieved (36-40). SRP is available for a new cycle while the signal sequence interacts with a 35 kDa integral glycoprotein of the ER membrane, the signal sequence receptor [41]. Translocation then takes place co-translationally, *i.e.* while translation is going on. The only additional component of the system known is signal peptidase, described as a six-subunit protein complex residing in the membrane which cleaves the signal peptide off the mature protein part at the luminal surface of the ER membrane [42]. During the process of translocation the amino-terminus of the signal peptide remains fixed at the cytosolic surface of the membrane, while the growing polypeptide chain traverses the membrane as a growing loop until the carboxy-terminus reaches the luminal surface [10, 43]. Translocation occurs through a membrane environment that is accessible to hydrophilic protein denaturants such as urea or alkaline pH [38].

Transport into the ER is not necessarily co-translational. In yeast (*Saccharomyces cerevisiae*) some proteins such as the precursor of the pheromone prepro-α-factor are post-translationally translocated as completed polypeptides (11). SRP or an SRP-equivalent has not been detected in *S. cerevisiae*. However, in another yeast species, *Schizosaccharomyces pombe*, a 7S RNA essential for cellular functions has been found which is similar to the 7S RNA of SRP [44]. Recent evidence indicates that cytosolic 70 kDa heat shock proteins and probably additional factors may take over certain functions of SRP in post-translational translocation by stabilizing a translocation-competent conformation of the precursor protein [45]. In the mammalian system a SRP- and docking protein independent pathway for translocation into the ER exists also, at least for small proteins such as prepro-mellitin [11]. Obviously, these proteins are so short that protein synthesis is terminated at the same time when the signal sequence emerges from the ribosome, and interaction with SRP does not take place. They are translocated post-translationally.

Early studies of protein export in *E. coli* have already established that the predominant mode of translocation is post-translational [1, 2]. M13 procoat was one of the first proteins whose assembly into the bacterial plasma membrane (corresponding to the inner membrane of Gram negative bacteria) was investigated in detail [2]. M13 coat protein is a small protein spanning the membrane with a single stretch. Its precursor contains a cleavable amino-terminal signal sequence (or leader peptide) which has the typical features of a type I signal. Membrane assembly of procoat turned out to be dependent on only one further membrane component, leader peptidase, a 37 kDaintegral membrane protein which cleaves the leader peptides of exported proteins at the periplasmic surface of the plasma membrane [46, 47]. In addition to ATP, the membrane potential difference across the plasma membrane was found necessary for translocation [48]. The same energy requirement has been found for a number of exported proteins [49-51].

### 3.1.2. The trigger factor-cycle

Unlike M13 procoat [52], however, many proteins including leader peptidase and proOmpA, the precursor of the outer membrane protein A, are dependent for export on additional proteinaceous factors in the cytosol or in the inner membrane. This has been demonstrated by the identification of a number of genes whose mutations affect protein export in *E. coli*. The development of an *in vitro* system for studying translocation of proteins (based on inverted *E. coli* inner membrane vesicles) enabled the biochemical analysis of these so-called Sec-gene products. The essential roles of two components, namely of the SecA gene product, a 94 kDa cytosolic factor with ATPase activity, and of the SecY gene product, a 42 kDa multi-spanning membrane protein, have been established most clearly (53, 54). In addition, a third component, *trigger factor*, was recently shown to be required for the translocation of proOmpA across the inner membrane [55].

Trigger factor was isolated as a 63 kDa cytosolic protein exploiting its stoichiometric binding to proOmpA [55-57]. Later it was found to be also present at ribosomes as a peripheral component of the large ribosomal subunit [58]. The present data strongly indicates that trigger factor fulfills a role similar to SRP in the eukaryotic system. The cyclic action of trigger factor during proOmpA translocation can be divided into several stages:
I) During or after translation newly synthesized proOmpA forms a 1:1 complex with trigger factor, whereby trigger factor stabilizes the translocation competent conformation of proOmpA;
II) at the plasma membrane proOmpA-trigger factor complex interacts with the secA protein which is reversibly bound to the membrane probably via the secY protein;
III) interaction with secA stimulates ATP hydrolysis by secA and proOmpA translocates across the membrane;
IV) trigger factor cycles back to the ribosome.

How ATP hydrolysis is coupled to translocation is still not clear. Interestingly, the function of trigger factor in stabilizing proOmpA can be substituted for by canine SRP [57], thus raising the possibility that trigger factor cycle and SRP cycle are variations of a common theme.

### 3.2. *Membrane integration of proteins*

The principles of protein translocation outlined above do not readily explain how membrane spanning proteins are integrated asymmetrically and with different topologies into the phospholipid bilayer. Apparently, the hydrophobic regions contained in the signal sequences of secreted proteins are not very different from the hydrophobic membrane-spanning segments typically found in integral membrane proteins (for review see Ref. 59). Indeed, an uncleaved signal sequence can serve as both, targeting signal and membrane anchor, at least if the hydrophobic stretch is longer than 15-20 amino acid residues [60]. In addition to uncleaved signals one more topogenic element, the *stop-transfer* sequence, was defined which functions in conjunction with the signal sequence. Stop-transfer sequences are continuous stretches of mostly hydrophobic residues which can stop the translocation of a polypeptide across the membrane if located at some distance behind a signal sequence.

It is important to note that–stop transfer sequences can function as internal (uncleaved) signal sequences [61]. Whether a hydrophobic segment acts as stop-transfer sequence or as signal sequence would only be determined by its relative position in the polypeptide chain. For example, a typical amino-terminal signal initiates the translocation of the polypeptide traversing the membrane as a loop until further translocation is halted by a stop-transfer sequence. The next hydrophobic segment of the polypeptide would then act as an internal signal sequence and so forth. Three different membrane topologies could be explained by combining cleaved or uncleaved signals with or without following stop–transfer sequences, namely single–spanning proteins with the amino–terminus either in the cytosol or the ER lumen (*i.e.* extracytoplasmic), or proteins spanning the membrane twice and having both termini in the cytosol. Multispanning proteins (*e.g.*, opsin which spans the membrane seven times, or the erythrocyte anion exchange protein band III with more than ten membrane spanning segments) would reach their final topology by successive insertion of internal signal sequences and stop-transfer sequences [10]. Experimental support for this model of *sequential insertion* [62, 63] recently came from studies using artificial polytopic proteins spanning the membrane up to four times [43]. These proteins were constructed by tandemly repeating portions of the cDNA coding for the human asialoglycoprotein receptor which is integrated into the plasma membrane by a single uncleaved signal/anchor domain.

Secreted proteins and proteins that are inserted into the membrane use the same translocation pathway, but only the interaction of the initial signal sequence with the translocation machinery appears to be SRP-dependent [43]. Assuming a proteinaceous channel through which protein translocation proceeds, one has to propose that internal signal sequences and stop-transfer segments must be able to laterally leave this channel to finally become embedded into the lipid bilayer. This then implies that the function of stop-transfer sequences, in addition to protein-lipid interaction, involves specific protein-protein interactions. Clearly, stop-transfer sequences must be defined functionally. A hydrophobic stretch of amino acids which is not recognized as a *stop-transfer signal* by the translocation machinery is completely transported across the membrane. On the other hand, it is not clear, whether all membrane-spanning sequences have to use the specific translocation apparatus to become inserted. The existence of so-called *embedding sequences* has been proposed, separate folded domains with a hydrophobic surface which would integrate spontaneously into the lipid bilayer without the help of a membrane protein [59, 62, 64]. This process would be triggered by the SRP-dependent interaction of the targeting sequence with the membrane. Incorporation of complete domains could be an attractive explanation in the case of complex membrane proteins (*e.g.* the channel-forming proteins phoE and ompF in the outer membrane of *E. coli* and the porin of the outer mitochondrial membrane) which in their mature parts lack continuous hydrophobic segments longer than five residues [65]. These proteins consist almost entirely of a number of β-pleated sheets which are membrane-spanning.

### 3.3. *Import of proteins into mitochondria*

Mitochondria, like chloroplasts, are assumed to have originated from prokaryotic ancestors by an endosymbiotic event, but have lost their autonomy during evolution. This is reflected by the fact that the mitochondrial genome codes only for a small percentage of the proteins located in mitochondria. Most mitochondrial proteins are coded for by nuclear genes, are synthesized on cytosolic polysomes as precursors and are post-translationally imported into the organelles (for reviews see Refs. 4 and 5). Mitochondria are surrounded by two membranes which are in close proximity at distinct sites. Only a few years ago it was demonstrated that protein import from the cytosol occurs at these so-called *translocation contact sites* [66]. The mechanism of protein translocation across two membranes in a single step is unique for mitochondria. Import into chloroplasts might follow a very similar mechanism. In mitochondria import requires energy in the form of ATP and the membrane potential difference across the inner membrane, whereas in chloroplasts ATP alone seems to be sufficient [5]. Protein import into mitochondria has been dissected into a number of distinct steps. The import pathways of most proteins follow general principles which will be summarized here.

#### 3.3.1. Targeting and recognition

Targeting information is in most cases contained in amino-terminal presequences of precursor proteins (see Section **2**) that are cleaved after translocation. However, uncleaved amino-terminal targeting sequences and internal targeting signals do occur [15, 67, 68]. In addition to the information for targeting to mitochondria, precursor proteins contain the complete information for their intramitochondrial sorting into one of the four sub-mitochondrial compartmentes: outer membrane, intermembrane space, inner membrane and matrix.

Precursors specifically interact with protease-sensitive receptors at the surface of the outer membrane [69-73]. These structures have not yet been identified. However, based on the differential protease sensitivity of binding sites and on experiments measuring competition for specific binding of different precursor proteins, it can be concluded that at least three classes of surface receptors exist [72, 73]. These types of receptors are specific for subclasses of precursor proteins. They might be clustered at contact sites between outer and inner membranes as recently proposed for a surface component involved in protein import into chloroplasts [74]. According to our recent working-hypothesis, precursors bound to different surface receptors are transferred to a *general insertion protein* (GIP) in the outer mitochondrial membrane which is believed to facilitate the membrane insertion of precursors destined to the different submitochondrial compartments [72, 73, 75).

Probably connected to *unfolding* of precursor proteins, this step or membrane insertion requires ATP [76] (see Section **4**). The interaction with GIP has been investigated for the precursors of the outer membrane protein porin and of the ADP/ATP carrier of the inner membrane [72, 75, 76]. GIP sites are saturable and have high affinity. Precursors bound to GIP are resistant to externally added protease (indicating insertion into the outer membrane has occurred) but are in a proteinaceous environment from which extraction by hydrophilic protein denaturants is possible. At the level of GIP the precursor of porin competes for the import of all other precursor proteins tested so far, except for the import of cytochrome *c* [72]. The route of apocytochrome *c* into the intermembrane space is exceptional in that it does not involve the receptor/GIP pathway [77, 78].

### 3.3.2. Translocation *via* contact sites

After interaction with GIP, precursors are either integrated into the outer membrane (as in the case of porin) or they are transported into translocation contact sites. Transfer into contact sites is dependent on the electrical component $\Delta\phi$ of the membrane potential across the inner membrane [79-81]. More specifically, $\Delta\phi$ (negative inside) appears to be required for the insertion into, or translocation across, the inner membrane of the positively-charged targeting sequences. The precise role of $\Delta\phi$ is unknown.

Evidence that translocation of proteins into mitochondria occurs at contact sites came from experiments in which precursor proteins spanning both mitochondrial membranes were accumulated as translocation intermediates [66, 82-84]. For example, precursor proteins that had antibodies bound to carboxy-terminal parts of the mature protein were arrested during translocation in a position reaching into the matrix with the amino-terminus (where proteolytic processing occurred) but with other parts of the molecule still being outside the mitochondrion where they were accessible to added protease [66, 82]. Thus, the two membranes had to be close enough together to be spanned by a single polypeptide chain. As demonstrated by immuno-gold labeling of translocation intermediates, these biochemically-defined translocation contact sites are identical with the morphologically-described sites of close contact between the two membranes [82]. Contact sites appear to be rather stable, independent of the energy state of mitochondria. They occur in limited number per mitochondrion [84]. By first accumulating saturating amounts of intermediates spanning contact sites further import of precursor proteins can be completely blocked. Saturability and the fact that spanning intermediates are accessible to hydrophilic denaturants [83] suggest that proteins are involved in the translocation of precursors.

### 3.3.3. Proteolytic processing and intramitochondrial sorting

During or after translocation, amino-terminal presequences of precursor proteins are specifically cleaved by the metal-dependent processing enzyme in the matrix [85-89]. Isolation of the enzyme from *Neurospora crassa* revealed that two components are required for activity: MPP, the mitochondrial processing peptidase proper, and PEP, the processing enhancing protein (57 kDa and 52 kDa, respectively) [87]. MPP is completely soluble in the matrix while at least part of PEP is peripherally associated to the inner surface of the inner membrane. PEP, which has no proteolytic activity on its own, strongly stimulates the activity of MPP. PEP may interact with the presequences of precursors as soon as they enter the matrix thus exposing the cleavage site to MPP. Sequence analysis of the two components showed considerable similarity suggesting a common origin of MPP and PEP [88]. Notably, the mitochondrial processing enzyme represents a new type of processing peptidase different from the membrane-integrated signal peptidase of the ER and the leader peptidase of *E. coli* [42, 46, 47].

Once translocated into the matrix and proteolytically processed many imported proteins have to undergo further sorting events and have to assemble into supramolecular complexes [90-92]. For example, proteins of the intermembrane space, such as cytochrome $b_2$ and cytochrome $c_1$, are re-translocated from the matrix back across the inner membrane to their target compartment [91]. This process is directed by a second targeting sequence which is linearly arranged behind the positively-charged mitochondrial targeting signal. This second targeting sequence contains a continuous stretch of hydrophobic residues and is structurally comparable to bacterial type signal sequences. It is cleaved by a second processing peptidase at the outer surface of the inner membrane. It was proposed that the export of proteins from the mitochondrial matrix follows a mechanism similar to bacterial protein export and has been conserved during the evolution of mitochondria from prokaryotic ancestors (*conservative sorting*) [5, 90, 91). Thus, the mitochondrial inner membrane is unique in that it contains two independent machineries translocating proteins in opposite directions. One evolutionary conserved component involved in protein import is the constitutively expressed intramitochondrial heat shock protein hsp60 [92]. This protein which forms a 14mer scaffold in the matrix is structurally and functionally related to the *E. coli* heat shock protein groEL and the chloroplast Rubisco binding protein [93]. In a conditional yeast mutant affecting hsp60 assembly of imported proteins into oligomeric complexes and the intramitochondrial sorting of cytochrome $b_2$ are defective [92].

## 4. Unfolding of precursor proteins

Results obtained with different membrane systems indicate that proteins have to be at least partially *unfolded* to be competent for membrane translocation. For example, the maltose-binding protein of *E. coli* was shown to be competent for export across the plasma membrane only as long as the precursor in the cytosol was in a loosely-folded, protease-sensitive, conformation [94]. Mitochondrial proteins could be trapped in contact sites spanning outer and inner membranes (see Section 3) thus demonstrating that at least part of the polypeptide chain had to be in an extended conformation during membrane translocation [66]. Import of an artificial precursor protein consisting of a mitochondrial presequence fused to dihydrofolate reductase (DHFR) was blocked by binding of methotrexate to the DHFR moiety [95]. Methotrexate, a strong folate antagonist, stabilizes the tightly folded state of DHFR. Unfolding by urea rendered the fusion protein translocation competent [96].

What are the physiological reactions that contribute to maintaining the translocation-competent, unfolded conformation of precursor proteins? It has been proposed that ATP-dependent *unfoldases* are present in the cytosol [97]. Such an enzyme has not yet been identified. However, reticulocyte lysates, used to synthesize precursor proteins *in vitro*, contain an ATP-dependent system that keeps precursors destined to the ER or to mitochondria in a loosely-folded state [35, 76]. Based on experimental evidence of several groups it is clear that constitutively expressed members of the 70 kDa family of heat shock proteins, together with additional activities, participate in these reactions [35, 45, 98-103]. Genetic deletion in yeast of three of the four 70 kDa heat shock proteins of the SSA-subgroup resulted in the accumulation of prepro-α-factor, (which is post-translationally translocated into the ER) and of a mitochondrial precursor protein in the cytosol [98]. Presumably, 70 kDa heat shock proteins bind to partially-unfolded proteins *via* hydrophobic interaction, and release them upon ATP hydrolysis [101, 102]. The first interaction might occur co-translationally. Precursors could undergo several cycles of binding to heat shock proteins until they would reach the translocation machinery of their target membrane. Possibly, the groEl homologous hsp60 in the mitochondrial matrix has a similar role in maintaining a conformation of imported precursor proteins which is competent for assembly with other subunits or for translocation from the matrix compartment across the inner membrane into the intermembrane space [92].

An alternative procedure to preventing misfolding of completed polypeptides by heat shock proteins and other conformation-preserving factors is to translocate precursors co-translationally. By its affinity for docking protein, binding of SRP to the signal sequence allows only for nascent chains to be targeted to the translocation machinery of the ER membrane. Arrest or slow down of translation after binding of SRP might be an efficient means to ensure correct intracellular sorting (see Section 3). In addition to its targeting function, SRP appears to play a role in preventing the folding of precursor proteins into a translocation-incompetent conformation. For example, SRP can substitute for trigger factor in the post-translational transport of proOmpA across the *E. coli* plasma membrane

[57]. It stabilizes prepro-α-factor for post-translational translocation across the ER membrane in yeast [103]. In certain cases translocation competence of secretory proteins is maintained only as long as the carboxy-terminus of a completed polypeptide chain is still attached to the ribosome. The term *ribosome-dependent* translocation was introduced to describe this requirement (11).

## 5. Energy requirements of protein translocation

What are the forces that drive the unidirectional transport of proteins across membranes? Originally, it was assumed that translocation was only possible while translation was going on. The polypeptide was believed to be pushed across the membrane as it was elongated by the ribosome machinery which derives its energy from the hydrolysis of ATP and GTP. We know now, however, that translocation is mechanistically independent of elongation. Another possible energy source of translocation was the proteolytic removal of presequences at the trans-side of the membrane. However, many proteins lacking cleavable presequences are translocated. Furthermore, preventing presequence cleavage by blocking the processing enzyme does not inhibit translocation.

In all systems studied, ribosome-indipendent membrane translocation of proteins requires the hydrolysis of nucleoside triphosphates (ATP or GTP) (for review see Ref. 5). In addition, transport across the bacterial plasma membrane and across the inner mitochondrial membrane is dependent on the potential difference across the respective membrane [48-51, 79-81]. The requirement for ATP has been related to *unfolding* or to maintaining a translocation competent conformation of precursor proteins (see Section **4**). It is not clear, whether hydrolysis of high-energy phosphate bonds plays a role in the mechanism of translocation itself. Artificially unfolded precursor proteins appear to have lost (or reduced) the ATP dependence of translocation [96,104]. Neither is the precise function of the electrochemical potential known. While in mitochondria only the electrical component $\Delta\phi$ and not $\Delta pH$ is required [81], translocation in bacteria seems to be dependent on the total promonmotive force [50, 51]. The polarity of the potential with respect to the direction of protein transport is opposite in the two systems. In case of mitochondria the membrane potential is only necessary for the insertion of the positively-charged targeting sequences into the inner membrane. It is not required for the translocation of the mature protein part [66, 76, 81].

In conclusion, our present knowledge is not sufficient to explain what drives the mature protein moiety of a precursor across the membrane. Neither is it known, how the unidirectionality of the process is determined. Having no alternative explanation at hand, it was proposed that tight folding of the protein occurring after translocation would be the indirect driving force for translocation. Prior to translocation the *unfolded* or nascent polypeptide chain would be on a higher level of energy where it is maintained by ATP hydrolysis. However, proteins do not immediately fold on the trans side of the membrane. To the contrary, their loosely-folded conformation appears to be preserved by interaction with heat shock proteins and maybe other components inside the organelles (*e.g.* the groEL homologue in mitochondria and chloroplasts, or the immunoglobulin heavy chain binding protein (BiP) of the ER lumen) [92, 105, 106]. This is most evident in cases where proteins have to undergo a second membrane translocation event (*e.g.* proteins of the mitochondrial intermembrane space or the thylakoid lumen in chloroplasts). Clearly, a number of open questions will find an answer only when the molecular mechanism(s) of protein translocation across membranes is understood in more detail.

**Acknowledgements**

I thank Dr. W. Neupert for numerous stimulating discussions and Dr. M.K. Hayer-Hartl for help in editing the manuscript.

**References**

[1] L.L. Randall, S.J.S.Hardy and I.R. Thom, *Annu. Rev. Microbiol.*, **41**, 507 (1987).
[2] W.T.Wickner, *Biochemistry,* **27**, 1081 (1988).
[3] P. Walter, and V. Lingappa, *Annu. Rev. Cell Biol.*, **2**, 499 (1986).
[4] N. Pfanner, F.U. Hartl and W. Neupert, *Eur. J. Biochem.*, **175**, 205 (1988).
[5] F.U. Hartl, N. Pfanner, D.W. Nicholson and W. Neupert, *Biochim. Biophys. Acta,* **988**, 1 (1989).
[6] T.H. Lubben, S.M. Theg, and K. Keegstra, *Photosynthesis. Res.,* **17**, 173 (1988).
[7] P.B. Lazarow and Y. Fujiki, *Annu. Rev. Cell Biol.*, **1**, 489 (1985).
[8] P. Borst, *Biochim. Biophys. Acta.*, **866**, 179 (1986).
[9] W.W. Just and F.U. Hartl, in *Peroxisomes in Biology and Medicine*, H.D. Fahimi and H. Sies (Editors), Springer-Verlag, Berlin (1987), pp. 402-416.
[10] W. Wickner and H.F. Lodish, *Science*, **230**, 400 (1985).
[11] R. Zimmermann and D.I. Meyer, *Trends Biochem. Sci.,* **11**, 512 (1986).
[12] K. Verner and G. Schatz, *Science,* **241**, 1307 (1988).
[13] R.D. Palmiter, J. Gagnon and K.A. Walsh, *Proc. Natl. Acad. Sci. USA,* **75,** 94 (1987).
[14] Y. Ebina, F. Kishi, F. Miki, H. Kagamiyama, T. Nakazawa and A. Nakazawa, *Gene,* **15,** 119 (1988).

[15] R. Zimmermann, U. Paluch, and W. Neupert, *FEBS Lett.,* **108,** 141 (1979).
[16] S.J. Gould, G.A. Keller and S. Subramani, *J. Cell Biol.*, **105**, 2923 (1987).
[17] S.J. Gould, G.A. Keller and S. Subramani, *J. Cell Biol.*, **107**, 897 (1988).
[18] M. Tropschug, D.W. Nicholson, F.U. Hartl, H. Köhler, N. Pfanner, E. Wachter and W. Neupert, *J. Biol. Chem.,* **263**, 14433 (1988).
[19] E.C. Hurt, B. Pesold-Hurt and G. Schatz, *EMBO J.,* **3,** 3149 (1984).
[20] A.L. Horwich, F. Kalousek, I. Mellmann and L.E. Rosenberg, *EMBO J.,* **4,** 1129 (1985).
[21] V.R. Lingappa, I. Chaidez, C.S. Yost and I. Hedgpeth, *Proc. Natl. Acad. Sci. USA,* **81,** 456 (1984).
[22] C. Watts, W. Wickner, and R. Zimmermann, *Proc. Natl. Acad. Sci. USA,* **80,** 2809 (1983).
[23] M.S. Briggs, L.M. Gierasch, A. Zlotnik, J.D. Lear and W.F. De Grado, *Science,* **228,** 1096 (1985).
[24] G. Von Heijne, *Eur. J. Biochem.,* **116,** 419 (1981).
[25] G. Von Heijne, *J. Mol. Biol.,* **192** 287 (1986).
[26] C. Milstein, G.G. Rowntee, T.M. Harrison and M. Mathews, *Nature (London),* **239,** 117 (1972).
[27] G. Von Heijne, *Eur. J. Biochem.,* **133,** 17 (1983).
[28] A. Ito, T. Ogishima, W. Ou, T. Omura, H. Aoyagi, S. Lee, H. Mihara and N. Izumiya, *J. Biochem,* **98,** 1571 (1985).
[29] G. Von Heijne, *EMBO J.,* **5,** 1335 (1986).
[30] G. Von Heijne, *Biochim. Biophys. Acta,* **947,** 307 (1988).
[31] E.C. Hurt, N. Soltanifar, M. Goldschmidt-Clermont, I.D. Rochaix and G. Schatz, *EMBO J.* **5,** 1343 (1986)
[32] A. Grossmann, S. Barthelt and N.H. Chua, *Nature (London),* **285,** 625 (1980).
[33] C. Dingwall and R.A. Laskey *Annu. Rev. Cell Biol.,* **2**, 367 (1986).
[34] K. Talmadge, S. Stahl and W. Gilbert, *Proc. Natl. Acad. Sci,* **77,** 3369 (1980).
[35] R. Zimmermann, M. Sagstetter, M.J. Lewis and H.R.B. Pelham, *EMBO J.,* **7,** 2875 (1988).
[36] G. Blobel and B. Dobberstein, *J. Cell Biol.,* **67**, 852 (1975).
[37] P. Walter, R. Gilmore and G. Blobel, *Cell,* **38**, 5 (1984).
[38] R. Gilmore and G. Blobel, *Cell,* **42**, 497 (1985).
[39] D.I. Meyer, E. Krane and B. Dobberstein, *Nature (London),* **297,** 647 (1982).
[40] R. Gilmore, P. Walter and G. Blobel, *J. Cell., Biol,* **95,** 470 (1982).
[41] M. Wiedmann, T.V. Kurzchalia, E. Hartmann and T.A. Rapoport, *Nature (London),* **328,** 830 (1987).
[42] F.A. Evans, R. Gilmore and G. Blobel, *Proc. Natl. Acad. Sci. USA,* **83,** 581 (1986).

[43] H.P. Wessels and M. Spiess, *Cell,* **55,** 61 (1988).
[44] P. Brennwald, X. Liao, K. Holm, G. Porter and J.A. Wise, *Mol. Cell Biol.,* **8,** 1580 (1988).
[45] R.J. Deshaies, B.D. Koch and R. Schekman, *Trends Biochem. Sci.,* **13,** 384 (1988).
[46] C. Zwizinski and W. Wickner, *J. Biol. Chem.,* **255,** 7973 (1980).
[47] P.B. Wolfe, W. Wickner and J. M. Goodman, *J. Biol., Chem.,* **258,** 12073 (1983).
[48] T. Date, J.M. Goodman and W. Wickner, *Proc. Natl. Acad. Sci. USA,* **77,** 4669 (1980).
[49] E.P. Bakker and L.L. Randall, *EMBO J.,* **3,** 895 (1984).
[50] B.L. Geller, N.R. Movva and W. Wickner, *Proc. Natl. Acad. Sci. USA,* **83,** 4219 (1986).
[51] L. Chen and P.C. Tai, *Proc. Natl. Acad. Sci. USA,* **82,** 4384 (1985).
[52] Y. Ohno-Iwashita and W. Wickner, *J. Biol. Chem.,* **258,** 1895 (1987).
[53] D.B. Oliver and J. Beckwith, *Cell,* **25,** 765 (1981).
[54] K. Ito, M. Wittekind, M. Nomura, K. Shiba, T. Yura, A. Miura and H. Nashimoto, *Cell,* **32,** 789 (1983).
[55] E. Crooke and W. Wickner, *Proc. Natl. Acad. Sci. USA,* **84,** 5216 (1987).
[56] E. Crooke, L. Brundage, M. Rice and W. Wickner, *EMBO J.,* **7,** 1831 (1988).
[57] E. Crooke, B. Guthrie, S. Lecker, R. Lill and W. Wickner, *Cell,* **54,** 1003 (1988).
[58] R. Lill, E. Crooke, B. Guthrie and W. Wickner, *Cell,* **54,** 1013 (1988).
[59] T.A. Rapoport and M. Wiedmann in *Current Topics in Membranes and Transport,* P.A. Knauf and J.S. Cook, (Editors), Academic Press, New York, (1985), Vol. **24,** pp. 1-63.
[60] J. Coleman, M. Inukai and M. Inouye, *Cell,* **43,** 351 (1985).
[61] M. Zerial, D. huyseboeck and H. Garoff, *Cell,* **48,** 147 (1987).
[62] G. Blobel, *Proc. Natl. Acad. Sci. USA,* **77,** 1496 (1980).
[63] D.D. Sabatini, G. Kreibich, T. Moromoto and M. Adesmik, *J. Cell Biol.,* **92,** 1 (1982).
[64] W. Wickner, *Annu. Rev. Biochem.,* **48,** 23 (1979).
[65] N. Overbecke, H. Bergmans, F. von Mansfeld and B. Lugtenberg, *J. Mol. Biol.,* **163,** 513 (1983).
[66] M. Schleyer and W. Neupert, *Cell,* **43,** 339 (1985).
[67] H. Arakawa, M. Takiguchi, Y. Amaya, S. Nagata, H. Hayashi and M. Mori, *EMBO J.,* **6,** 1361 (1987).
[68] R.A. Stuart, W. Neupert and M. Tropschug, *EMBO J.,* **6,** 2131 (1987).
[69] C. Zwizinski, M. Schleyer and W. Neupert, *J. Biol. Chem.,* **258,** 4071 (1983).
[70] C. Zwizinski, M. Schleyer and W. Neuperr, *J. Biol. Chem.,* **259,** 7850 (1984).

[71] R. Pfaller and W. Neupert, *EMBO J.*, **6,** 2635 (1987).
[72] R. Pfaller, H.F. Steger, J. Rassow, N. Pfanner and W. Neupert, *J. Cell Biol.*, **107**, 2483 (1988).
[73] R. Pfaller, N. Pfanner and W. Neupert, *J. Biol. Chem.*, **264**, 34 (1989).
[74] D. Pain, Y.S. Kanwar and G. Blobel, *Nature (London),* **331,** 232 (1988).
[75] N. Pfanner and W. Neupert, *J. Biol. Chem.*, **262**, 7528 (1988).
[76] N. Pfanner, M. Tropschug and W. Neupert, *Cell,* **49**, 815 (1987).
[77] D.W. Nicholson, H. Köhler and W. Neupert, *Eur. J. Biochem.,* **164**, (1987).
[78] D.W. Nicholson, C. Hergersberg and W. Neupert *J. Biol. Chem.*, **263**, 19034 (1989).
[79] M. Schleyer, B. Schmidt and W. Neupert, *Eur. J. Biochem.,* **125**, 109 (1982).
[80] S.M. Gasser, G. Daum and G. Schatz, *J. Biol. Chem.*, **257**, 13034 (1982).
[81] N. Pfanner and W. Neupert, *EMBO J.,* **4**, 2819 (1985).
[82] M. Schwaiger, V. Herzog and W. Neupert, *J. Cell Biol.,* **105**, 235 (1987).
[83] N. Pfanner, F.U. Hartl, B. Guiard and W. Neupert, *Eur. J. Biochem.,* **169**, 289 (1987).
[84] J. Rassow, B. Guiard, V. Herzog, F.U. Hartl and W. Neupert, Submitted.
[85] P. Böhni, S. Gasser, C. Leaver and G. Schatz in *The Organization and Expression of the Mitochondrial Genome,* A.M. Kroon and C. Saccone (Editors), Elsevier/North-Holland, Amsterdam. (1980), pp. 423-433.
[86] B. Schmidt, E. Wachter, W. Sebald and W. Neupert, *Eur. J. Biochem.,* **144**, 581 (1984).
[87] G. Hawlitschek, H. Schneider, B. Schmidt, M. Tropschug, F.U. Hartl and W. Neupert, *Cell,* **53**, 795 (1988).
[88] R.A. Pollock, F.U. Hartl, M.Y. Cheng, J. Ostermann, A. Horwich and W. Neupert, *EMBO J.,* **7**, 3493 (1988).
[89] C. Witte, R.E. Jensen, M.P. Yaffe and G. Schatz, *EMBO J.* **7**, 1439 (1988).
[90] F.U. Hartl, B. Schmidt, E. Wachter, H. Weiss and W. Neupert, *Cell,* **47**, 939 (1986).
[91] F.U. Hartl, J. Ostermann, B. Guiard and W Neupert, *Cell,* **51**, 1027 (1987).
[92] M.Y. Cheng, F.U. Hartl, J. Martin, R.A. Pollock, F. Kalousek, W. Neupert, E.M. Hallberg, R.L. Hallberg and A.L. Horwich, *Nature (London),* **337**, 620 (1989).
[93] S.M. Hemmingsen, C. Woolford, S.M. van der Vies, K. Tilly, D.T. Dennis, C.P. Georgopoulos, R.W. Hendrix and R.J. Ellis, *Nature (London),* **333**, 330 (1988).
[94] L.L. Randall and S.J.S. Hardy, *Cell,* **46**, 921 (1986).
[95] M. Eilers and G. Schatz, *Nature (London),* **322**, 228 (1986).
[96] M. Eilers, S. Hwang and G. Schatz, *EMBO J.,* **7**, 1139 (1988).
[97] J.E. Rothman and R.D. Kornberg, *Nature (London),* **322**, 209 (1986).

[98] R.J. Deshaies, B.D. Koch, M. Werner-Washburne, E.A. Craig and R. Shekman, *Nature (London),* **332,** 800 (1988).

[99] W.J. Chirico, M.G. Waters and G. Blobel, *Nature (London),* **332**, 805 (1988).

[100] H. Murakami, D. Pain and G. Blobel, *J. Cell Biol.,* **107,** 2051 (1988).

[101] H.R.B. Pelham, *Cell,* **46**, 959 (1986).

[102] H.R.B. Pelham, *Nature (London),* **332**, 776 (1988).

[103] D.I. Meyer, *Trends Biochem. Sci.,* **13**, 471 (1988).

[104] N. Pfanner, R. Pfaller, R. Kleene, M. Ito, M. Tropschug and W. Neupert, *J. Biol. Chem.,* **263**, 4049 (1988).

[105] S. Munro and H.R.B. Pelham, *Cell,* **46**, 291 (1986).

[106] C.K. Kassenbrock, P.D. Garcia, P. Walter and R.D. Kelly, *Nature (London),* **333**, 90 (1988).

# ELECTRICAL CURRENTS INDUCED BY ION PUMPS ON BLACK LIPID MEMBRANES

ERNST BAMBERG, KLAUS FENDLER

*Max-Planck Institut für Biophysik*
*Kennedyallee 70*
*D-6000 Frankfurt 70*

Contents



*Bioelectrochemistry III*
Edited by G. Milazzo and M. Blank
Plenum Press, New York, 1990

## 1. Introduction

Charge translocation in light-driven and ATP-driven ion-pumps has been studied on planar lipid membranes. The method consists of the adsorption of membrane fragments containing the ion-pump to a planar lipid film. As examples the light-driven pump Bacteriorhodopsin and the $Na^+K^+$-ATPase from pig kidney have been investigated. In a short circuit experiment the displacement and pump currents were detected *via* the capacitive coupling between the membrane fragments and the underlying planar film.

Bacteriorhodopsin was activated directly by light. The pump activity of the $Na^+K^+$-ATPase was initiated by use of a photolabile non-hydrolysable ATP derivative (caged ATP). After an U.V. flash the inactive ATP analog is converted to normal ATP [1], so that the $Na^+K^+$-ATPase is activated in a synchronised manner by the light induced concentration jump of ATP. The time needed for photolysis of the caged ATP at physiological pH values is in the millisecond range [2].

On the example of bacteriorhodopsin-containing purple membranes associated with a planar lipid membrane the composed membrane system is described in terms of an equivalent circuit diagram. Pump currents were measured under stationary conditions and the kinetics of these currents were determined and related to the photocycle of bacteriorhodopsin obtained by flash photolysis. The spectroscopic behaviour of bacteriorhodopsin is reviewed by Stoeckenius and Bogomolni [3]. In a second series of experiments the electrical pump currents of the $Na^+K^+$-ATPase, which is an enzyme of the $E_1$-$E_2$ type, were studied. The currents can be described within the consecutive model of the Albers post-reaction-scheme (see Fig. 12). During the phosphorylation of the enzyme 3 $Na^+$ are transported outward and 2 $K^+$ are transported into the cell during dephosphorylation.

## 2. Description of the compound membrane and stationary currents

After light-excitation, bacteriorhodopsin, the light-driven $H^+$ pump from *Halobacterium halobium* [4] returns to the ground state *via* a series of intermediates and thereby translocates protons through the membrane (Fig. 6A). The kinetics of this light-driven proton pump have been studied so far mainly with purple-membrane sheet in aqueous suspension [5]. From these studies the spectral characteristics as well as the lifetimes of a number of intermediates in the photocycle could be obtained. In order to understand the pumping mechanism, the spectroscopic transitions have to be correlated with the elementary steps of proton translocation through the bacteriorhodopsin molecule. Information on proton transfer kinetics may be obtained by studying photoelectric transients in oriented purple-membrane samples. Orientation has been achieved by a number of different methods, for instance by attaching purple membranes or bacteriorhodopsin-containing vesicles to lipid-impregnated filters, black lipid membranes [6–8], or to thin Teflon films [9, 10]. Another method consists of orienting purple-membrane sheets in water by an electric field [11].

When a purple membrane suspension is added to one aqueous side of the black film, photosensitivity develops in the course of minutes. A schematic representation of the experimental situation is given in Fig. 1. Records of the short-circuit photocurrent obtained about 1 h after addition of the bacteriorhodopsin are shown in Fig. 2. The first measurement (Fig. 2A) was carried out in the absence of protonophore. Under these conditions the black film has an extremely low conductivity on the order of 10 nS $cm^{-2}$)

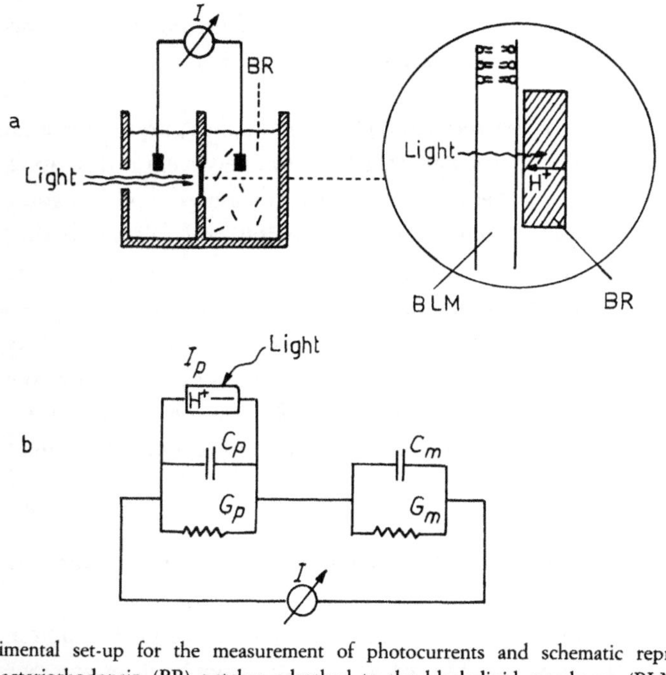


Fig. 1a – Experimental set-up for the measurement of photocurrents and schematic representation of the arrangement of bacteriorhodopsin (BR) patches adsorbed to the black lipid membrane (BLM). (b) Equivalent circuit for the BR-BLM system.

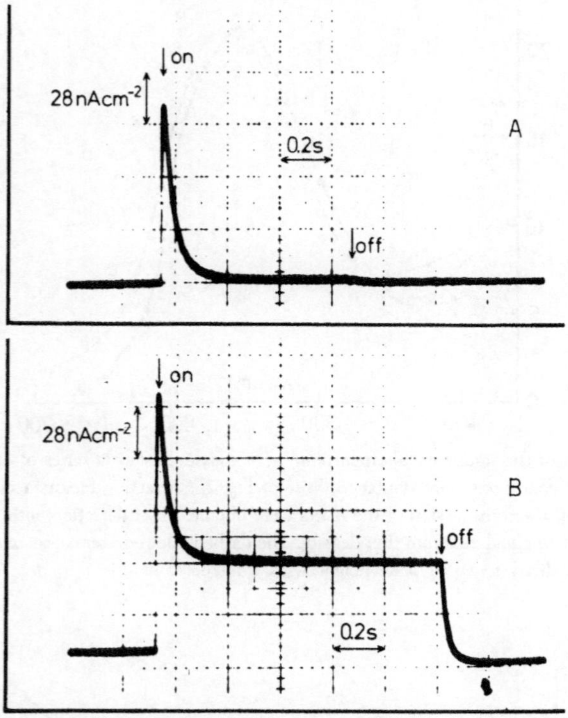


Fig. 2 - Short-circuit photocurrent after addition of purple membrane fragments to one aqueous compartment. The aqueous phase contained 0.1 *M* $MgCl_2$ and 0.5 m*M* Tris, pH 7.0. The light was filtered through a K4 filter (Balzers, $\lambda_{max}$ 550 nm).The area of the black film was 1.8 $mm^2$ A) Photocurrent in the absence of protonophore. The sign of the photocurrent corresponds to a proton transfer towards the bacteriorhodopsin-free side B) After addition of protonophore to the bacteriorhodopsin-free compartment. After switching off the light the current falls to a slightly negative value and approaches zero level (horizontal trace at left) within a few seconds.

and is virtually impermeable to protons. After switching on the light the current rises within less than 20 ms and thereafter declines with a time constant of the order of 50 ms to a low level. The sign of the photocurrent was always the same, corresponding to a proton transfer toward the bacteriorhodopsin-free side. In the second part of the experiment (Fig. 2B) a protonophore was added to the membrane system. It is seen from figure 2B that in the presence of the protonophore a large stationary photocurrent is observed after the decline of the early transient. The highest (quasi)-stationary photocurrents at light intensity of about 2 W/$cm^2$ were of the order of 0.1 μA/$cm^2$. If a continuous monolayer of purple membrane would be present on the planar film, the total charge corresponding to 1 $H^+$ per bacteriorhodopsin would be about 1.4 μC $cm^{-2}$. If a current of 100 nA $cm^{-2}$ flows for several minutes, the transferred charge is far in excess of 1.4 μC $cm^{-2}$. This means that a continuous working of the proton pump is observed in these experiments.

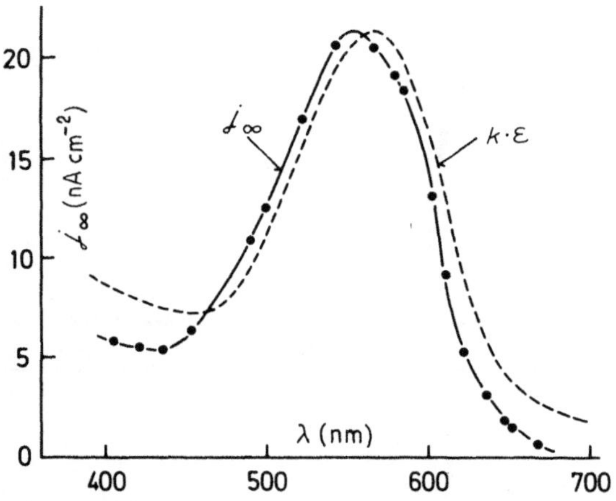


Fig. 3 - Action spectrum of the stationary photocurrent $j_\infty$, as measured with a series of narrow-band interference filters. The experimental conditions were similar to those of Fig. 2. Correction factors accounting for the emission spectrum of the lamp and the transmission of the filters were obtained by calibration with a bolometer. The action spectrum was normalized to equal quantum flux density. The dashed line represents the extinction coefficient of the purple membrane, normalized to equal peak height with a factor *K*.

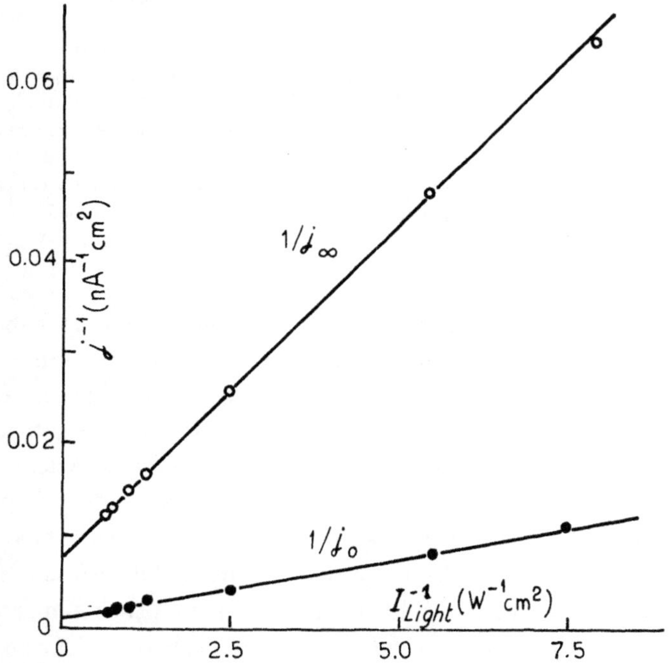


Fig. 4 - Reciprocal values of the initial current $j_o$ and of the stationary current $j_\infty$ as a function of reciprocal light intensity $I_{light}$. All measurements have been performed with the same membrane in the presence of protonophores; the aqueous phases contained 0.1 *M* $MgCl_2$ and 0.5 m*M* Tris, pH 7.0.

In Fig. 3 the action spectrum of the stationary photocurrent $j_\infty$, as obtained from measurements with a series of narrow-band interference filters, is compared to the optical absorption spectrum of the purple membrane. The action spectrum is slightly blueshifted (by about 15 nm), but otherwise closely agrees with the absorption spectrum. Both the initial current $j_0$ (which is observed after switching on the light) as well as the stationary current $j_\infty$ saturate with increasing light intensity. This is shown in Fig. 4 where the reciprocal values of $j_0$ and $j_\infty$ are plotted as a function of reciprocal light intensity. Both $j_0$ and $j_\infty$ may be represented by the equation:

$$j = j_{sat} \frac{I_{light}}{I_{light} + I_{light,\,1/2}} \tag{1}$$

where $j_{sat}$ is the saturation current and $I_{light,\,1/2}$ is the halfsaturation intensity which is approximately the same for $j_0$ and $j_\infty$ ($I_{light,1/2} = 0.5$ W cm$^{-2}$).

The observation that the stationary photo-current strongly increases upon addition of protonophores (Fig. 2) argues against the possibility that the purple membrane is incorporated into the black film. The most likely interpretation of the findings represented in Fig. 2 consists in the assumption that the purple-membrane sheets are attached to the black lipid film in a preferential orientation (Fig. 1a). If the proton permeability of the underlying black film is low, only a transient capacitive photocurrent occurs which creates an electric potential drop between the intermediate layer and the aqueous phase. On the other hand, after incorporation of proton-permeable channels or protonophore into the black film, a permanent photocurrent can flow between the aqueous phases. This method may be analyzed on the basis of an equivalent circuit (Fig. 1b) similar to that used by HERRMANN and RAYFIELD [12] for the interpretation of their experiments with purple membrane vesicles. Light absorbed by bacteriorhodopsin drives a pump current $j_p$ which tends to build up a potential difference $U_p$ across the purple membrane. In general, the pump current $j_p$ depends on potential difference $U_p$. As a first approximation, $j_p$ may be represented as a linear function of $U_p$ [12]:

$$j_p = j_{p,o} \left(1 - \frac{U_p}{U^*}\right) \tag{2}$$

where $j_{p,0}$ is the pump current for $U_p = 0$ and $U^*$ is a constant. (The term $U_p/U^*$ also accounts for gradients of proton activity built-up by the pump which should be roughly proportional to $U_p$). Under short-circuit conditions the relation $U_p = U_m$ holds, where $U_m$ is the potential difference across that part of the black lipid film which is covered by the purple membrane. $G_m$ and $C_m$ are the conductivity and the specific capacitance of the covered part of the black film, and $G_p$ and $C_p$ are the corresponding values of the purple membrane (undert short-circuit conditions where the total potential difference across the membrane vanishes, the uncovered parts of the membrane may be omitted in the circuit analysis). $G_p$ not only contains the conductivity across the purple membrane sheet but also accounts for any leakage pathway within the contact layer between black film and purple membrane. Simple circuit analysis may be used to calculate the time course of the externally measured current $j$. The result reads [13]:

$$j\,(t) = j_\infty + (j_0 - j_\infty)\ exp\ \left(\frac{t}{\tau}\right) \tag{3}$$

$$j_0 = j_{p,0}\ \frac{C_m}{C_m + C_p} \tag{4}$$

$$j_\infty = j_{p,0}\ \frac{G_m}{G_m + G_p + (j_{p,0} / U^*)} \tag{5}$$

where the quantity $\tau$ is given by equation (6):

$$\tau = \frac{C_m + C_p}{G_m + G_p + (j_{p,0} / U^*)} \tag{6}$$

and all other quantities $j_0$, $j_\infty$, $G_m$, $G_p$, $C_m$, $C_p$ are referred to unit area of the black film, $G_m$ and $G_p$ represent the conductivities of the black lipid membrane and the purple membrane, respectively, and $C_m$ and $C_p$ are the corresponding capacitances.

After switching off the light the time course of the current is given by $\tau_{off}$:

$$\tau_{off} = \frac{C_m + C_p}{G_m + G_p} \tag{7}$$

Implicit in the derivation of equations (3)-(10) is the assumption that the intrinsic time-constants of the pump are small compared with the time constants $\tau$ and $\tau_{off}$ of the equivalent circuit. As the turnover rate of the pump is of the order of 100 $s^{-1}$ [14] this assumption seems reasonable within the time scale of the experiments represented in Fig. 2.

Equations (3)-(6) approximately account for the observed photoresponse of the system. The undoped planar film has a low conductivity $G_m$ and therefore, according to equation (5), the stationary membrane current $j_\infty$ should be small. The conductivity $G_m$ can be increased by addition of the uncoupler carbonylcyanide-*p*-trifluoromethoxyphenyl-hydrazone (FCCP), which acts as a proton carrier. From previous studies [15] it is known that the FCCP-induced proton conductivity of the bilayer is proportional to the aqueous FCCP concentrations. According to equation (5), the quantity $1/j_\infty$ should be a linear function of $1/G_m$ (and therefore of $1/c$). A linear relationship between $1/j_\infty$ and $1/c$ was obtained indeed (data not shown), [13].

According to equation (6) the time constant $\tau$ for the decay of the photocurrent should depend on the pump current $j_{p,0}$ and therefore on the light intensity $I_{light}$. The relationship between $j_{p,0}$ and $I_{light}$ may be expressed by (compare equation 1):

$$j_{p,o} = j_{sat,p,o} \frac{I_{light}}{I_{light} + I_{light,\,1/2}} \tag{8}$$

where $j_{sat,\,p,Q}$ is the value of $j_{p,0}$ in the limit $i_{light} \rightarrow \infty$. Equation (6) may be then written as follows:

$$\frac{1}{\tau} = \frac{G_m + G_p}{C_m + C_p} + \frac{j_{sat,p,o}/U^*}{C_m + C_p} \bullet \frac{I_{light}}{I_{light} + I_{light,\,1/2}} \tag{9}$$

Figure 5 shows that the reciprocal decay time $1/\tau$ tends to saturate with increasing light intensity $I_{light}$, as predicted by equation (9).

## 3. Kinetics of the photocurrents

From Fig. 5 the value of the *RC* time $\tau_{off}$ of the compound membrane system can be read to be about 0.25 s. Optimal adjustment of the measuring system yields a time resolution of about 1 μs [16]. Therefore relaxations of the photocurrent should be observable in a time window of 1 μs – 0.25 s. The time-dependent photocurrent after excitation with a short laser flash is given by:

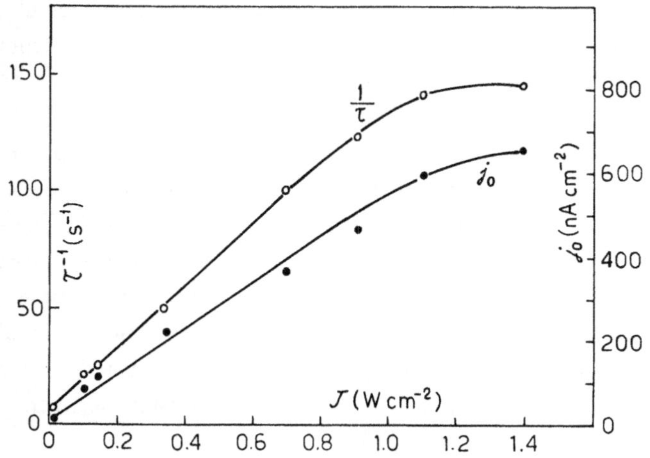


Fig. 5 - Reciprocal decay time $1/\tau$ of the photocurrent as a function of light intensity $I_{light}$. The aqueous phase contained 0.1 *M* $MgCl_2$ and 0.5 m*M* Tris, pH 7. In addition, the corresponding values of the initial current $j_o$ are plotted which have been measured in the same experiment.

$$j\,(t) = \frac{C_m}{C_p+C_m}\,\Sigma_t\, j_{p,o,i}\, exp(-\frac{t}{\tau_i}) \qquad (10)$$

where $j_{p,0,i}$ is the amplitude of the exponential component $\tau_i$ of the current generated by the proton pump at zero potential difference across the purple membrane.

The time course of the photocurrent is shown in Fig. 6b. The polarity of the current is defined as indicated in Fig. 1, *i.e.*, $j > 0$ corresponds to a motion of positive charge towards the supporting membrane. During the first 10 µs after the flash a negative current is observed which rises steeply towards positive values and thereafter declines to zero. (The finite slope of the falling phase of $j(t)$ within the first 5 µs is an artifact caused by the time constant of the circuit which was about 3 µs in this case). The shape of $j(t)$ for $t > 5$ µs can be represented with high accuracy by an expression consisting of a sum of four exponential terms:

$$j\,(t) = \sum_{i=1}^{4} a_i\, exp\,(-\frac{t}{\tau_i}) \qquad (11)$$

where $a_i$ is the amplitude of the exponential component $\tau_i$ of the measured current. Using three instead of four exponential terms does not adequately fit the experimental $j(t)$ curves, *i.e.* with a three-exponential fit the deviation between theoretical and experimental curves was consistently larger than the experimental scatter. On the other

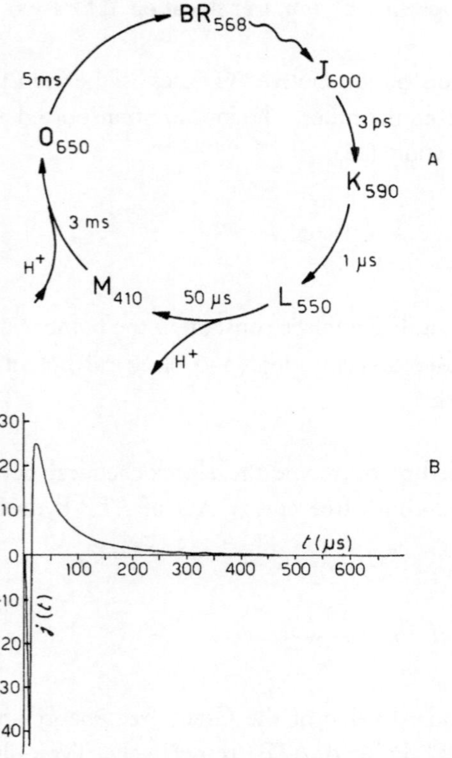


Fig. 6A- Photocycle of bacteriorhodopsin with the intermediates BR, K, L, M and O. The subscript denotes the absorption wavelength of the particular intermediate. The lifetimes of the intermediates vary according to the conditions. Typical values are given in the figure. For the M decay, which has to be described by 2 time constants of similar value only an overall lifetime is given.

B) Photocurrent $j(t)$ after a 10-ns laser flash at 575 nm. The signal represents an average over 3000 single excitations. A qualitatively identical signal but superimposed with considerable noise is obtained in a single-flash experiment. The energy of the laser flash was 2 μJ, corresponding to 4 mJ/cm$^2$. The temperature was $T$ = 22 °C. For times > 10 μs a highly accurate fit of the experimental $j(t)$ curve is obtained with $\tau_1$=3.9 μs, $\tau_2$=21.0 μs, $\tau_3$=64.4 μs, $\tau_4$=420 μs, $a_1$=−708 nA, $a_2$=56.8 nA, $a_3$=17.6 nA, $a_4$=1.23 nA. The fitted curve (not shown) is almost indistinguishable by eye from experimental curve for $t$ > 10 μs (apart from the noise components).

hand, introducing further terms does not significantly improve the fit. The analysis was therefore carried out with four exponential functions throughout. In almost all experiments the calculated fit curve was identical with the experimental curve for times $t$ > 10 μs within the width of the drawing line in a plot such as Fig. 6B. Comparing these results with the photocycle of Bacteriorhodopsin (Fig. 6A) a good agreement with the optically observed transitions K→ L→M→ BR was obtained. (Note that 2 time-constants are assigned to the L→ M transition [16]).

## 4. Electrogenic properties of ion translocating ATPases

The basic function of ion-motive ATPases is the transport of charge from one compartment of the cell to the other. The ions are transported against an electrochemical potential given by equation (12):

$$\Delta\tilde{\mu} = RT \ \text{In} \frac{c'}{c''} + n \ \mathsf{F} \ (\phi' - \phi'') \tag{12}$$

where $R$ is the gas costant, $\mathsf{F}$ Faraday's constant, $n$ the number of transported charges and $c'$, $c''$ *and* $\phi'$, $\phi''$ are the concentrations and electrical potentials, respectively, at both sides of the membrane.

The energy needed to overcome the electrochemical potential is provided by the hydrolysis of ATP. The GIBBS free energy $\Delta G$ of ATP hydrolysis is given by equation (13):

$$\Delta G = \Delta G_o + RT \ \ln \frac{c_D c_P}{c_T} \tag{13}$$

where $\Delta G_o$ is the standard value of the GIBBS free energy and $c_D$, $c_P$, and $c_T$, are the concentrations of ADP, Pi, and ATP, respectively. Typically in an animal cell $\Delta G$ amounts to 60 kJ/mol corresponding to about 600 meV/molecule [17].

The equilibrium potential, or reversal potential, of an ion translocating ATPase is defined as the potential difference at which the free energy of ATP hydrolysis and the work required for the reversible stoichiometric transfer of ions are equal and opposite [18]. The NaK-ATPase for example exchanges 3 $Na^+$ ions against 2 $K^+$ ions. With typical ion gradients of 10 and 30 for $Na^+$ and $K^+$ and a GIBBS free energy of ATP hydrolysis of 600 meV/molecule, equation 12 yields a reversal potential difference of 240 mV.

While the reversal potential only establishes the thermodynamic limit for the function of the ion pump, more detailed information may be obtained from the steady state current *versus* potential difference behaviour of the transport system. In particular predictions about the number of potential-difference-sensitive steps are possible under certain conditions [19].

Since the reaction cycle of an ion translocating ATPase consists of many partial reactions each of which may be electric-potential-sensitive, current *versus* potential

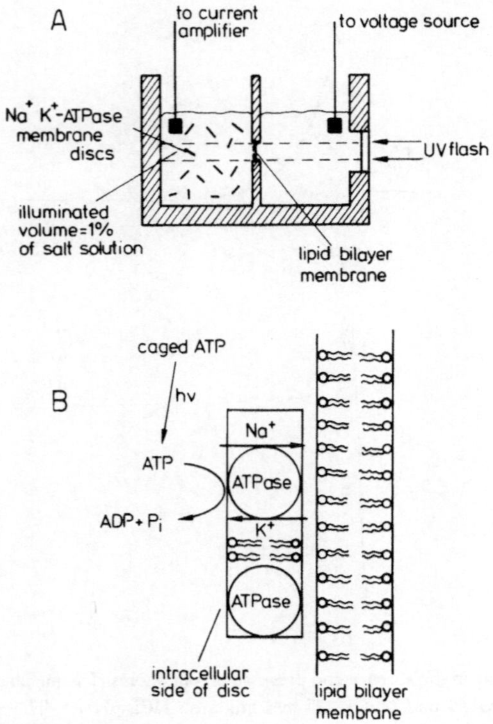


Fig. 7 - Schematic representation of the bilayer set-up.
A) Teflon chamber with black lipid membrane (BLM) and adsorbed ATPase membrane fragments.
B) Proposed sandwich-like arrangement of discs and underlying lipid membrane. The two membranes are capacitively coupled.

difference curves may be complicated. For the same reason, interpretation in terms of an explicit kinetic scheme is not straightforward. For the $Na^+K^+$-ATPase a variety of models has been proposed to predict its current *versus* potential difference behaviour (see *e.g.* Refs. 17, 20-22). Other ATPases have attracted much less attention since the experimental evidence is still very poor.

## 5. Pump currents of the $Na^+K^+$-ATPase

Adsorption of membrane fragments containing the ATPase to a BLM offers the possibility to monitor directly the charge translocation taking place during the ATPase pumping cycle. Pump currents were initiated by a light-induced concentration jump of ATP *via* the conversion of caged ATP into its active form. With this method the electrogenic transport activity of the $Na^+K^+$-ATPase could be detected with high sensitivity (0.5 pA) and high time resolution (1 ms). The experimental situation is shown schematically in Fig. 7. 15 min after addition of ATPase and caged ATP, a current could

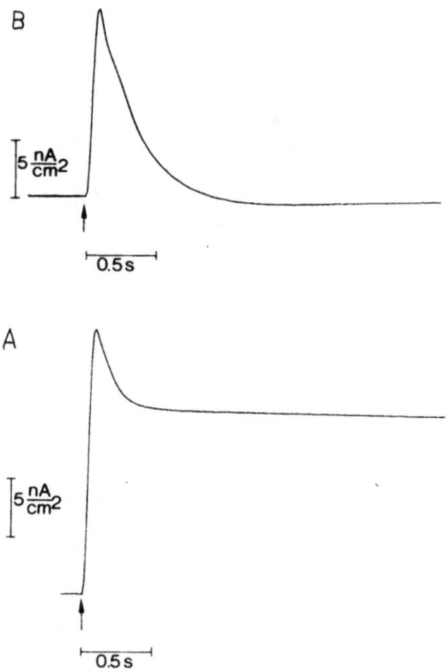


Fig. 8 - Short circuit currents in the absence and presence of ionophores. The membrane bathing solution contained 130 m*M* NaC1, 20 m*M* KC1, 3 m*M* $MgC1_2$, 25 m*M* imidazole-HCl, pH 7.5. The arrows indicate opening of the shutter, which was opened for 125 ms.

A) 100 μ*M* caged ATP, no addition of ionophores.

B) Same membrane system after addition of monensin and 1799 to a concentration of 10 μ*M* and 0.6 μ*M*, respectively.

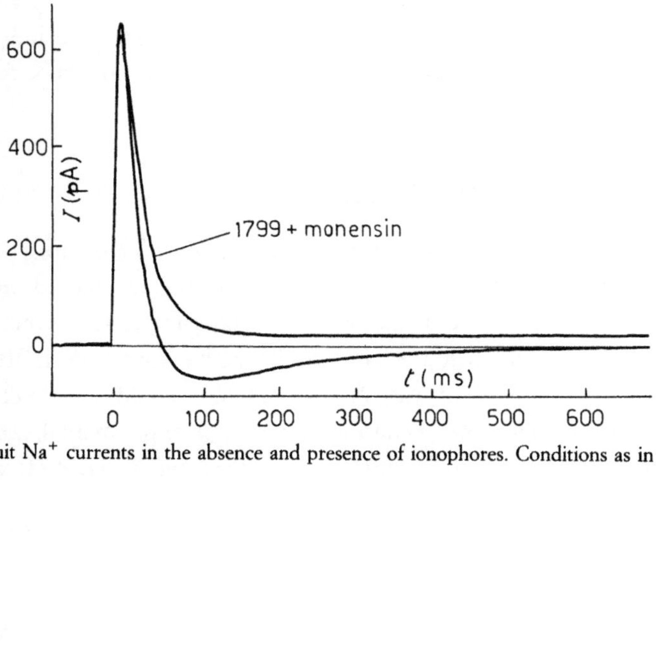


Fig. 9 - Short circuit $Na^+$ currents in the absence and presence of ionophores. Conditions as in Fig. 8 but without KC1.

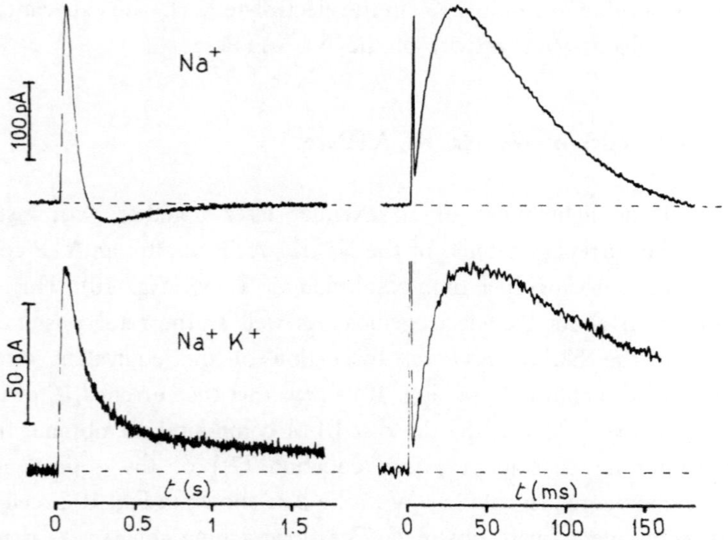


Fig. 10 - Short circuit currents without and with $K^+$. The signal was recorded simultaneously with two different sample rates (left hand side and right hand side). Conditions: 3 m*M* $MgC1_2$, 25 m*M* imidazole-HC1 at pH 6.2, 7 μ*M* ATP released per flash, 130 m*M* NaC1 and 20 m*M* KC1 (lower traces).

be induced by an UV light pulse. In the absence of ionophores the current is transient, whereas the permeabilisation of the underlying lipid bilayer yields a sustained current (Fig. 8) which decays slowly to zero due to the consumption and the dilution of the liberated ATP into the non-irradiated volume. Because of their insensitivity to UV light a carrier system consisting of monensin as an alkali metal ion/$H^+$ exchanger together with the protonophore 1799 was used for these experiments.

The sign of the current was always the same and corresponded to a positive charge motion towards the protein-free side of the BLM. The discs are preferentially oriented with the extracellular side facing the planar bilayer. $Mg^{2+}$ was necessary for the appearance of pump currents. If not otherwise indicated, in all experiments a concentration of free $Mg^{2+}$ of 3 m*M* was used. Together with additional control experiments [23] these experiments show that the observed current reflects the pumping activity of the $Na^+K^+$-ATPase.

The appearance of stationary $Na^+$ pump currents can prove that the enzyme acts as an electrogenic $Na^+$ pump. They can be detected on the sandwich-like structure by increasing the conductivity of the supporting lipid bilayer with an appropriate carrier system. A stationary $Na^+$ pump current was obtained indeed after the addition of monensin and the protonophore 1799 to the membrane system (Fig. 9). It can be shown that this effect is not due to residual $K^+$ in the electrolyte [24]. The experiment therefore demonstrates the electrogenic activity of the $Na^+$-ATPase.

## 6. Transient kinetics of the $Na^+K^+$-ATPase

Using a 10 ns light pulse of an excimer laser together with fast recording instrumentation the current generated by the $Na^+K^+$-ATPase after an ATP concentration jump could be measured with a time resolution of 1 ms (Fig. 10). This experiment provides information about the electrogenicity as well as the rate constants of partial reaction steps of the $Na^+K^+$-ATPase. Inspection of the equivalent circuit of the disc-BLM compound membrane (see Fig. 10) shows that the current $I_p(t)$ of the pump is distorted by the network formed by the disc-BLM compound membrane. In particular the characteristic time constant $\tau_{off}$ [see equation (7)] of the equivalent circuit is introduced in the measured signal [16]. With the data shown in Fig. 10 excellent fits with 3 exponential components were obtained. The slowest time constant $\tau_3$ determined by this way has to be assigned to the characteristic time constant $\tau_{off}$ of the equivalent circuit [25]. This is further supported by the experiment shown in Fig. 9. If the membrane conductivity is increased by the addition of ionophores, the slow negative component $\tau_3$ of the $Na^+$ current disappears.

The fast two time-constants $\tau_1=10$ ms and $\tau_2=83$ ms (values given for pH 6.2 and saturating ATP concentration) reflect charge translocations during the reaction cycle of the enzyme [25]. Similar results have been recorded in the absence of $K^+$ [26,27]. Because $\tau_3$ is much larger than $\tau_1$ and $\tau_2$, distortion of the pump current by the network is small and can be neglected. The two intrinsic time constants of the $Na^+K^+$-ATPase found in the photoinduced current suggest a two-step first-order reaction sequence with an electrosilent transition followed by an electrogenic one. The current transient generated by this model is given by equation (14):

$$I(t) = \frac{A}{\tau_2 - \tau_1}\left[\exp\left(-\frac{t - t_D}{\tau_2}\right) - \exp\left(-\frac{t - t_D}{\tau_1}\right)\right] \tag{14}$$

where the adjustable parameter $A$ was taken to be the same at all ATP concentrations. The delay time $t_D$ was introduced to account for the kinetics of ATP release from caged ATP. This function was fitted to the data at different ATP concentrations (Fig. 11). The saturating behaviour of $1/\tau_2$ shows that substrate binding is involved in this step and that it consequently has to precede the reaction characterized by $\tau_1$. The situation is, however, complicated by the fact that caged ATP is a competitive inhibitor of the $Na^+K^+$-ATPase [24,28].

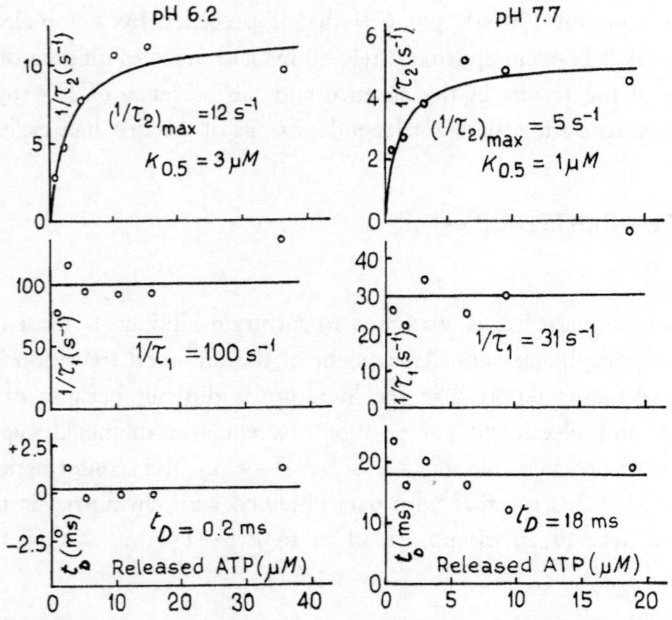


Fig. 11 - ATP dependence of the fit parameters of equation 14. The ATP dependence of the inverse time constants $1/\tau_1$ and $1/\tau_2$ and of the delay $\tau_D$ (see equation 14) of the UV flash induced current was measured at pH 6.2 and pH 7.7. The electrolyte contained 130 m*M* NaC1, 20 m*M* KC1, 3 m*M* $MgC1_2$ different amounts of caged ATP and 25 m*M* imidazole at pH 6.2 and pH 7.7, respectively. For $1/\tau_2$ the solid line represents a fit of a MICHAELIS MENTEN type concentration dependence to the data. For $\overline{1/}\ \tau_1$ and $t_D$ the solid line is a horizontal line at the average value $\overline{1/\tau_1}$ and $t_D$ respectively.

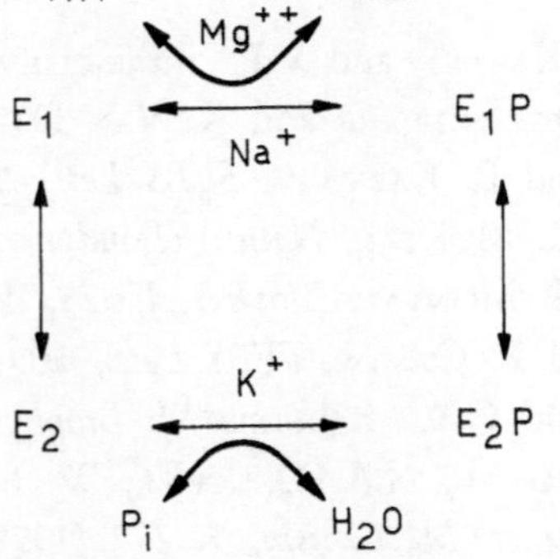


Fig. 12 - Reaction cycle of the $Na^+K^+$ATPase (ALBERS-POST scheme). $E_1$ and $E_2$ correspond to the two conformations of the enzyme, EP, to the phosphorylated form.

The arguments given above support a kinetic model consisting of a fast electrogenic step with rate constant 100 $s^{-1}$ (pH 6.2) that is preceded by a non-electrogenic slow process which takes place in approximately 80 ms and involves binding of the substrate. The similarity of the results in the absence and the presence of $K^+$ suggests that the current is generated during the $Na^+$ dependent steps of the first half cycle (see Fig. 12):

$$E_1 + ATP \rightarrow E_1.ATP \rightarrow E_1P \rightarrow E_2P$$

As already discussed $\tau_2$ is attributed to substrate binding. $\tau_1$ must then be due to reaction of the phosphoenzymes. Assignment of the observed transition by comparison with the rate constants reported in the literature is difficult because of differences in enzyme origin and electrolyte composition. Recent measurements suggest, that the electrogenic reaction step of the $Na^+K^+$-ATPase is the conformational transition $E_1P \rightarrow E_2P$ [29,30]. This together with data obtained with chymotrypsin-treated enzyme [26] argues in favor of an assignment of $\tau_1$ to $E_1P \rightarrow E_2P$.

## References

[1] J.H. Kaplan, B.I.I.I. Forbush and J.F. Hoffman, *Biochemistry,* **17,** 1929 (1978).
[2] J.A. McCray, L. Herbette, T. Kihara and D.R. Trentham, *Proc. Natl. Acad. Sci. USA,* **77,** 7237 (1980).
[3] W. Stoeckenius and R.A. Bogomolni, *Ann. Rev. Biochem.,* **52,** 587 (1982).
[4] D. Oesterhelt and W. Stoeckenius, *Nature (London),* **233,** 149 (1971).
[5] W. Stoeckenius, R.H. Lozier and R.A. Bogomolni, *Biochim. Biophys. Acta,* **505,** 215 (1979).
[6] L.A. Drachev, D. Kaulen and V.P. Skulachev, *FEBS Lett.,* **87,** 161 (1978).
[7] M.C. Blok, K.J. Hellingwerf and K. Van Dam, *FEBS Lett.,* **76,** 45 (1977).
[8] Z.N. Dancshazy and B. Karvaly, *FEBS Lett.,* **72,** 136 (1976).
[9] H.W. Trissl and M. Montal, *Nature (London),* **266,** 655 (1977).
[10] F.T. Hong and M. Montal, *Biophys. J.,* **25,** 465 (1979).
[11] L. Keszthelyi and P. Ormos, *FEBS Lett.,* **109,** 189 (1980).
[12] T.R. Herrmann and G.W. Rayfield, *J. Biophys.,* **21,** 111 (1978).
[13] E. Bamberg, H.J. Apell, N.A. Dencher, W. Sperling, H. Stieve and P. Läuger, *Biophys, Struct. Mechanism,* **5,** 277 (1979).
[14] R.H. Lozier, R.A. Bogomolni and W. Stoeckenius, *J. Biophys.,* **15,** 955 (1975).
[15] F.S. Cohen, M. Eisenberg and S. McLaughlin, *J. Membr. Biol.,* **37,** 361 (1978).
[16] A. Fahr. P. Läuger and E. Bamberg, *J. Membr. Biol.,* **60,** 51 (1981).

[17] P. de Weer, *Fortschr. Zool.,* **33,** 387 (1986).
[18] P de Weer, D.C. Gadsby and R.F. Rakowski, *Annu. Rev. Physiol.,* **50,** 225 (1988).
[19] A.V. Lafaire and W. Schwarz, *J. Membr. Biol.,* **91,** 43 (1986).
[20] J.B. Chapman, E.A. Johnson and J.M. Kootsey, *J. Membr. Biol.,* **74,** 139 (1983).
[21] J.A. Reynolds, E.A. Johnson and C. Tanford, *Proc. Natl. Acad. Sci. USA,* **82,** 6869 (1985).
[22] P. Läuger and H.J. Apell, *Biochim. Biophys. Acta,* **945,** 1 (1988).
[23] K. Fendler, E. Grell, M. Haubs and E. Bamberg, *EMBO J.,* **4,** 3079 (1985).
[24] G. Nagel, K. Fendler, E. Grell and E. Bamberg, *Biochim. Biophys. Acta,* **901,** 239 (1987).
[25] K. Fendler, E. Grell and E. Bamberg, *FEBS Lett.,* **224,** 83, (1987).
[26] R. Borlinghaus, H.J. Apell and P. Läuger, *J. Membr. Biol.,* **97,** 161 (1987).
[27] H.J. Apell, R. Borlinghaus and P. Läuger, *J. Membr. Biol.,* **97,** 179 (1987).
[28] B. Forbush III in *The sodium pump,* I.M. Glynn and C. Ellory (Editors), Comp. of Biologists Ltd, G.B (1985), p. 599
[29] R. Goldshlegger, S.J.D. Karlish and A. Rephaeli, *J. Physiol.,* **387,** 331 (1987).
[30] M. Nakao and D.C. Gadsby, *Nature (London),* **323,** 628 (1986).

# THE MEASUREMENT OF SURFACE POTENTIALS AND TRANSMEMBRANE POTENTIALS IN CELLS AND ORGANELLES

HAGAI ROTTENBERG

*Pathology Department, Hahnemann University*
*Philadelphia, PA. USA*

Contents



*Bioelectrochemistry III*
Edited by G. Milazzo and M. Blank
Plenum Press, New York, 1990

## 1. Introduction

Most living cells and organelles carry discrete charges on their membrane surface, which result in the presence of surface potential differences (shortly called *surface potentials*) between membrane surface and the contacting bulk medium. In addition, the selective active transport of ions and the resulting ion concentration differences lead to the formation of transmembrane potential differences (shortly called *membrane potentials*). Both surface potentials and membrane potentials play important roles in metabolism, control and signaling processes in cells and organelles. Thus, the ability to measure these various potentials and to distinguish between them, is crucial to the understanding of many cellular and organelles's processes. In this Chapter I shall review briefly the most useful methods available at present, and discuss their principles and their applications.

## 2. The measurement of surface potential

Biological membranes carry discrete electrical charges at their surface. These charges are carried by acidic phospholipids, gangliosides and by membrane proteins, most of which are also negatively charged. Another significant contribution to the membrane surface charge, which tends to reduce the net negative charge, arises from the surface binding and adsorption of various charged molecules and ions (mostly divalent cations). These discrete charges induce electrostatic potentials in the aqueous phase adjacent to the membrane surface [1]. Biological fluids and the cytoplasm contain relatively high concentrations of mono– and divalent cations, which would largely screen these charges and limit the effective surface potential to a distance of only a few nm. Nevertheless, this potential is believed to play a role in regulating a variety of important membrane processes [1–3]. Because of the great complexity and severe limitations imposed by the nature of biological membranes, a careful selection of an appropriate technique and a prudent interpretation of the results is necessary [4].

To begin with, the distribution of phospholipids, proteins, gangliosides and other charged species between the two membrane surfaces is asymmetric. Hence, the density and type of charges on one surface of the membrane are, in general, different from those on the opposite surface. It is, therefore, necessary to select methods that allow an estimation of the surface potential on each separate membrane surface. Secondly, in some biological membranes, protein clustering, aggregation and lateral segregation of phospholipids may result in non–uniform distribution of charges. In such cases, a method which yields a measurement of an average delocalized surface potential may not be informative. Thirdly, the charges on biological membranes may be located at various distances from the phospholipid headgroups, which form the surface separating the membrane from the aqueous phase. Gangliosides and membrane proteins project into the aqueous phase up to a distance of several nm. A charge located at a distance from the phospholipid headgroup produces a potential which would be sensed differently by a probe located near the phospholipids headgroups or on an adjacent site. For membranes in which a major fraction of their surface is occupied by proteins (*e.g.*, the inner membrane of the mitochondria), the definition of a surface becomes blurred and the measured potential differences would strongly depend on the localization of the probes. Finally, it must be remembered that in intact biological membranes, various processes, such as enzymatic reactions, active ion transport and the existence of membrane potentials and pH differences may greatly distort the information obtained in measurements of surface potential. This may pose a severe limitation since it is often desired to measure changes in surface charge in relation to membrane processes.

### 2.1. *Electrophoretic mobility*

This method, which depends on the measurement of the velocity of migration of charged particles in an electric field, is quite useful in model systems of uniform phospholipid particles. However, the method is of limited use with intact biological membranes. This is because the charge distribution and charge location in relation to the membrane solution interface are not sufficiently uniform and the surface not sufficiently smooth to satisfy the assumptions of the HELMHOLTZ-SMOLUCHOWSKI theory upon which the calculation of the potential is based. The calculated $\zeta$ potential from measurement of electrophoretic mobility of mitochondria is compared with other methods in Table 1.

Most other methods for the measurement of surface potential (and the calculation of surface charge density) depend on the potential–induced accumulation of probe ions near the charged surface. These methods are more applicable to biological membranes, but careful choice of the appropriate method is required. The most common difficulty in applying these methods to biological membranes is the rather high permeability of most biological membranes to charged molecules. A common error is to assume that a probe that was shown to be relatively impermeant in liposomes is similarly impermeant in all membranes. In general biological membranes are, orders of magnitude, more permeable to

Table 1 – Surface potentials in rat liver mitochondria

| Method | Probe | External pH | $\phi_s$(mV) | Reference |
|---|---|---|---|---|
| Electrophoretic mobility | | 7.4 | −21 | 12 |
| Fluorescence | ANS | 7.2 | −21 | 10 |
| Phosphorescence | $Tb^{3+}$ | 7.2 | −5 to −7 | 16 |
| e.s.r. | $Cat_{12}$ | 7.2 | −20 | 21 |
| e.s.r. | $Cat_{12}$ | 7.2 | −15 | 19 |

ions and charged amphiphilic molecules than liposomes. This is probably the result of the effect of integral membrane proteins on the membrane dielectric constant, which reduces the potential barrier in the core of the membrane, thus facilitating charge transfer [5]. This increased permeability is particularly problematic if it is desired to measure the surface potential in a system where membrane potentials, pH differences or other ion concentration differences exist or are induced during the measurements. In such cases, the measured response is often a result of the latter processes, which may affect both the probe distribution across the membrane and its binding to the membrane. Several cases of such complications are described below.

### 2.2. *Methods for measuring directly the accumulation of probe ions near the membrane surface*

The electrostatic potential difference, $\Delta\phi_s$ arising from the membrane surface charge leads to accumulation of ions of opposite charge according to the induced field. When the ions are at electrochemical equilibrium their excess concentration near the surface is related to the surface potential by the BOLTZMANN relationship:

$$\frac{c_s}{c_0} = \exp\left(-\frac{z\mathsf{F}\Delta\phi_s}{RT}\right) \tag{1}$$

where $c_o$ is the concentration in the bulk phase, $c_s$ the concentration at the surface, $z$ the ion charge, $\phi_s$ *the surface potential,* F the faraday constant, $R$ the gas constant and $T$ the absolute temperature. Because the direct measurement of $c_s/c_0$ is usually quite difficult, these methods are not widely used.

### 2.2.1. Resonance raman dye probes

In this method the accumulation of a positively-charged styryl dye (WW-638), which aggregates when present at high concentration near the membrane surface, is estimated from the aggregation-induced changes in its resonance raman spectrum. The method has been used to estimate the surface charge of purple membranes and liposomes reconstituted with bacteriorhodopsin [6]. Although the dye was shown to be relatively impermeant in liposomes, its permeability in intact biological membranes is unknown and should be assessed before application (see below).

### 2.2.2. Fluorescent acridines

The fluorescence of positively-charged acridines is quenched when aggregated near the surface. This method is in principle identical to the method described above. 9–Amino–acridine quenching was used to probe surface potential in chloroplasts [7,8]. This probe should be used with the utmost caution. Firstly, the probe permeates the membrane readily in its neutral form and distributes according to the pH gradient [9]. The accumulation of the probe in a compartment of low pH also results in fluorescence quenching. Thus, the finding that the addition of salts to a chloroplast suspension increases fluorescence, may be due, in part, to the effect of salts on the membrane DONNAN–potential, which is responsible for the existence of a substantial pH gradient in chloroplasts even in the dark [9]. Also, it is now recognized that 9–amino–acridine binds to membranes, which also induces quenching and modulates the surface charge, thus further complicating the interpretation of these results.

## 2.3. *Methods which depend on the enhanced membrane binding of a charged probe due to its potential–induced accumulation adjacent to the membrane surface*

These methods are based on the same principle as the methods described above (Section 2.2.) except that instead of a direct determination of the probe accumulation near the surface, one measures the enhancement of probe binding that results from the increased accumulation. When the binding characteristic is analyzed, the surface potential can be estimated from its effect on the apparent dissociation constant of the probe.

$$K_d = K_{0,d} \; exp \; (\frac{zF\phi_s}{RT}) \tag{2}$$

hence

$$\Delta\phi_s = \frac{RT}{zF} \ln \frac{K_d}{K_{0,d}} \tag{3}$$

where $K_d$ is the apparent binding constant in the presence of surface charge ($\phi_s$). Depending on the membrane polarity (which is usually negative, $-\Delta\phi_s$), and the probe charge (either $+z$ or $-z$), one would get enhancement of binding (lower $K_d$) or attenuation (high $K_d$), respectively. This analysis holds for those probes that bind to saturable binding sites. In practice it is not always necessary to determine the values of $K_d$. Provided that this value is known approximately, and that the assay is carried out at a concentration well below the value of $K_d$, the extent of binding is proportional to the $K_d$. Hence, $\Delta\phi_s$ can be calculated from the extent of binding ($m$) in the presence and absence ($m_0$) of surface potential:

$$\Delta\phi_s = \frac{RT}{zF} \ln \frac{m}{m_0} \tag{4}$$

Of course, the latter equation can also be used with probes which do not show saturation binding, since in this case the extent of binding is proportional to the partition coefficient. Commonly, the determination of probe binding in the addition of high concentrations of salts which efficiently screen the surface charge ($m_0$ or $K_{0,d}$) is accomplished by the addition of high concentrations of salts which efficiently screen the surface charge and abolish the surface potential. However, with many biological membranes this technique presents a problem, since many ions, particularly cations, bind to the membrane surface, and thus directly modulate the surface charge. If the binding parameters of the screening ion are known, a correction can be made. A simple, but not a perfect, alternative way to overcome this problem is to perform a *null point* titration of surface potential [10].

Another common complication in the application of these methods to biological membranes is the fact that the binding of the charged probe itself modulates the potential [11]. This can be taken into account when calculating charge density since the binding is usually measured. However, it is preferable, whenever possible, to choose a probe that can be measured with high sensitivity at surface concentration which is negligible when compared to the membrane charge density.

### 2.3.1. Nuclear magnetic resonance

In this technique, one measures the potential–induced binding of paramagnetic transition metal ions to the phospholipids headgroups from their effects on the $^{31}P$–n.m.r. signal. The limitations of this technique, when applied to intact biological membranes, are considerable. Firstly, the narrow $^{31}P$–n.m.r. signal is obtained only in small sonicated vesicles. Secondly, some of the ions used are either transported by the $Ca^{2+}$ carrier (*i.e.*, manganese) or strongly inhibit various membrane processes, which may lead to further complications.

### 2.3.2. Fluorescent and phosphorescent probes

In this method the binding of charged probes to the membranes is detected by the fluorescence (or phosphorescence) of the bound probe. To be useful it should be possible to distinguish clearly between the bound and free species. A large difference in quantum efficiency, emission wavelength, or both, enables one to monitor changes in the amount of bound probe, free probe, or both.

(*a*) ANS and TNS. These anionic fluorescent molecules (1–anilinonaphtalene–8–sulfonate and 2–(*p*–toluidinyl)–napthalene–6–sulfonate) are the most widely used probes for surface potential determination [10–14]. The hydrophobic anion fluorescence is greatly enhanced and exhibits a strong blue shift on binding to hydrophobic sites, thus enabling easy determination of the amount of bound probe. Although these probes are also widely used in biological membranes, there are many complications which often result in erroneous determination. Unlike artificial phosholipid membranes, there are high–affinity sites for probe binding which are located on membrane proteins. These sites, which often contribute most of the fluorescence at low probe concentrations, may be located at various distances from the membrane surface, either in the membrane core or projecting into the aqueous bulk. Moreover, these sites are often heterogeneous, with different affinities and extremely different quantum yields. Some may not contribute at all to the fluorescence but strongly affect the binding parameters [10]. Finally, although these probes are relatively impermeant in liposomes they are quite permeant in biological membranes, and therefore respond to the membrane potential. Indeed, these same probes are used to estimate the membrane potential in vesicular membrane systems having a positive membrane potential [10,15].

(*b*) $Tb^{3+}$. This phosphorescent cation of the lanthanide series can also be used to measure surface potentials [16]. It binds with high affinity to high affinity $Ca^{2+}$ binding sites. This probe also binds to membrane proteins so that the same reservations discussed above in regard to ANS apply here. However, in mitochondria, at least, there are fewer types of binding sites and these sites are easily separated. The protein–bound probe is estimated from the phosphorescence which is excited by fluorescence energy transfer from the protein. With this probe it is, therefore, possible to measure surface potentials at specific sites. This may result in more interesting and biologically relevant information. The main drawback of this probe is its avidity in forming insoluble complexes with a variety of anions, particularly inorganic phosphate and phosphate esters. This can present a serious problem, as it severely limits the choice of medium composition. In mitochondria, the problem is particularly acute since the matrix contains large amounts of phosphate which slowly leaks out during the course of the measurement. We have solved this problem by briefly treating the mitochondria with N–ethyl–maleimide [(NEM) (40 nmole/mg protein, 1 min at room temperature)] to inhibit phosphate transport. Alternatively, this problem can be minimized by performing the assay quickly at low temperatures (below 10 °C), or by depleting the mitochondria of phosphate before the determination. The problem does not exist with submitochondrial particles or other membrane preparations that do not leak phosphate. In mitochondria there are two sites for $Tb^{3+}$ binding. One with high affinity (1.5 μ*M*) and low capacity and the other with relatively lower affinity (29 μ*M*) and higher capacity. We found the low affinity site more suitable for surface potential determination. The sites are easily distinguishable because of different phosphorescence decay times. With a commercial phosphorescence spectrometer (Perkin-Elmer LS5), it is possible to select a window in which only the phosphorescence of the slow decay site is detected. The lower potential difference measured, compared with other probes, probably indicates that the $Tb^{3+}$ is bound to a protein which projects out of the phospholipid surface.

### 2.3.3. e.s.r. probes

In this method a charged aliphatic compound, which is labelled with a spin moiety (usually doxyl), is used to measure the free and bound species from the differences in their e.s.r. spectra. The most serious problem in applying these probes to biological systems is the difficulty in finding a probe which would be sufficiently hydrophobic to allow detection of the signal from the bound species at practical membrane concentrations and at the same time be impermeant. Since permeability increases with partition coefficient and the permeability of biological membranes is much greater than the one of liposomes, it has not been possible, as yet, to identify with certainty such probes. The widely used $Cat_{12}$ probe (4–N–N,N–dimethyl–N–dodecyl)–ammonium–2,2,6,6–tetramethyl–piperidine–1–oxyl), which is sufficiently impermeant in liposomes [17], was found to be permeant in red blood cells [18] and even more so in mitochondria [19,20]. Thus its use for determination of putative potential–induced changes in surface charge

[21,22] could not be justified, since the probe clearly responds to the induced membrane potential [19]. Another drawback of the spin probe technique is its relatively low signal sensitivity as compared to fluorescent probes. Finally, the nitroxy group is reduced quickly [23] if the membrane contains even very weak redox activity. The latter problem can be partially solved by lowering the temperature, and the addition of specific enzyme inhibitors and strong oxidants such as ferricyanide. An advantage of this probe over the above fluorescent probes is that it partitions into the lipid phase, thus allowing a determination of the surface potential adjacent to the phospholipid headgroup.

#### 2.3.4. Surface pH indicators

This method is based on the fact that the existence of a surface charge would change the concentration of protons near the surface. A pH indicator (which binds protons) and is located at or near the surface) can report these changes. 4–Heptadecyl–4–hydroxycumarin (p*K* 8.65) was used for the determination of surface potential in liposomes [24,25] and stomatitis virus [26].

To demonstrate the various results that can be obtained by the use of these various methods in a single membrane preparation, we show in Table 1 estimates of surface potential in rat liver mitochondria obtained by different methods. As discussed above, the various estimates which result from different locations of the probe may be useful in establishing the heterogeneous distribution of surface charge on the mitochondrial membrane surface. Thus, the significantly lower potential difference detected by $Tb^{3+}$ suggests a protein binding site which projects out of the membrane surface. Alternatively, lateral protein and lipid segregation may lead to uneven distribution of charges at the surface. These alternatives are being explored by various means and may lead to better understanding of the topology of the mitochondrial inner membrane. The application of these probes to the measurements of mitochondrial surface charge in the presence and absence of membrane potential, suggests that there is no significant change in surface charge due to energization. The reported changes [12,21] were shown to be due to the effect of membrane potential on the transport and distribution of surface charge probes [27].

## 3. The measurement of membrane potentials

The most direct measurement of the transmembrane electrical potential difference is obtained from the use of microelectrodes. The potential difference between two reference electrodes located at the bulk solutions bathing the two sides of the membrane is, by definition, the membrane potential. However, the application of microelectrodes to organelles, small cells and vesicles is severely limited due to size limitation. Thus, the methods used in these systems are still mostly indirect.

### *3.1. The equilibrium distribution of permeant ions*

The most widely used techniques for estimation of $\Delta\phi$ are based on a very simple and fundamental thermodynamic principle. For a vesicular system at equilibrium, the electrochemical potential different of permeant ions vanishes, *i.e.*:

$$\Delta\tilde{\mu}_i = 2.3\ RT \log \frac{a_{in}}{a_{out}} + z\mathsf{F}\Delta\phi = 0 \tag{5}$$

and therefore:

$$\Delta\phi = -\frac{2.3\ RT}{z\mathsf{F}} \log \frac{a_{in}}{a_{out}} \tag{6}$$

where, for monovalent ions at 25 °C the term $2.3\ RT/(z\mathsf{F})$ equals 59 mV.

To estimate $\Delta\phi$, using equation (6), it is necessary to determine the activities of an equilibrated ion on both sides of the membrane. It is difficult to determine ion activities inside vescicles when the activity in the surrounding medium is higher. Therefore, when the membrane potential is negative, it is necessary to choose as the test ion a cation (which accumulates in the vesicle). Similarly, an anion should be used when the membrane potential is positive. The potential–induced accumulation of the proper test ion compensates for the much smaller internal volume of cells and vesicles in the suspension. The available methods for $\Delta\phi$ determination differ from each other by the choice of the test ion and the technique which is used to determine its concentration difference. In fact, it is now clear that with few exceptions most of the potential sensitive optical probes depend on the same fundamental mechanism, namely, the establishment of electrochemical equilibrium distribution of a test ion.

#### 3.1.1. The distribution of lipophilic ions

In biological membranes, most ions permeate through specific transport systems. The mechanism of these transport systems often involves coupling between the movement of ions, either as co–transport with other ions, or as exchange for other ions. Also, there are cases of coupling of the ion transport to enzymatic reactions (pumps). Obviously, in these cases the difference of the electrochemical potential of the test ion at steady state does not necessarily vanish, and hence the apparent equilibrium distribution is not equal to the membrane potential. To apply equation (6) correctly the rate of independent electrogenic transport of the test ion must be much faster than the rate of other transport processes.

A transport mechanism which often insures a true equilibration, involves a partition of the ion into the membrane lipid phase, diffusion in the membrane, and dislocation. Fast equilibration of ions across the membrane by this mechanism requires that the ion be lipophilic, *i.e.*, the ion must have a high partition coefficient between water and lipids. In addition a high diffusion coefficient in the membrane is necessary. Organic ions with a diffuse charge distribution are often sufficiently lipophilic to provide rapid equilibration by this mechanism. This is particularly true in biological membranes (in contrast to artificial phospholipid membranes) since the increase of the membrane dielectric constant by membrane proteins and the presence of mobile charged species reduces the potential barrier for ion diffusion across the membrane [5].

Several synthetic lipophilic ions in which the charge is shielded by a hydrophobic shell are highly permeant [28]. The rate of membrane diffusion of lipophilic ions may be increased by addition of trace amounts of a lipophilic ion of the opposite charge [28]. Another strategy to insure true electrochemical equilibration of ions is to load the membrane with an ionophore (*i.e.*, a lipophilic ligand of a specific ion) which results in fast diffusion of the lipophilic ion complex and net transport of the test ion across the membrane [29]. The classical example of this method is the use of valinomycin to increase the elctrogenic transport of $K^+$ or $Rb^+$. When $K^+$–valinomycin transport rate far exceeds other transport pathways (with excess valinomycin), potassium may reach electrochemical equilibrium even when other $K^+$ transport pathways exist which are not–electrogenic. The advantage of $K^+$ (or $Rb^+$) –valinomycin method is that the content of the test ion (*i.e.*, the cation–valinomycin complex) in the membrane is usually negligible in comparison with the content of free ion in the bulk phases. Hence, there is no need to correct for the amount of bound ions when measuring the ion distribution, and the membrane potential can be calculated accurately from measuring the concentration of $Rb^+$ (or $K^+$) in the medium and the vesicles.

The main drawback of this method is that the potassium concentration in the medium must be kept to a very low level, preferably below 50 $\mu M$, in order to reduce the magnitude of the $K^+$ current at steady state which affects the measured potential difference. This may be too restrictive for many physiological experiments. Another disadvantage of this method is that it is necessary to determine accurately the internal volume of the vesicles [30]. These measurements are often tedious and not very accurate.

Because of the above–mentioned disadvantages, many investigators have preferred the use of synthetic lipophilic ions, particularly $TPMP^+$ and $TPP^+$ for the measurement of $\Delta\phi$. However, these ions are potent inhibitors of membrane enzymes and transport systems which limit the use of these probes in physiological and biochemical experiments. This problem can be partially averted by using a very low concentration of the probe ion, preferably less than 3 $\mu M$. In addition, a considerable fraction of these ions is bound to the membrane. Hence, when estimating the internal concentration from the measured uptake of the ion, it is necessary to correct for the amount of bound ions.

A reliable correction procedure must depend on an adequate physical model. Several attempts have been made to describe lipophilic cation binding and distribution [31–34]. The most successful model assumes independent binding sites on the internal and external membrane surfaces which are equilibrated with their respective bathing solutions [35]. This model was applied for the estimation of membrane potential from the spectra of lipophilic cations [35] and from the distribution of radioactively labeled lipophilic cations [36]. One of the unexpected advantages of the use of lipophilic ions that bind extensively to the membrane [*e.g.*, $TPP^+$) is that it is not necessary to determine the exact value of the internal volume because most of the accumulated ion is bound internally. Since the binding is proportional to the internal *concentration* and not the internal free ion *content*, the volume (which determines the content) is less important.

### 3.1.2. Methods of determination of lipophilic ion distribution

There are two basic methods for the determination of these concentration differences. In the first method, the net uptake of the test ion by the vesicles is determined from the disappearance of the ion from the suspending medium. In the second method, the vesicles are separated from the suspending medium and the ion content of both the vesicles and the medium is determined.

The advantage of the net uptake method is that it does not disturb the equilibrium. With a separation there is often a loss of ions from the vesicles during the separation. Also, provided that the test ion equilibrates quickly and that a fast detection of the external concentration is possible (see below) the net ion uptake method can be used for kinetics studies. The disadvantage of the net uptake method is that since the medium volume is much larger than the vesicle volume, the change in the external concentration of the test ion is very small compared to the change of the internal concentration. Hence, the sensitivity of this method is quite limited and the method is reliable only when there are relatively large potential differences.

In contrast, the separation method is very sensitive and can reliably detect potential differences of a few mV. However, care must be taken that no significant amount of ion is lost during separation, and it is often difficult to find a reliable separation method, particularly for vesicles of small size.

*(a)* Selective ion electrodes. In this method a selective electrode is inserted into the vesicle suspension and the net uptake is determined from the change in external concentration as detected by the electrode. To calculate the concentration difference, the amount of ions taken by the vesicle is divided by the total internal volume of the vesicles in the suspension, which must be determined separately. Also, depending on the kind of the test ion, it is usually necessary to correct for binding, as described below. There are many commercial ion selective electrodes of good quality. In addition, it is possible to construct such an electrode for any desirable test ion from either planar phospholipid membranes or ion-exchange membranes [37].

*(b)* Flow dialysis. Another variant of this method is the use of flow–dialysis [38]. Here, the concentration of the test ion in the suspension is estimated from the rate of diffusion of the test ion across a dialysis membrane, since, to a first approximation, the rate is proportional to the concentration. The rate of diffusion is estimated from collecting fractions of a solution flowing past a dialysis membrane which contains the vesicle suspension. Usually, in this method, the ion is radiolabeled and the content of the dialysate is estimated by scintillation counting. In principle, other detection methods, *i.e.*, optical absorbance, fluorescence, *etc.,* can be used with the appropriate probe. With most probes the response is slow, hence the method is primarily used to estimate steady-state potential differences.

*(c)* Vesicle separation techniques. In these methods the test ion is usually radioactively labelled, although optical absorption, or fluorescence detection may be used with the appropriate ion. The separation can be accomplished by fast centrifugation or filtration. Centrifugation and filtration through a layer of silicon oil is effective for large organelles and cells. With vesicles that can be sedimented in a few minutes, either by a conventional microfuge (EPPENDORF or BECKMAN) or by airfuge (BECKMAN), we find that centrifugation (without filtration through silicon oil) is quite adequate. The filtration through silicon oil may help to reduce the loss of ions from the pellet; however, the passage of the vesicles through the oil causes osmotic imbalance and often results in shrinking, which may lead to artificially increased internal concentrations and overestimation of the potential difference.

When using the centrifugation method without silicon oil, the pellet must be separated from the medium immediately after centrifugation. With the microfuge we use narrow 400 $mm^3$ ($\mu l$) tubes, in which a very small area of the pellet surface is in contact with the supernatant. The inclusion of a radiolabel marker for external volume (*i.e.*, sucrose or other sugar) is often necessary.

### 3.1.3. The determination of $\Delta\phi$ from the binding of charged lipophilic e.p.r. probes [35]

This method is, in principle, identical to the method described above, except for the detection method. Since both the amounts of bound probes and free probes can be estimated from the e.p.r. spectra without separation of vesicles, it combines some of the advantages of the electrode and of the centrifugation techniques. However, because of the difficulty in estimating small amounts of bound probes in the presence of high amounts of free probes and *vice versa,* it is necessary to adjust the probe/membrane ratio carefully for each value of $\Delta\phi$. Also, the weakness of the e.p.r. signal requires relatively high concentrations of both membranes and probe and the probes are susceptible to reduction by membrane enzymes. A detailed description of the method is found elsewhere [35].

## 3.2. *Optical probes*

Many probes for the determination of membrane potential from changes in optical signals (absorption or fluorescence) have been introduced in recent years [39]. Several of these probes are potentially very useful since they often offer both high sensitivity to small changes of membrane potential and fast response. However, there are two general drawbacks in working with these probes, both of which result from the fact that the precise mechanism of the response of these probes is not always known. Firstly, in general, it is not possible to calculate the magnitude of $\Delta\phi$ directly from the response since the mechanism of the response is unknown. The solution for this problem is to calibrate the signal, either by imposing known diffusion potential differences or by calibrating against the ion distribution method. This procedure is not always successful since results often depend on the particular conditions of the calibration procedure and are not always applicable to other conditions.

Second, there are factors other than $\Delta\phi$ that also affect the probe signal. Often the probe responds to changes in surface charge, surface pH, solvent effects, quenchers, *etc.* Since it is seldom possible to properly correct for all of these effects, there is often uncertainty even in regard to qualitative $\Delta\phi$ estimates with these probes. While the precise mechanism, which is responsible for the optical response is not known, the underlying mechanism of the *slow* dyes, which are used in bioenergetics, is believed to be the same as that of the lipophilic ion response as described above [39]. With few exceptions (*e.g.*, the "515" shift of chlorophylls and carotenes in photosynthetic membrane), most optical probes used in bioenergetics are permeant ions. Hence, the reponse is the result of the potential–driven redistribution of the probe between the internal and external solution and the inner and outer membrane surface.

In principle, the distribution of the probe between the membrane surfaces and compartments should be related to $\Delta\phi$ by the same relationships derived for lipophilic ions [35,36]. The added complication, which usually precludes a fully quantitative description, is the often unknown relationship between the probe concentration and the optical signal in each compartment. The *fast* probes, mostly used in neurophysiology, are impermeant and their response depends on a different mechanism [38].

In some cases the probe fluorescence characteristics are such that the free probe is not fluorescent and only the bound probes contribute to the fluorescence. In such cases the potential difference, could be calculated from the fluorescence by equations similar to those used for the determination of $\Delta\phi$ from the signal of e.p.r. probes [35]. For example, ANS would have been a good candidate for such a method, if not for the fact that a considerable fraction of the bound probe is not fluorescent [10]. More commonly, the probe fluorescence or absorption is modulated when the probe aggregates either in the compartment of high concentration, or on the membrane surface where it is concentrated. However, without exact knowledge of the aggregation equilibrium constants and the fluorescence characteristic of the various species, it is hard to give satisfactory, fully quantitative description of such processes. Moreover, the extent of aggregation will depend sharply on dye/membrane ratios requiring a precise calibration for each dye/membrane ratio. Thus, at present, practical application of these probes largely depends on proper calibration procedures.

### 3.2.1. Calibration by the imposition of known diffusion potentials

In this method a series of measurements in which diffusion potential differences (shortly called *diffusion potentials*) are artificially imposed, are used to construct a calibration curve. The most common procedure is to use saturating amounts of valinomycin to make membrane fully permeable to potassium. The potential difference is imposed by using various concentrations of a potassium salt in the suspending medium. If the initial internal potassium concentration is known and provided that valinomycin–induced potassium permeability is far greater than the permeability to all other ions in the suspension, the potential difference can be calculated from the NERNST equation:

$$\Delta\phi = -59 \log \frac{[K^+_{in}]}{[K^+_{out}]} \tag{7}$$

where $\Delta\phi$ results to be expressed in m.V.

While this calibration procedure is widely used, it is not always adequate. Firstly, in most biological membranes (unlike phospholipid vesicles), the membrane is relatively permeable to several other ions, such as $C1^-$, $H^+$, $Ca^{2+}$, $Na^+$, *etc.* While it is sometimes possible to eliminate some of these by choosing a proper medium and using chelators (*e.g.*, EDTA), it is often impossible to eliminate all relevant permeant ions. For istance, in mitochondria, even at neutral pH, the permeability to $H^+$ is sufficient to collapse the valinomycin–induced potassium potential difference within a few seconds. A proper calculation of the diffusion potential requires consideration of all permeant species, as expressed in the GOLDMAN equation:

$$\Delta\phi = -59 \log \frac{P^+ c^+_{in} + P^- c^-_{out}}{P^+ c^+_{out} + P^- c_{in}} \quad \text{(mV)} \tag{8}$$

where $P^+$ stands for the permeability coefficient of the cation, $c^+$ for its concentration while $P^-$ and $c^-$ represent the respective quantities for the anion.

In mitochondria, adequate description of the potential difference is obtained (in the presence of EGTA) by considering only potassium and proton permeabilities:

$$\Delta\phi = -59 \log \frac{P_{K^+} [K^+_{in}] + P_{H^+} [H^+_{in}]}{P_{K^+} [K^+_{out}] + P_{H^+} [H^+_{out}]} \quad \text{(mV)} \tag{9}$$

In calibrating the membrane potential (as in measuring the potential differences) it is important to remember that most probes respond also to the surface charge [10, 40]. Hence a proper calibration requires changing the potassium concentration while keeping the ionic strenght high and constant (This can be accomplisched by including a high concentration of $Mg^{2+}$ in the medium).

Another problem with this calibration procedure is that the potential differences obtained in the experimental system are often considerably higher than the actual potential differences that can be reliably obtained with this calibration. This means that the experimental points would fall outside the calibration curve. Since the calibration curve is often not linear and tends to saturate at high potential differences, it is not possible to estimate correctly these potential differences from such calibration curves.

### 3.2.2. Calibration against the ion distribution method

Optical probe signals have been calibrated against microelectrode measurements and against the distribution of radiolabeled ions. In general, this is a reliable method of calibration. One selects general conditions that are expected to yield potential difference values over a wide range and measures the potential difference by a reliable method in parallel with the optical signal. Besides the obvious precaution that the method of calibration itself must be as reliable as exists, it is important to remember that this calibration is only valid for the conditions in which the calibration curve was constructed. Often, changing the condition (*i. e.*, salt, pH, temperature, substrates, dye/membrane ratio) may yield entirely different calibration curves. Here, we review briefly the most widely used optical probes in bioenergetics.

(*a*) ANS. This is the oldest and widely used probe of *energization* in bioenergetics [41]. It is now clear that the energy–induced response is a reflection of the membrane potential [10]. Because the probe is negatively charged, it accumulates in vesicles of positive potential difference (*e.g.*, submitochondrial particles, chromatophores, *etc.* which enhances the internal binding and fluorescence. The enhancement is a sensitive measure of the potential difference in this system. In systems of negative potential difference, such as mitochondria, the generation of potential difference causes efflux of dye which leads to dissociating of internal bound dye and quenching. This response it not as sensitive, since most of the dye is bound externally to begin with. The dye also responds to both surface charge and pH, and caution must be exercised to eliminate interference from these side effects [10]. The fluorescence has been calibrated by external diffusion potential and by $Rb^+$ distribution. There is a considerable difference between the calibration curve of $\Delta\phi$ generated by the substrates ATP and succinate [10].

(*b*) Carbocyanine dyes. Many different positively charged dyes of this class have been used for monitoring changes of $\Delta\phi$ [39]. The choice is dictated sometime by their absorption and emission characteristics. The most commonly used dye in systems with negative potential differences is the positively charged dis–$C_3$–5. Usually, the dye is calibrated with potassium–valinomycin diffusion potential, using the NERNST equation, and assuming linear response in the entire range. This is inadequate, as discussed above. A more adequate calibration is obtained by the use of $Rb^+$ distribution. Here, too, the response is not linear with succinate as a substrate but is fairly linear with ATP. This dye, too, was shown to respond to surface charge and pH and the same precautions must be taken, as discussed above.

(*c*) Oxanol–IV. This negatively charged dye accumulates in positively charged vesicles where it binds to the internal membrane surface, resulting in shifts in its absorbance and fluorescence [42]. Thus, its response is in essence identical to that of other lipophilic ions. It is widely used in photosynthetic membrane vesicles, and sub-mitochondrial particles where the membrane potential is positive.

(*d*) Safranine. This positively charged dye accumulates in negatively charged vesicles and organelles where it aggregates, leading to shifts in its absorption spectra [43]. While it has been used fairly widely, the high dye concentrations required for accurate measurement may produce various artifacts [41].

**Aknowledgements**

Supported by PHS Grants CM-28173 and AA0-7238.

## References

[1] B.H. Honig, W.L. Hubbel, and R.F. Fleweling, *Annu. Rev. Biophys. Biophysic. Chem.,* **15**, 163 (1986).
[2] J. Barber, *Biochim. Biophys. Acta,* **594**, 253 (1980).
[3] C.M. Armstrong, *Physiol. Rev.,* **61,** 44 (1981).
[4] H. Rottenberg, *Methods Enzymol.* **171,** 364 (1989).
[5] J.P. Dilger, S.G.A. McLaughlin, T.J. Mc.Intosh and S.A. Simon, *Science,* **206**, 1196 (1979).
[6] B. Ehrenberg, Y. Berezin, *Biophys. J.,* **45**, 663 (1984).
[7] W.S. Chow, and J. Barber, *Biochim. Biophys. Acta,* **591**, 82 (1980).
[8] R.W. Mansfield, H.Y. Nakatani, J. Barber, S. Mauro and R. Lannoye, *FEBS Lett.,* **137** 133 (1982).
[9] H. Rottenberg, T. Grunwald and M. Avron, *Eur. J. Biochem.,* **25,** 54 (1972).
[10] D.E. Robertson, and H. Rottenberg, *J. Biol. Chem.,* **258**, 11039 (1983).
[11] S. McLaughlin and H. Harary, *Biochemistry,* **15**, 1941 (1976).
[12] T. Aiuchi, N. Kamo, K. Kurihara and Y., Kobatake, *Biochemistry,* **16**, 1626 (1977).
[13] D.H. Haynes and H. Staerk, *J. Membr. Biol.,* **17**, 313 (1974).
[14] R. Gibrat, C. Romieu and C. Grignon, *Biochim. Biophys. Acta,* **736**, 196 (1983).
[15] A.A. Jasaitis, V.V. Kuliene, and V.P. Skulachev, *Biochim. Biophys. Acta,* **234**, 177 (1971).
[16] K. Hashimoto and H. Rottenberg, *Biochemistry,* **22**, 5738 (1983).
[17] D.J. Castle and W.L. Hubbel, *Biochemistry,* **15**, 4818 (1976).
[18] G.S.B. Lin, R.I. Macey and R.J. Mehlhorn, *Biophys. Acta,* **732**, 683 (1983).
[19] K. Hashimoto, P. Angiolillo and H. Rottenberg, *Biochim. Biohys. Acta,* **764**, 55 (1984).
[20] L. Wojtczak and A. Szewczyck, *Biochem. Biophys. Res. Commun.,* **136**, 941 (1986).
[21] A.T. Quintanilha and L. Packer, *FEBS Lett.,* **78**, 161 (1977).
[22] A.T. Quintanilha and L. Packer, *Arch. Biochem. Biophys.,* **190**, 206 (1978).
[23] A.T. Quintanilha and L. Packer, *Proc. Natl. Acad. Sci. USA,* **71**, 570 (1977).
[24] P. Fromherz, *Biochim. Biophys. Acta,* **323**, 326 (1973).
[25] M.S. Fernandez, *Biochim. Biophys. Acta,* **646**, 23 (1981).
[26] R. Pal, W.A. Petri Jr., Y. Barenholz and R.R. Wagner, *Biochim. Biophys. Acta,* **727**, 185 (1983).

[27] H. Rottenberg, in *Ion Interactions in Energy Transfer Biomembranes.* G.C. Papageorgiou, J. Barber and Papas, (Editors), Plenum, New York (1986), p. 29.

[28] L.L. Grinius, A.A. Jasaitis, T.P. Kadziauskas, E.A. Liberman, V.P. Skulachev, V.P. Topali, L.M. Tsotine, M.A. Vladimirov, *Biochim. Biophys. Acta,* **216**, 1 (1970).

[29] P.W. Reed, *Methods Enzymol,* **55**, 435 (1979).

[30] H. Rottenberg, *Methods Enzymol.,* **172**, 63 (1989).

[31] A. Zaritzky, M. Kihara, R.M. NcNab, *J. Membr. Biol.,* **63**, 215 (1981).

[32] T.S. Lolkema, A. Abbing, K.S. Hellingwerf and W.N. Koning *Eur. J. Biochem.* **130**, 287 (1983).

[33] R.S. Richie, *Prog. Biophys. Mol. Biol.,* **43**, 1 (1987).

[34] M. Demura, N. Kamo and Y. Kobatake, *Biochim. Biophys. Acta,* **812**, 377 (1985).

[35] D.S. Cafiso and W.L. Hubbel *Biochemistry,* **17**, 187 (1978).

[36] H. Rottenberg, *J. Membr. Biol.,* **81**, 127 (1984).

[37] N. Kamo, M. Muratsugu, R. Hongoh and Y. Kobatake, *J. Membr. Biol.,* **49**, 105 (1979).

[38] S. Ramos, S. Schuldiner, and H.R. Kaback, *Methods Enzymol.* **55**, 68 (1979).

[39] A.S. Waggonner, in *The Enzymes of Biological Membranes,* A.N. Matonesi, (Editor), Plenum, New York (1985) Vol. **1**, P. 313.

[40] K. Masamoto, K. Matsura, S. Itoh, M. Nishimura, *Biochim. Biophys. Acta,* **638**, 108 (1981).

[41] A. Azzi, *Biochem. Biophys. Res. Commun.,* **37**, 254 (1969).

[42] C.L. Bashford and J.C. Smith, *Methods Enzymol.,* **55**, 569 (1979).

[43] K.F.O. Akerman and M.F.K. Wikstrom, *FEBS Lett.,* **68**, 191 (1976).

[44] V.G.R. Valle, L. Pereiza–da–Silva, and A.E. Vercesi, *Biochem. Biophys. Res. Commun.,* **135**, 189 (1986).

# FACTORS CONTROLLING THE ION CONDUCTANCE OF CHANNELS

MARCO COLOMBINI

*Department of Zoology*
*Laboratories of Cell Biology*
*University of Maryland*
*College Park, MD 20742 U.S.A.*

Contents



*Bioelectrochemistry III*
Edited by G. Milazzo and M. Blank
Plenum Press, New York, 1990

# 1. Introduction

## 1.1. *Definition of a channel*

Ion flow through membranes is not favored because it involves movement of charge through a low dielectric constant medium. The strong interaction between ions and the highly polarizable fluid, water, means that a great deal of energy is needed to remove ions from water. The low polarizability of the lipid environment in the membrane does not interact strongly with ions and therefore acts as a good barrier to ion flux. For these reasons, ion flux through membranes must be facilitated by a transport process or some polar defect in the membrane.

Of the transport systems that facilitate ion flux through membranes, channels can be defined as structures that induce a continuous water-filled pathway (the aqueous pore) through the membrane. This definition excludes structures that bind ions and physically translocate them through the membrane, *i.e.* the carriers. It may turn out that this distinction may blur as more is learned about the molecular mechanism by which carriers translocate ions. Indeed, at least one channel (VDAC in mitochondria) has regions in its primary sequence that have homology to carriers (ADP/ATP exchanger in mitochondria) [1,2]. However, the definition does limit channels to structures that:
1) can carry charge across membranes; and
2) allow ions to flow in some continuous way (at least for some period of time) through the membrane.

## 1.2. *Classification*

The channels can be further classified based on their effective pore size (*i.e.* the narrowest portion of the pore). Those with a narrow pore tend to interact more intimately with the transported substance and therefore (in general) display a higher degree of selectivity. Ions probably interact weakly with the walls lining the pore of the larger channels (Fig. 1) thus resulting in lower selectivity. Indeed, a layer of water of hydration may partially insulate the ion from the walls of the large pore. Although this classification is somewhat arbitrary, it is useful in providing a framework for what follows.

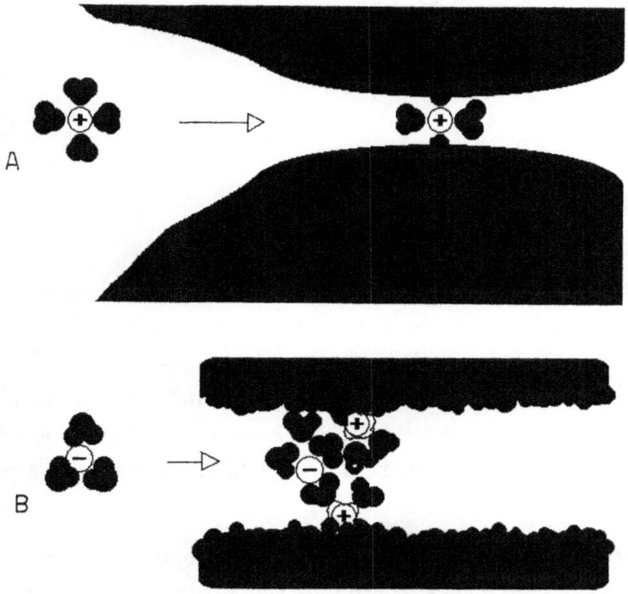


Fig. 1. - Permeation of ions through channels with a narrow (A) or relatively wide (B) pore size. In the channel with the narrow constriction (A), the ion must exchange some of its waters of hydration for ligating groups at the narrow portion of the channel. In the pore without the narrow constriction (B), the ion may not come into intimate contact with the walls of the pore. It must always be surrounded by water.

Selectivity can be based on the size and/or the charge of the translocated species. It is easy to see that large channels will allow large ions to cross the membrane and that their large size should result in little ion-channel interactions and therefore low selectivity. However, the electric fields emanating from charges can extend for relatively long distances and may induce selectivity in even large channels.

## 2. Barriers to ion flow

Ion flux through the channel can be divided into 5 stages (Fig. 2).

1. Diffusion through the unstirred layer (on side 1): ions flow from the bulk phase to the mouth of the channel.
2. Entry (into the channel proper): from the channel's mouth (which may contain a vestibule or funnel-shaped access region) to the inside of the channel proper.
3. Translocation: through the channel.
4. Exit: from the channel proper to the channel's mouth on side 2.
5. Diffusion through the unstirred layer (on side 2): ion flow from the channel's mouth to the bulk phase on side 2.

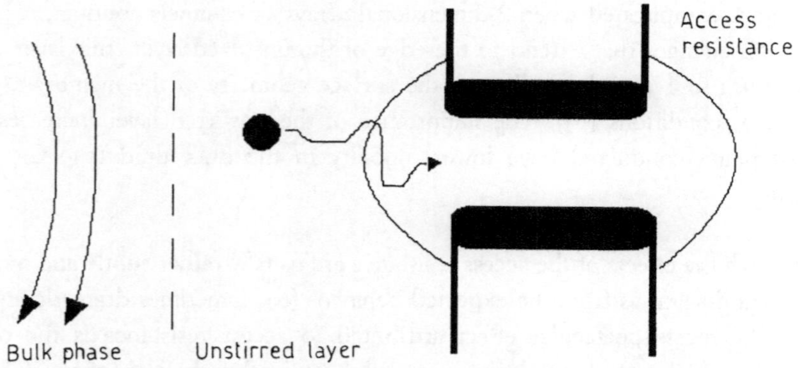


Fig. 2. - Barriers to the flow of ions through a channel. On leaving the bulk phase, the ion must diffuse through the unstirred layer which may easily be 4 orders of magnitude thicker than the membrane. If the flux through the pore is fast, some of the electric field may extend out from the mouth of the channel. This access resistance region is indicated by the two paraboloid curves on either side of the channel.

## 2.1. *Access to the mouth of the channel*

For a single channel, processes 1 and 5 are, in principle, the same. It is through these processes that ions from the bulk phase access the channel. Usually only a small portion of the unstirred layer significantly inhibits the flow of ions. The access resistance can be estimated by integrating the resistance of the medium to a circular disk at the mouth on one side of the channel [3, 4].

$$R_{access} = \frac{\varrho}{4r} \tag{1}$$

where $\varrho$ is the resistivity of the medium and $r$ is the radius of the pore's mouth. For a symmetrical pore the total access resistance would be twice this value. Obviously, the importance of this resistance depends on its magnitude relative to the resistance of the channel proper. If the channel behaved like a cylinder of bulk solution, the access resistance would make the channel effectively longer by 1.6 $r$ [4]. The channel could be more or less conductive than an equivalent cylinder of bulk solution resulting in a greater or lesser effect of the access resistance.

The above treatment ignores convection and stirring because most of the access resistance occurs well within the unstirred layer next to the membrane. The situation becomes more complicated when 2-dimensional arrays of channels conduct, resulting in concentration changes that extend to the edge of the unstirred layer (this layer can vary from a few μm to 1 mm depending on the surface geometry of the membrane and the environmental conditions [5]). The importance of the unstirred layer increases further when ions being conducted have lower mobility in the bulk fluid (*e.g.* $Ca^{2+}$ in the cytoplasm).

Although the effects of the access resistance are usually rather subtle and most often result in small deviations from the expected behavior [6], sometimes dramatic effects are observed. The most spectacular effect attributed to access resistance is the observed ultra-steep potential difference dependence of the mitochondrial channel, VDAC [7]. Large, highly-negatively charged polymers are capable of making VDAC channels at least an order of magnitude more potential difference dependent than any channel observed to date as shown by conductance potential difference plots in Fig. 3. The untreated channels are steeply potential difference dependent (x). Each channel closes at both positive and negative applied voltages [8, 9]. The addition of 25 m*M* dextran sulfate (8 kDa) results in a large increase in potential difference dependence (+). The evidence strongly favors the

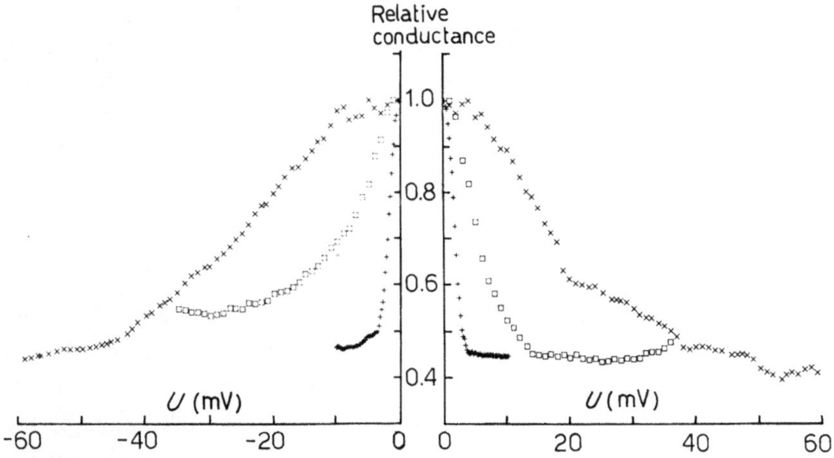


Fig. 3. - Dextran sulfate induces ultra-steep voltage dependence in VDAC. The channels were solubilized from the mitochondrial outer membrane of *N. crassa* and inserted into a planar phospholipid membrane made from soybean phospholipids. Both aqueous phases consisted of 1*M* LiCl, 5 m*M* $CaCl_2$. The data were collected either in the absence (x) or the presence of 6.25 μM (□) or 25 m*M*(+) dextran sulfate. Experiments were performed under voltage-clamp conditions (*i.e.* the voltage was held constant and any current changes were monitored) (from Mangan and Colombini [7]).

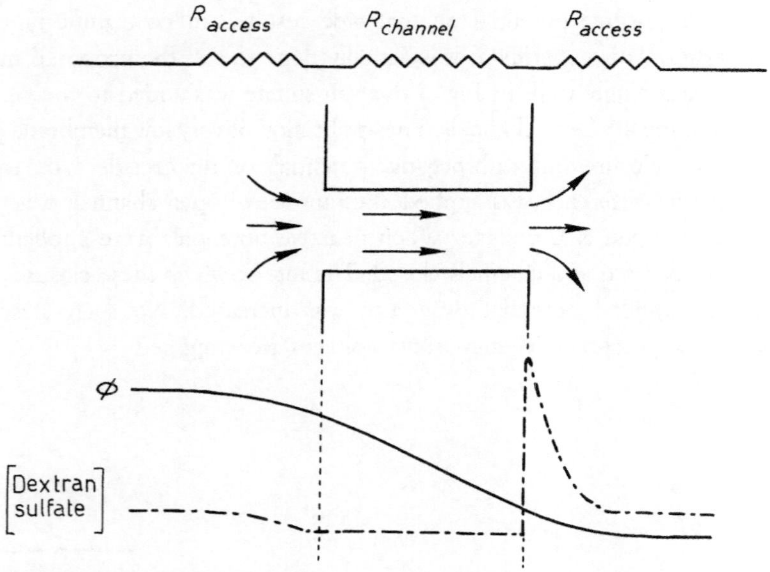


Fig. 4. – Mechanism for the ultra-steep potential difference dependence. The channel and the access-resistance regions are depicted in the top of the figure as resistors in series. The arrows indicate the flow of positive ions through a channel in a membrane. The solid curve shows the proposed potential profile through the channel. Note that some of the potential change occurs outside the mouth of the channel. This results in a region of positive charge on the right side of the channel. Dextran sulfate (broken line) accumulates at the mouth of the pore in the positively charged region and interacts with the channel.

mechanism shown in Fig. 4. According to this mechanism, the applied potential difference across the membrane results not only in an electric field in the channel proper, but also in the access resistance region (presumably because of the high conductance of the channel proper). Thus, the potential in the negative side of the membrane is less negative in a small region next to the mouth of the channel. An impermeant polyanion present in the bulk phase should partition into this region and accumulate there. Naturally, the larger the charge on the polyanion the more the accumulation, or, more precisely, the greater the probability of finding the polyanion at the mouth of the channel. If the proximity of the polyanion affects the channel, as it does in the case of VDAC, then a potential difference dependent change in the channel would result. In the case of VDAC, the polyanion increases VDAC's probability of being in the closed conformation therefore an ultra-steep potential difference dependence is the result.

Of all the experimenal evidence that implicates the access resistance in this phenomenon, the most convincing is the asymmetry of the effect. If the polyanion is added only to one side, the effect only occurs if the side to which the polyanion was added was made negative. A positive potential on that side resulted in no significant change in VDAC's properties. This experiment was actually done to test the proposed mechanism and it worked exceedingly well. In Fig. 5 dextran sulfate was added to the cis side of a membrane containing 4 VDAC channels. The application of very low membrane potentials resulted in channel closure only with negative potentials on the cis side. Note that, at the instant the potential difference was applied, the number of open channels was 4 because this is the number open at 0 voltage. When negative potentials were applied, stepwise current changes occurred and channels closed. The rate at which these closures occurred increased as the applied potential difference was increased. No such closures were observed with low potentials when positive voltages were applied.

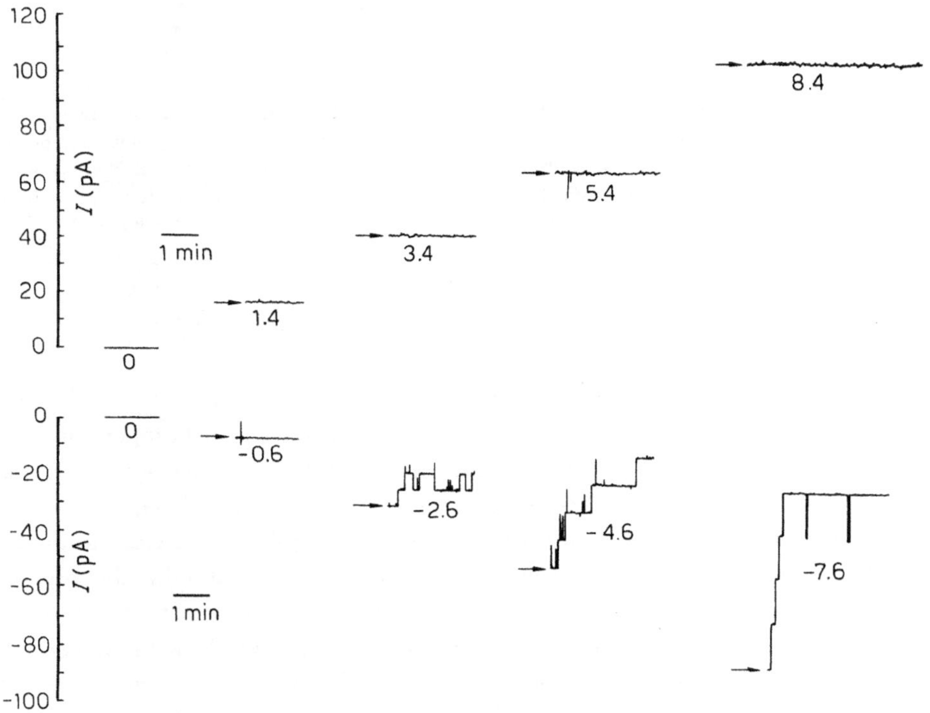


Fig. 5. – Asymmetry of the effect of dextran sulfate (8 kDa) on the potential difference dependence of VDAC. The experimental system was as described in Fig. 3. Four channels were inserted into the membrane. Dextran sulfate was added to one aqueous phase (designated the cis side) to a final conc. of 125 $\mu M$. The upper traces show the current responses to the indicated applied potentials (positive on the cis side). The lower traces show the current responses to the indicated negative potentials on the cis side. In between the applied potentials, the voltage was returned to zero where any channels that may have closed rapidly open. The arrows indicate the level of the instantaneous current, *i.e.* prior to the closure of any channels (from MANGAN and COLOMBINI, [7]). Figures given under each graph represent the applied potential difference $U$ in mV.

*2.2. Entry into and exit from the channel*

The entry and exit processes need not be identical if the channel is not symmetrical. However, in principle, they can be considered to be equivalent. These processes may involve:
*a)* the exchange of ligand groups around the ion from water or hydroxyl ions to polar groups on the walls of the pore (as indicated in Fig. 1A),
*b)* steric effects which must be met for the ion (especially complex ions) to enter the pore (*i.e.* the physical size of the permeant as compared to the size of the narrowest part of the channel),
*c)* ion-induced conformational changes in the pore (the ion or any permeant can exert force on the channel causing some degree of distortion),
*d)* ion-induced conformational changes in the water within the channel (the ion will polarize water within the channel and this may influence its ability to permeate), and
*e)* repulsive or attractive forces with ions already present in the channel (whether mobile or fixed charges). These steps may contribute to or determine the channel's selectivity and permeability.

For large pores, these processes might be approximated by liquid-liquid junctions (appendix to Ref. 10). They may involve little dehydration of the mobile ion (Fig. 1B). The hydrated ion moves from one environmental condition to another. The environment may differ in electrical potential (resulting from fixed charges within the pore[1], hydrostatic pressure[2], viscosity (probably incrased due to water of hydration), and ionic strength (due to the presence of fixed charges and their counterions within the pore). These factors will influence the ion's partitioning between the two spaces.

For small pores, entry and exit may be dominated by the rates and energies associated with ligand exchange and the ion-induced changes in the channel and its contents. Some of the ion's water of hydration must be shed (Fig. 1A) and replaced by interactions with polar groups in the channel. The energies of dehydration and binding have been shown to be important in determining the selectivity of channels with small pores [14]. Small ions are favored by high field-strength binding sites while large ions are favored by low field-

1. Electroneutrality can be violated within a channel because the dimensions are comparable to the DEBYE length [11]. Thus an electrical potential can exist within the channel resulting from the presence of a net charge.

2. The hydrostatic pressure within the channel is likely to be rather different from that in the bulk phase. Since water must be in equilibrium between the bulk phase and the channel interior, any difference in the concentration of solutes between the channel interior and the bulk will make the water concentration either more dilute or more concentrated. This can only be compensated for by an increase or decrease in the hydrostatic pressure within the channel. The hydrostatic pressure could influence the entry and exit of solutes. It should also apply stress on the walls of the channel. This issue is not simple because some of the water is probably tightly associated with the walls of the channel [12]. Purposely changing the hydrostatic pressure can lead to interesting phenomena [13].

strength sites (for alkali metal ions, one of the 11 EISENMAN selectivity sequences is generally observed). In the large pores, the hydrated ion may enter with its tightly bound shell of electrons with only little distortion. The more intimate the interactions between the ion and the walls of the channel, the more effect the ion will have on the channel. Therefore one might expect that the entry and exit processes may be more complex and thus slower for small as compared to large pores.

### 2.3. *Translocation within the channel*

The translocation process is the motion of the ion through the channel from one location where it can leave the pore to the other (Fig. 6). In principle, these locations could be one and the same thus eliminating this step to translocation (Fig. 1A). However, there is evidence for such a translocation step in many channels. Translocation may involve:
*a)* exchange of ligands around the ion,
*b)* movement through some intermediate binding/low energy sites,
*c)* steric constraints,
*d)* movement of part of the ion's electric field through the low dielectric constant medium in the membrane, and
*e)* movement of the ion away from counterions in the medium.

Points *d* and *e* are critical and basic to the problem of ion translocation through a channel. The removal of some of the ion's high dielectric constant environment is a step that requires a great deal of energy. Most of this energy must be compensated for by interaction with the walls of the channel. When an ion translocates through a channel, it is

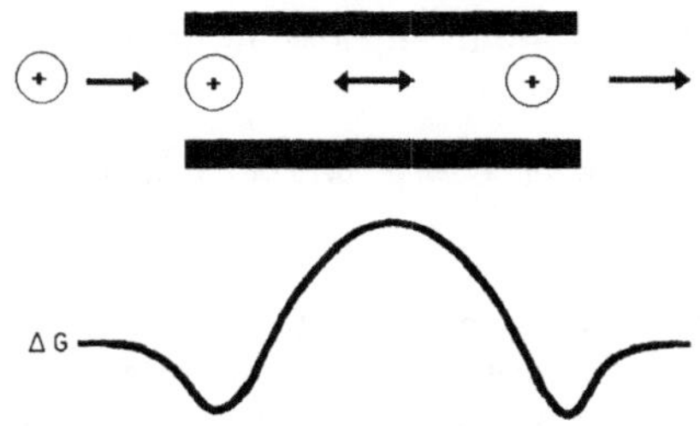


Fig. 6. – The process of translocating ions through a channel may involve overcoming an energy barrier in the middle of the channel. In the upper figure the ion is shown entering the channel, translocating through the channel, and exiting the channel. The lower figure indicates how the free energy of the ion may change as it moves through the channel.

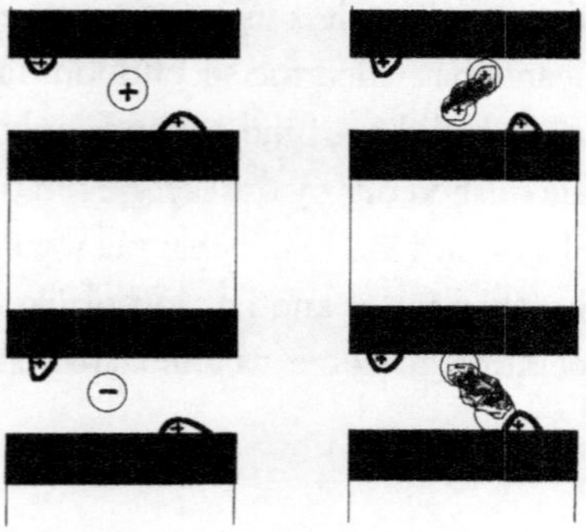

Fig. 7. – Four ions of different sizes and charge are shown moving through the same kind of channel. The positive charges adjacent to the walls of each channel are fixed charges (part of the protein). The sphere around the permeating ion is meant to indicate that the electric field produced by the ion extends beyond the ions's physical size.

often moved away from counterions for half of its journey. Energy is needed to separate charge and this is exactly what is happening in the case of a channel with no net charge in the channel, this so-called image force can be represented as an energy barrier in the middle of the pore (Fig. 6). The energy wells depicted in the figure represent a net preference for interacting with the walls of the channel. This favorable interaction helps to effectively reduce the magnitude of the barrier in the middle of the channel.

For small ions moving through large pores, translocation may resemble movement through bulk solution with little interaction with the walls lining the pore (Fig. 1B, or even a larger pore relative to ion size). However, for large ions, the size of the structure may force the ion to interact more intimately with the walls of the same pore. In Fig. 7, the circle around the permeating ion is meant to indicate how the electric field around the ions extends out and results in interaction with charges and other groups on the walls of the pore. While small ions could move through with little interaction, larger charged molecules may be forced to interact.

For small channels, the interaction may be extensive but binding must be weak enough to allow ions to flow through the channel. Strong binding results in long residency times within the channel and therefore small rates ot ion flow. It is the binding that confers the high selectivity of these pores.

If an aqueous pore through a channel were large enough to allow a variety of cations to cross, one might expect that those cations that can be bound would be slowed down and therefore would be less permeable than the ions that can be bound more poorly. However, a rather simple mechanism (Fig. 8) allows the ions that are more tightly bound to be translocated more rapidly. Two binding sites in series together with electrostatic repulsion between their bound ions means that when the second ion binds, it tends to force the first ion out of the channel. An ion that did not bind or binds only weakly could not stay on the binding site long enough to be pushed out by the second ion nor could it effetively push out a bound ion. The presence of a bound ion in the channel would act as an effective barrier to the translocation through the channel of another ion of like sign. This type of selectivity mechanism seems to be operating in some calcium channels [15].

## 3. Channel gating

### *3.1. General*

Most biological channels exist in multiple conformational states differing in their ability to conduct ions. Channel gating is the process by which channels change their ability to conduct ions by undergoing transitions from one conformational state to another. In most cases, these transitions are controlled by environmental signals: the membrane potential, the concentration of a particular chemical, the tension of a membrane, etc. The signal results in shifting the relative energy levels of the channel's conformational states and therefore in large changes in the probability of finding a channel in a particular conformational state. For example, the electric potential difference-gated gap junctions of amphibian blastomeres are normally open in the absence of an electric field [16]. When an electric field is applied, the channels close. The electric field is causing the closed state of the channel to become the low-energy state.

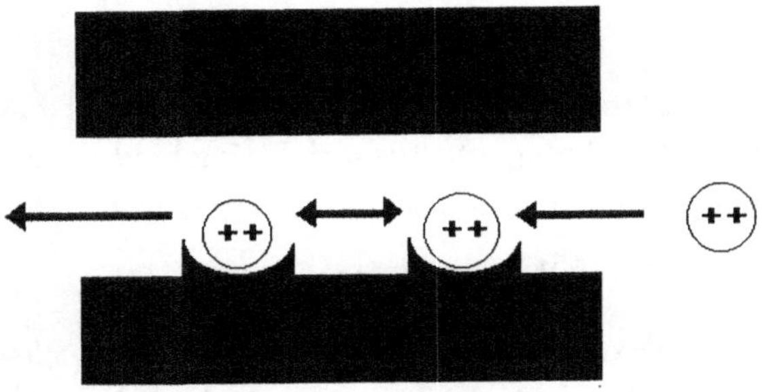


Fig. 8. – A channel that possesses two binding sites for divalent ions. Permeation occurs by one ion entering the channel and binding to an unoccupied site. This ion can now repel another located at the second site and thus induce the second ion to move on.

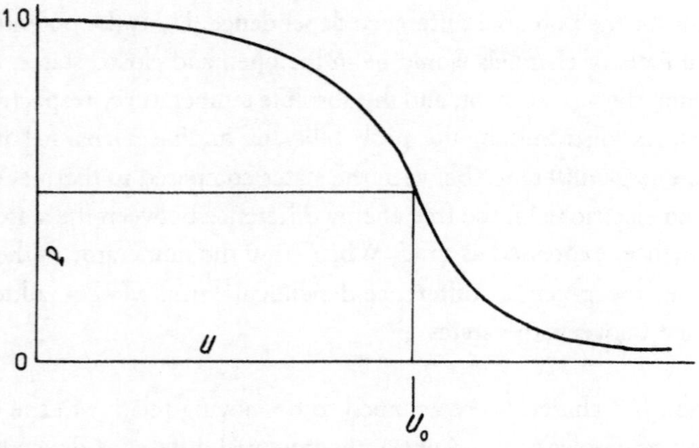


Fig. 9. – An illustration of how the applied potential difference dependent properties of a channel can be quantitated in terms of two parameters, $n$ and $U_O$ where $P$ = Probability of being open = $1/[1+\exp\{(n\mathrm{F}U-n\mathrm{F}U_o)/RT\}]$
$n$ – measure of the steepness of the potential difference dependence – mininum number of charges on the sensor
$U$ – applied potential difference
$U_o$ – location of the switching region – potential difference at which half the channels are closed.

### 3.2. *Electric potential difference*

Focusing on potential difference-gating, the probability of finding a channel in a given state, say the open state, would depend on the transmembrane potential difference $U$ (Fig. 9). It has been assumed for a long time that voltage-dependent channels must have some domain, called the *sensor*, that allowed the channel to detect and respond to the electric field. The sensor's response to the electric field would change the overall energy level of the channel resulting in the conformational change in the channel. In order for a protein domain to act as a sensor, it was felt that it could either be a large dipole that aligned with the electric field or be a charged entity that moved relative to the field. The conformational change, that results in a modification of the ability of the channel to conduct ions,must be tightly coupled to the motion of the sensor. In this way, a potential difference-dependent conformational change could take place.

Expressing these ideas in a quantitative manner [8, 17, 18] one obtains:

$$\frac{N_o}{N_c} = \exp\left[-\frac{n\mathrm{F}(U-U_o)}{RT}\right] \qquad (2)$$

where $N_o$ and $N_c$ are the number of channels in the open and closed states respectively; $n$ is the number of charges that would have to experience the entire potential difference, $U$, in order to account for the potential difference dependence; $U_o$ is the potential difference at which equal numbers of channels would be in the open and closed states; F, $R$, and $T$ are *Faraday*'s constant, the gas constant, and the absolute temperature, respectively. According to this BOLTZMANN distribution, the probability of finding a channel in a given state depends on the energy difference between the states compared to thermal energy ($RT$). In the absence of an electric field, the free energy difference between the states is some value and, for convenience, expressed as $n\mathsf{F}U_o$.When $U$=0 the numerator in the exponential is $n\mathsf{F}U_o$. Otherwise, the potential difference-dependent term, $n\mathsf{F}U$ is added to yield the energy difference between the states.

In equation (2), charges were assumed to be moving relative to the electric field. If the reorientation of dipoles were occuring, the potential difference-dependent term would be the difference between the product of the electric field at the dipole and the component of the dipole moment in the direction of the field, in the two states of the channel. Otherwise the concepts are very similar. It is remarkable that potential difference-gated channels work at all. In order for the membrane potential to be able to dramatically change the probability of finding a channel in a given state, the potential difference-dependent energy change must be large enough to overcome the energy difference between the channel's conformational states, say an open and a closed state. Since these states have very different abilities to conduct ions or molecules through the membrane, it would not be surprising that their energy levels were different by the equivalent of a few hydrogen bonds, say 50 kJ/mole. Now let us consider what that would mean to the channel's ability to be potential difference gated. If the sensor consisted of a charged domain, the maximum effect of the potential would be achieved when the domain's motion resulted in the charges being translocated all the way through the membrane, a remarkable feat considering the apolar nature of the membrane. If the sensor consisted of 4 charges moving all the way through the potential difference across the membrane, a 125 mV membrane potential would be needed to just cancel an energy difference of 50 kJ/mole (see Fig. 10). Such a large potential would correspond to an electric field of a quarter of a million volts per cm. This force is close to the maximum that a membrane can withstand and would still be insufficient to do more than make channels be equally probably in the open and closed conformation.

The situation can be improved either by making the channel more potential difference-dependent (move more charge through the membrane) or by fine tuning the channel so that the energy of the open conformation is almost identical to that of the closed conformation. This fine tuning does occur. The estimated energy differences between these conformations for the mitochondrial channel, VDAC, is less than the energy of one hydrogen bond, about 8 kJ/mol [8, 19]. In this way, the application of just 20 mV can result in a very large change in the probability of finding a channel open and therefore in the permeability of the membrane which contains these channels. In porin from *E.coli*, the high

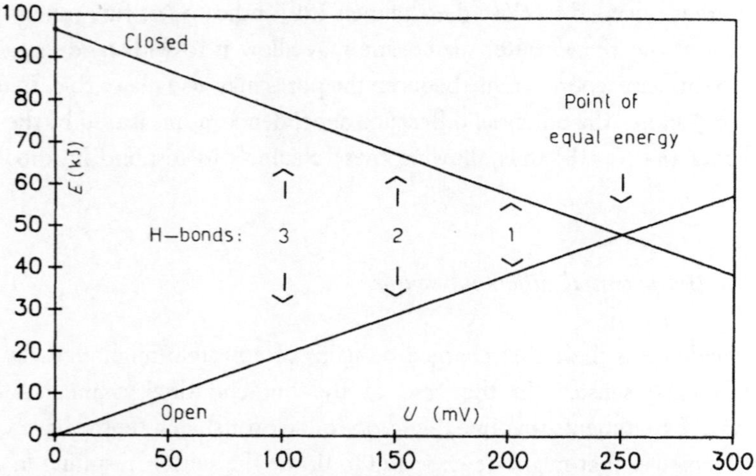


Fig. 10. – A theoretical calculation of how the energy level (*E*) of the open or closed states of a channel can vary with the potential difference across the membrane. The upper curve indicates the energy level of the closed state while the lower curve represents the energy level of the open state. Where the lines intersect, the two states have equal energy. The numbers in the center indicate the energy difference between the states in terms of H-bonds ( *i.e.* 1, 2, or 3 H-bonds).

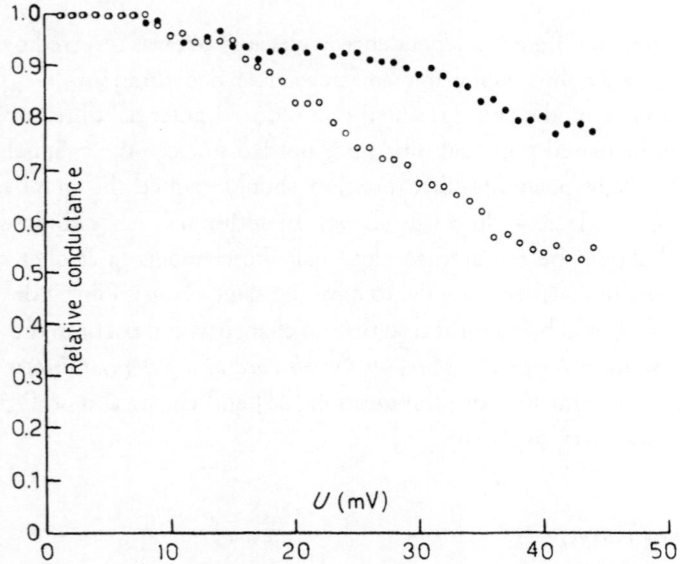


Fig. 11. – Partial titration of the potential difference sensor. The experimental conditions were similar to those described in Fig. 3 except that 5 m*M* MES buffer was added and the VDAC used was purified from rat liver mitochondria. The open circles were recorded at pH 6.2, while the closed circles were recorded after the pH was raised to 10.7 by addition of the buffer, CAPS, to both aqueous phases (from Bowen *et al.* [19]).

potential difference needed to close the channels [20] indicate less fine tuning. However, the ruggedness of the *E. coli* outer membrane may allow it to withstand large potentials differences. In this case, cooperativity between the pores may also play a role. In the sodium channel in squid axons, the potential difference dependence, as measured by the parameter $n$, is a bit larger ($n$=6) [18] thus allowing these channels to respond in a lower voltage range.

### 3.3. *The electric potential difference sensor*

What evidence is there that charged domains of a protein might actually serve as a potential difference sensor? In the case of the mitochondrial channel, VDAC, the straight-forward experiments that one might do to look for such a sensor have worked out very well. pH changes that might be expected to titrate the sensor, resulting in a structure of reduced charge, were, in fact effective at reducing the potential difference – dependence of the channels (Fig. 11) [21]. A p*K* of 10.6 was estimated for the charges on the sensor. Another way of neutralizing the charged sensor is by chemical modification. Since amino groups were suspected, succinic anhydride was used to convert some of them to carboxyl groups. This resulted in a loss of potential difference dependence that was even more pronounced than that shown in Fig. 11 [19]. The following observations strengthened the conclusion that the sensor was really being neutralized as opposed to many alternative explanations:

a) the loss of potential difference dependence at elevated pH was reversed by returning the pH back to neutrality thus ruling out an irreversible denaturation;

b) partial titration or modification resulted in a reduced potential difference dependence coupled with an increased potential difference needed to close the channels. The second argument is especially powerful since titration should reduce the number of effective charges resulting in a decrease in $n$ (see above). In addition $U_0$ should increase because a higher potential should be required to close half the channels (a smaller charge on the sensor should require a higher potential to have the same result). Finally the decrease in $n$ and increase in $U_0$ should be reciprocal so that no change would occur in the value of $n\mathrm{F}U_0$ *(the energy difference between the states in the absence of a potential difference).* The fact that this did indeed occur for both the titration [21] and chemical modification (Fig. 12) [19] of VDAC was very gratifying.

### 3.4. *Gating mechanisms*

Many mechanisms can be conceived whereby the permeability of a channel can be altered by a conformational change. When restrictions based on the know structure of proteins and the properties of the membrane environment are imposed, fewer models are

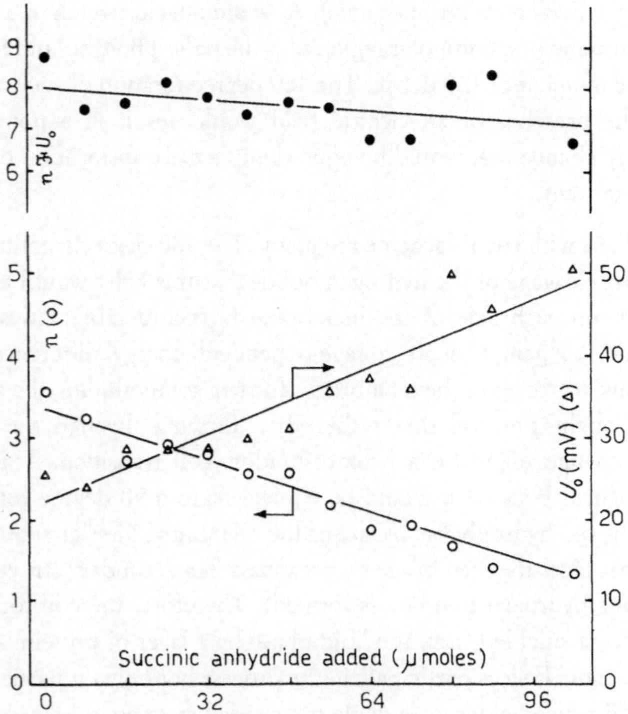


Fig. 12. – Changes in the voltage-dependence parameters of VDAC as a function of succinic anhydride addition. The experimental conditions were similar to those in Fig. 3 except that KCl was used instead of LiCl and 50 m*M* MOPS was present (pH=7.2). Rat liver VDAC was used in this experiment. Results such as those shown in Fig. 3 were collected and fitted to equation 2 to obtain the values of $n$ and $U_o$. This was done after sequential additions of succinic anhydride. (from Doring and Colombini [19]).

likely. If further requirement is imposed that the gating be coupled to the motion of a potential difference sensor and the process accounts for the observed steepness in potential difference dependence, the models that remain are fewer still.

In order to account for the strong potential difference dependence of the gating process in potential difference-gated channels, the conductance change must be linked to either the movement of many charges along the electric field or the alignment of dipoles with the field. If dipoles were involved, a great many of them would have to act in concert to account for the potential difference dependence. Of the secondary structures formed by proteins, the one with the strongest dipole moment is the alpha helix. The amide linkages of the protein backbone are aligned so that their dipole moments (3.5 debye) are parallel to the helix axis. The result is the effective separation of one unit of charge (the N terminus being positive and the C terminus, negative). A 34 aminoacid stretch of alpha helix would span a 5 nm membrane (the non-polar region plus the polar phospholipid head groups) and have a net dipole moment of 120 debye. The 180 degree rotation of such a structure in the membrane in the presence of an electric field would result in a potential difference dependent energy change that would be equivalent to the translocation of a single charge through the membrane.

The problems with such a scheme are many. The 180 degree rotation of such a helix would require the cleavage of the hydrogen bonds that this helix would have formed with the water phases on each side of the membrane (a considerable activation energy). In addition, in order to gain enough voltage-dependent energy difference, multiple long helices would have to rotate in the membrane (further compounding the activation energy problem). One might propose that rather than flipping through the membrane, the conformational change might be a helix-to-random-coil transition. This would require twice as many helices because it would be equivalent to a 90 degree rotation. However, more importantly, the hydrophobic nature of the membrane interior requires that the part of the protein that faces the membrane be organized in a secondary structure (this ensures that the backbone hydrogen bonding is formed). Therefore, the coil form would not be likely, unless it were shielded from the lipid phase by a layer of protein. Another solution might be to have more helices participate in the process but only tilt partially (Fig. 13). The tilted state should have a wider pore while the aligned state would have a narrower pore. This would be a natural coupling between the motion of the sensor and the permeability change. Indeed, this type of motion has been proposed for the gating process induced by $Ca^{2+}$ in rat liver gap junctions (not potential difference-gated) [22]. However, a serious

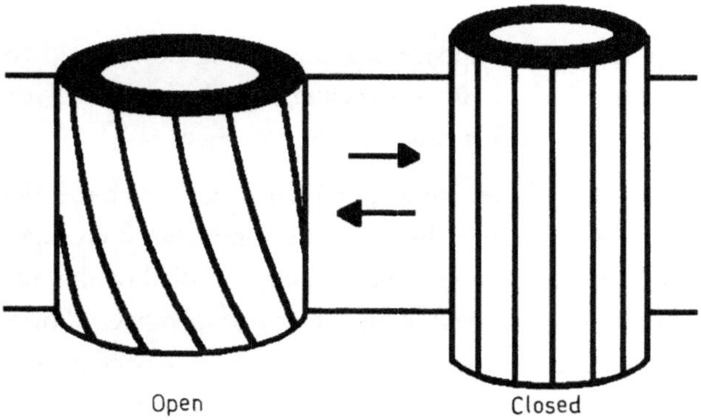


Fig. 13. – A conformational change in a channel formed by a bundle of alpha helices. When the helices become aligned, perpendicular to the plane of the membrane, the channel becomes a little longer and its pore size is reduced.

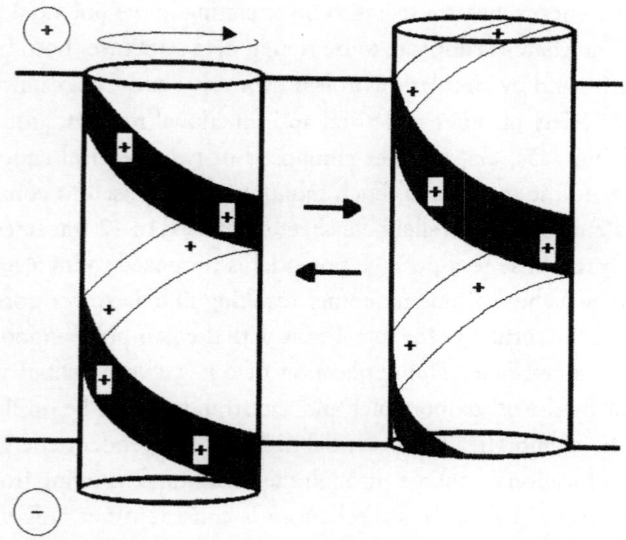

Fig. 14 – A model for potential difference gating in the sodium channel (adapted from Ref. 23). An amphipathic alpha helix spans the membrane in association with the rest of the protein (not shown). It is mainly hydrophobic except that every third amino acid side chain is positively charged. On the left, the membrane potential is keeping the helix down (more into the cytoplasm). When the potential is eliminated, the helix screws upward (toward the extracellular space) resulting in a net movement of positive charge across the membrane.

problem with applying this mechanism to potential difference-gated channels is the requirement that the helices are all aligned in the same direction. Aligning dipoles in parallel is unfavorable because the strong dipole moment would cause the dipoles to repel each other.

Although one cannot rule out the possibility that the potential difference sensor might be a dipole, the movement of net charge seems much more likely. From the aminoacid sequence of the potential difference-gated $Na^+$ channel, several investigators have noticed the presence of sequences that are highly non – polar except that every third (sometimes fourth) aminoacid is either a lysine or arginine [23]. They proposed [23] that these stretches form alpha helices that slide or screw through the membrane in the presence of an electric field (Fig. 14). Since there are 4 such sequences in 4 homologous regions on the quarter-million molecular weight polypeptide, each need only move a short distance for the overall potential difference dependence to be very steep. Similiar sequences (with high homology) are found in the related potential difference-gated $K^+$ and $Ca^{2+}$ channels. Finally, the 4 homologous domains are reminiscent of the gating particles of Hodgkin and Huxley [18].

A rather different mechanism seems to be operating in the potential difference-gated ch nnel, VDAC. The channels appear to be rich in beta structure (both by looking at the primary sequence (z) and by circular dichroism (MANGAN and COLOMBINI, unpublished). For this and for a variety of other structural and functional reasons, an unusual model is being considered (Fig. 15), *i.e.*, a dimer composed of two identical subunits oriented in opposite directions in the membrane. Each monomer is proposed to consist of one alpha helix flanked by a curving antiparallel beta sheet composed of 12 transmembrane strands. Channel closure in response to a positive potential is proposed to involve the sliding of 6 strands out of the beta sheet of one monomer resulting in a narrower pore. The extruded strands would sit on the surface of the membrane with the non-polar amino acid side chains embedded into the membrane. The application of a negative potential would cause the same response but by the other monomer and the strands would lie on the surface of the opposite side of the membrane. The potential difference dependent energy change would arise from the translocation of charge through the membrane resulting from the extrusion of the 6 transmembrane chains. This mechanism is consistent not only with the primary sequence but also with experimental findings including measured changes in pore diameter [24] and internal volume [13].

## 4. Channel blockage

Many agents have been described which interfere with channel function by blocking the flow of ions through channels. A particularly interesting aspect is the potential

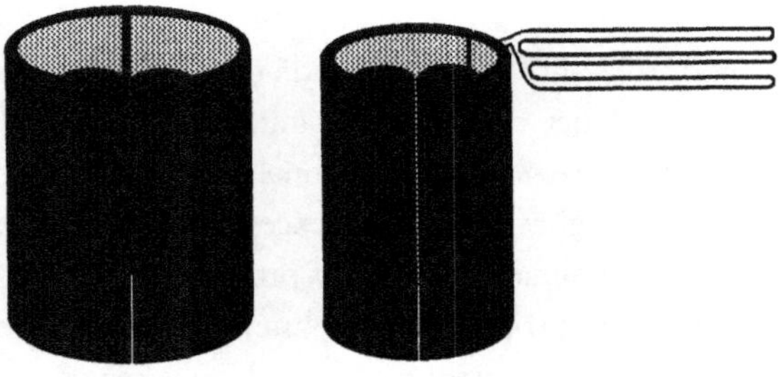


Fig. 15 – A model for potential difference gating in the mitochondrial channel, VDAC. The channel consists of two monomers oriented in opposite directions in the membrane. Each monomer consisting of an amphipathic alpha helix and a 12 stranded antiparallel beta sheet. Channel closure occurs when a potential difference is applied and 6 of the beta strands are pulled out of the channel and lie on the surface of the membrane.

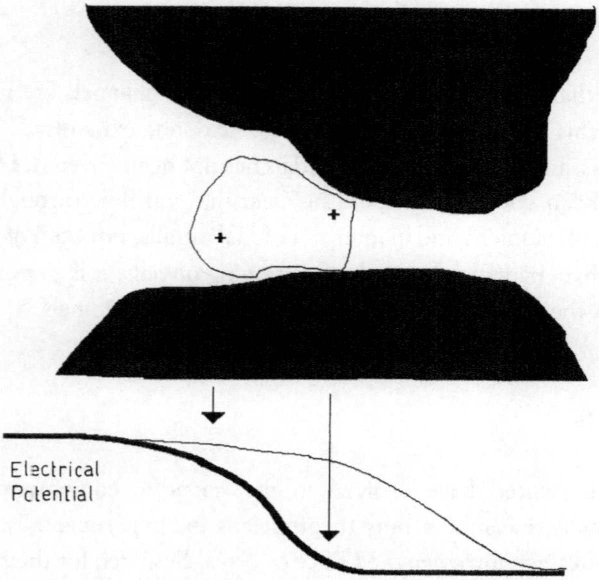


Fig. 16. – Potential difference dependent block of a channel. The doubly-charged blocker changes the electrical potential profile through the channel when it blocks the channel. The thin line is the profile before blockage while the thick line is the profile after blockage. The charge on the right side of the blocker experiences a large potential change when the blocker enters the channel. By contrast, the charge on the left side experiences only a small potential change.

difference dependence of certain blockers. A charged substance that can penetrate part of the way through the channel (Fig. 16) is a good candidate for such blocking action. The electric field can either enhance the blocking or relieve it depending on the direction of the field and the charge on the blocker.

The degree of potential difference dependence of the blocking action depends on the charge of the blocker, $z$, and the fraction of the membrane potential, $d$, through which the charge moves before its motion ceases due to steric factors. Thus the potential difference-dependent component of the process is $z\mathsf{F}dU$. By knowing $z$ and measuring the potential difference dependence of the block (in a similar way to measuring the voltage dependence of channels) one can calculate $d$. Unfortunately, without knowing what the electrical potential profile is within the channel before and after blockage, it is difficult to make any interpretation from the value of $d$. As shown in Fig. 16, the potential profile through the channel changes as the blocker interferes with ion flow.

## 5. Conclusion

The factors that influence the flow of ions through channels are many and diverse. The coverage in this chapter has been selective and not exhaustive. The influence of phenomena such as surface and dipole potentials has not been covered. However, from the factors that have been covered, it should be clear that ion flow through channels is rich in both interesting phenomena and in interpretational pitfalls. For both of these reasons and for the vital role that channels play in the functioning of cells, it is important to gain even more insight into the process of ion permeation through channels.

## Acknowledgements

The ideas presented have evolved in the scientific community because of the contributions of many researchers, both theoreticians and experimentalists. I am grateful to them. I also gratefully acknowledge the Office of Naval Research for their financial support through grant # N00014-85-K-0651.

## References

[1] M. Forte, H.R. Guy and C.A. Mannella, *J. Bioenerg. Biomembr.*, **19**, 341 (1987).
[2] R. Kleene, N. Pfanner, R. Pfaller, T.A. Link, W. Sebald and W. Neupert, *EMBO J.*, **6**, 2627 (1987).
[3] J.E. Hall, *J. Gen. Physiol.*, **66**, 531 (1975).
[4] B. Hille, *Ionic Channels of Excitable Membranes*, Sinauer Assoc. Inc. Sunderland, MA, (1984) p. 184.
[5] J.A. Schafer and T.E. Andreoli in *Membrane Physiology*, 2nd, Edition, T.E. Andreoli, J.F. Hoffman, D.D. Fanestil and S.G. Schultz (Editors) Plenum Medical Book Co., New York, (1987), Chap. 11, p. 185.
[6] O.S. Anderson, *Biophys. J.*, **41**, 147 (1983).
[7] P.S. Mangan and M. Colombini, *Proc. Natl. Acad. Sci. USA*, **84**, 4896 (1987).
[8] S.J. Schein, M. Colombini and A. Finkelstein, *J. Membr. Biol.*, **30**, 99 (1976).
[9] M. Colombini in *Ion Channel Reconstitution*, C. Miller (Editor), Plenum Publishing Co., New York (1986), Chap. 22, p. 533.
[10] D.M. Adelsberger-Mangan and M. Colombini, *J. Membr* Biol. **98**, 157 (1987).
[11] A. Finkelstein and A. Mauro in *Cellular Biology of Neurons*, E. R. Kandel (Editor), American Physiological Society, Bethesda MD, (1977), Vol. **1**, p. 161.
[12] M.S. Prouty, A.N. Schechter and Parsegian, *J. Mol. Biol.*, **184**, 517 (1985).
[13] J. Zimmerberg and V.A. Parsegian, *Nature (London)*, **323**, 36 (1986).

[14] G. Eisenman and S. Krasne in *MTP International Reviews of Science*, Biochemistry Series, C.F. Fox (Editor), Butterworth, London, (1975), Vol. **2**, p. 27.
[15] R.W. Tsien, P. Hess, E.W. McCleskey and R.L. Rosenberg, *Annu. Rev. Biophys. Chem.*, **16**, 265 (1987).
[16] D.C. Spray, R.L. White, A. Campos de Carvalho, A.L. Harris and M.V.L. Bennett, *Biophys. J.*, **45**, 219 (1984).
[17] G. Ehrenstein, H. Lecar and R. Nossal, *J. Gen. Physiol,* **55** 119 (1970).
[18] A.L. Hodgkin and A.F. Huxley, *J. Physiol.*, **117**, 500 (1952).
[19] C. Doring and M. Colombini, *J. Membr. Biol.,* **83,** 81 (1985).
[20] H. Schindler and J.P. Rosenbusch, *Proc. Natl. Acad. Sci. USA*, **75**, 3751 (1978).
[21] K.A. Bowen, K. Tam and M. Colombini, *J. Membr. Biol.*, **86**, 51 (1985).
[22] P.N.T. Unwin and P.D. Ennis, *Nature (London)*, **307**, 609 (1984).
[23] W.A. Catterall, *Science*, **242**, 50 (1988).
[24] M. Colombini, C.L. Yeung, J. Tung and T. König, *Biochim. Biophys. Acta*, **905**, 279 (1987).

# MOLECULAR ASPECTS OF THE NEUROTRANSMISSION BY THE ACETYLCHOLINE RECEPTOR SYSTEM

EBERHARD NEUMANN

*Faculty of Chemistry, University of Bielefeld,*
*P.O. Box 8640, D-4800 Bielefeld 1, F.R.G.*

Contents



*Bioelectrochemistry III*
Edited by G. Milazzo and M. Blank
Plenum Press, New York, 1990

## 1. Introduction

The nicotinic acetylcholine receptor (AcChR) is the key protein in the rapid signal transmission between cholinergic nerves and the innervated target cells of vertebrate muscles and of the electric organs of many electric fishes [1,2]. The trans-synaptic signal transfer (see Fig. 1), finally resulting in muscle contraction or in electric organ discharge, is initiated by the neuronally triggered release of the neurotransmitter acetylcholine (AcCh). According to the historical concept by NACHMANSON [1] this cationic activator molecule binds to the AcCh receptor and causes a conformational change. Thereby a pathway (channel) for the flow of $Na^+$ and $K^+$ ions is opened that causes depolarisation of the postsynaptic membrane. Rapid depolarisation, in turn, triggers the action potentials that are necessary for the muscle cell contraction or the discharge of the electric organ electrocytes.

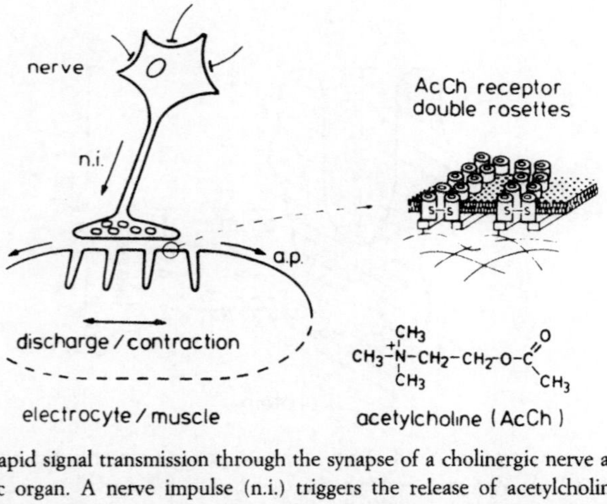


Fig. 1 - Scheme for the rapid signal transmission through the synapse of a cholinergic nerve and a muscle cell or an electrocyte of an electric organ. A nerve impulse (n.i.) triggers the release of acetylcholine ($AcCh^+$), stored in synaptic vesicles. AcCh-binding to the AcCh receptors (AcChR), organized in double rows in the postsynaptic membrane of the target cell (muscle or electrocyte), finally causes action potentials (a.p.) leading to muscle contraction or electrocyte discharge.

Frequently it is stated that the nicotinic AcChR is the best characterized receptor-channel. However, many fundamental molecular aspects of the cholinergic neurotransmission are unresolved, even on the level of the AcChR channel protein. For example, it is still not clarified whether the physiological elementary signal, *i.e.* the smallest observed single channel event of cholinergic cell membranes, is due to one receptor molecule or rather due to the receptor dimer. It is also not known what is the functional significance of the receptor organisation in double rows [3], see Fig. 1. Furthermore, the hysteretic binding of AcCh to *Torpedo californica* (*T. cal.*) receptors remains to be a challenge not only for physical chemical interpretation but also for a possible role in synapse function [4,5]. Finally, it has been recently discovered that *T. cal.* AcChR has lipid kinase activity and autophosphorylation capacity [6]. Again, it is not known on what functional level, activation and channel opening or rather desensitization, this newly discovered enzymatic activity, is relevant.

## 2. The various AcChR species and channel activities

The AcChR of electric fish can be isolated in two macromolecular forms, depending on the experimental conditions of isolation [7-9]. When sulfhydryl-alkylating agents are present in the first tissue homogenization step the dimer (D), consisting of two disulfide-linked monomers (M), is the predominant species. The monomeric form prevails when reducing agents such as dithiothreitol (DDT) are used in the homogenization medium. As outlined in the various summaries (see, *e.g.*, Ref. 9), the monomer or light (L) form has a subunit stoichiometry of two α-chains, one β-, one γ- and one δ-chain ($\alpha_2\beta\gamma\delta$). The relative

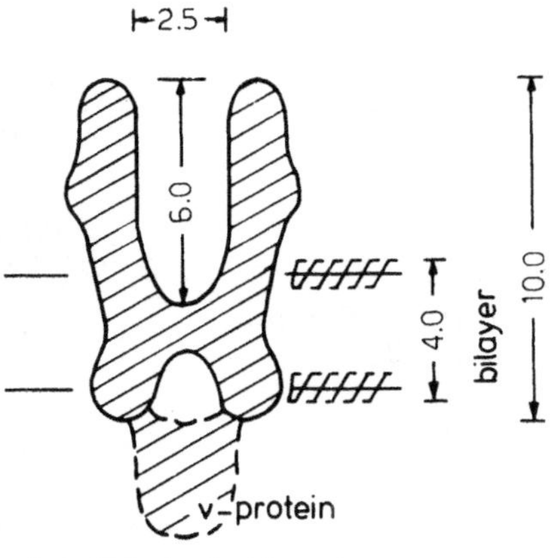


Fig. 2 - Tube structure of the *Torpedo* AcChR, sketched from electronmicroscopic data by TOYOSHIMA and UNWIN, [Nature (London) (1988)]. Axial section through the cylindrically averaged structure, showing the funnel-like channel mouths and the channel part in relation to the lipid bilayer and peripherical protein (v-protein or 43 kDa protein) associated with the proper receptor on the cytoplasmic side. The length dimensions are given in nm. The AcCh binding sites are presumed to be on top of the receptor tube facing the synaptic gap.

molar mass of the monomer is $M_r$ 290 000; the dimer or heavy (H) form is twice as large. Figure 2 depicts the tube structure of the *Torpedo* AcChR; some molecular dimensions of the monomer fragment of the presumably dimeric AcChR are shown [10].

### 2.1. *AcChR oligomers in solution*

In the course of studies to reconstitute the isolated and purified AcChR proteins in lipid bilayers it was found that the receptor monomers exhibit an inherent tendency to specific non-covalent association in vesicular lipid bilayers as well as in solution [11, 12]. In particular the reversible formation of stable dimers according to $2M \rightleftarrows M_2 \rightleftarrows D$ *i.e.* with reformation of the δ-δ sulfide bond, readily occurs in the presence of phospholipids [12, 13]. Non-covalent and covalent [13] aggregations of monomeric receptors to dimers (and higher oligomers [14]) might be one of the reasons why, until recently [11], no characteristic functional differences between monomers (M) and dimers (D) have been observed.

Due to the intrinsic tendency to non-covalent dimer formation in the presence of phospholipids many monomer preparations may have actually dealt with dimers of the type $M_2$. Although phospholipids are not a necessary condition to form $M_2$, the stability of the non-covalent dimer strongly increases in the presence of lipids ($K_d \approx 10^{-10}$ *M*) compared to that in the absence of lipids ($K_d \approx 10^{-6}$ *M*); see Ref. 12.

### 2.2. *Single channel events*

The functional reconstitution of the isolated and purified AcChR proteins from *T. cal.* electric tissue involved reincorporation of the channel proteins into lipid vesicles, formation of monolayers and finally of planar bilayers from these vesicles. For the protein reincorporation into vesicles, the conventional cholate dialysis sometimes caused perturbed and irregular channels and receptor aggregations. Provided the planar bilayers are adjusted to cohesive pressures of 30 and 32 mN/m, 16 mole per cent cholesterol of the total lipid, both the dimer and the monomer exhibit single channel events when agonists, such as AcCh, carbamoylcholine or suberoyldicholine, are added (Fig. 3). As depicted in Fig. 3, the most striking difference between the channel events is that the maximum channel conductance (≈20 pS) of monomers (M) is about half as large as that (≈40 pS) of dimers (D). Furthermore, the amplitude and time characteristics of the channels formed by the isolated dimers (≈40 pS) in lipid bilayers are almost quantitatively the same (≈40±10 pS) as for unextracted receptors (microsacs) from a different *Torpedo* species: (*Torpedo marmorata* [15]).

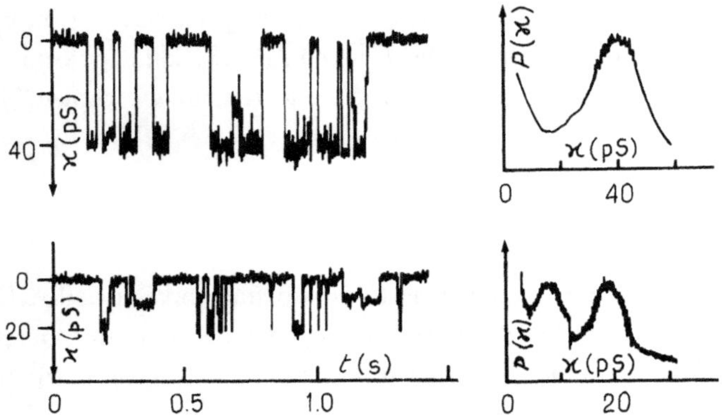


Fig. 3 - Conductance ϰ *versus* time traces and channel histograms *P* (ϰ) (probability of occurrence) of reconstituted *T. cal.* AcChR in planar lipid bilayer (16 mole % cholesterol), 5 min after addition of suberoyldicholine (0.5 μ*M*); 20 °C, 0.5 *M* NaCl. Upper trace: the native dimer (D) at 100 mV; lower trace: monomer (M) at 150 mV [11].

## 2.3. *Effect of reducing agents*

When the disulfide reducing agent DDT was added to a planar bilayer containing dimers (D), the initial dimer channel events started to change: at first conductance levels characteristic of the monomeric receptor occurred at the expense of dimer events; after about 10 min the channel events became undefined and within 1 h the channel activity disappeared. It thus seems that disulfide reduction first abolishes one of the two channels within a dimer and with progressing time the second one also disappears. These observations may relate to inhibitory effects of DDT on ligand binding and on receptor activity [16, 17]. In view of the dimer conductance level being twice as large as that of the monomer the dimer appears to cause a *double-channel* with apparent *single-channel* characteristics.

## 2.4. *AcChR channel oligomerization*

The isolated and purified AcChR monomers (M) and dimers (D) reconstituted in planar bilayers may give rise to a synchronized opening and closing of up to several, dimeric channel units caused by protein association [11]. For instance, when the cohesive pressure of a bilayer with monomer channels was raised from 32 to 40 mN/m by the addition of $CaCl_2$ to a final concentration of 0.5 m*M*, large conductance peaks developed at the expense of the monomer levels; they are grouped in activity clusters of multiple dimer levels. Most remarkably, only even multiples of the monomer level (20 pS) are populated corresponding well to the dimer level (40 pS) and multiples thereof (80, 120 and 160 pS); peaks at uneven multiples do not occur.

The specific channel oligomerization in multiples of dimers corresponds well with the oligomerization of receptor proteins in solution where, after crosslinking, the dimer complexes $(M_2)_n$, or $D_n$, n = 1,2,3.., may be identified [12, 13]. The channel characteristics of the dimer $M_2$ is hardly distinguishable from that of the (native) dimer D. The data of SCHINDLER *et al.* [11] have been recently criticized by SCHÜRHOLZ [18]. Indeed, contrary to frequent statements, it is still very difficult to reconstitute the isolated AcChR.

### 2.5. *Double channel and pair cooperativity*

Recalling the oligomerization tendency of the isolated AcChR monomers in solution and in the planar lipid bilayer it appears that contact formation between the monomers alone leads not only to stable complexes but also to pronounced changes in the channel characteristics, especially in the channel conductance. The monomer-monomer interactions within the dimer in the lipid phase do apparently lead to the reestablishment of the δ–δ disulfide bridges [13].

The factor of 2 in the conductance difference between monomers and dimers and the occurrence of multiples of dimers in the protein complexes as well as in the channel conductances can be consistently interpreted in terms of a simple model: synchronous opening and closing of two monomeric channels within a dimer (D and $M_2$). Therefore, the operationally defined *dimer-channel* with apparent *single-channel* characteristics has to be visualized as two cooperative monomeric channels that act fully in concert. The fact that the same channel conductances were observed with isolated dimers and with unextracted microsacs [15] suggests that the basic functional units of the observable AcChR channels in the biomembrane are, in reality, synchronized double-channels.

At present it is not known if the principle of cooperative double-channel gating applies to other cholinergic systems as well. Remarkably corresponding patch-clamp data for single-channel conductances [19-20] are closer correlated to the channel conductance of the reconstituted *Torpedo* dimer than to that of the monomers. However, the comparison is obscured by the different ionic conditions applied besides possible species-dependent differences.

Since the AcChR in the native postsynaptic membrane can be visualized in double rows of dimers [3], it appears likely that the oligomeric channel cooperativity (consistent with the receptor oligomerization in solution) is the functional counterpart of the native organization in double rows. This feature and the enhancement of long range cooperativity by the intact disulfide bridge between the monomers of the D-species together with the effect of reducing agents on the ratio of monomers and dimers in the protein isolation [11], suggest that the disulfide-bridged *dimer D is the native and functionally operative channel unit.*

*2.6. Gating scheme*

If the *dimer-channel* is indeed the functional conductance unit in the biomembrane, the HILL coefficient of about 2 for the channel activation (for review see Ref. 9) may be interpreted in a different manner compared to previous analyses. The dimer channel activation may result from the successive association of two agonist molecules (A) with the two monomeric low affinity conformer $R_l$ of the dimer species $R_l \cdot R_l$; *i.e.* each monomer channel within the dimer binds only one agonist molecule. The functionally relevant step is then the concerted transition of the complex $AR_l \cdot AR_l$ to the conducting conformer $AR_l^* \cdot AR_l^*$

$$
\begin{array}{rcl}
2\ A + R_l\ R_l \rightleftarrows R_l \cdot AR_l + A & \rightleftarrows & AR_l \cdot AR_l \\
 & & \uparrow\downarrow \\
 & & AR_l^* \cdot AR_l^*
\end{array}
$$

The HILL coefficient of 2 is thus related to the highly cooperative, practically synchronous transition of the receptor dimer, in its low affinity state for agonist binding. The occupation of only one of the α-subunits within a monomeric part of the dimer compares well with the channel activation by irreversibly binding agonists. It is sufficient for channel activation when only one *irreversible agonist* binds per monomer after reduction of disulfide bridges in the vicinity of a binding site for AcCh on one of the two α-subunits [21-23] (see also Ref. 24). The covalent agonist labeling of a second α-subunit requires much higher concentration of irreversible agonists [25].

There are still other arguments in favor of the activation scheme in terms of two agonist molecules and the dimer. Recent dialysis studies have confirmed that the final equilibrium state for the binding of $^3$H AcCh to membrane-bound receptors (microsacs) and to purified dimers is associated with only *one* AcCh molecule bound per monomer ($K \approx 5$ n*M*, 4 °C, *T. cal.*), independent of whether detergent is present or not; at equilibrium, there is no cooperativity to this final high affinity state [4, 26]. If the channel activation would require two agonist molecules per receptor monomer, then the final slow phase of receptor inactivation (associated with a HILL coefficient of one) leading to one agonist bound per monomer would have to release one of the two previously bound agonist molecules. However, the sequence:

$$2\ A + R_l = R_lA_2 = R_l^*A_2 = R_hA + A$$

appears very unusual. Furtheron, according to NEUBIG *et al.* (1982) the functional unit of $^{23}Na^+$ efflux gating in *Torpedo* membranes (microsacs) comprises two α-bungarotoxin (α-Bgt) sites [27]. Since in the *absence* of detergents there is only one α-Bgt molecule bound with high affinity per monomeric AcChR [4, 5] the functional unit of ion transport gating must be the dimer (D).

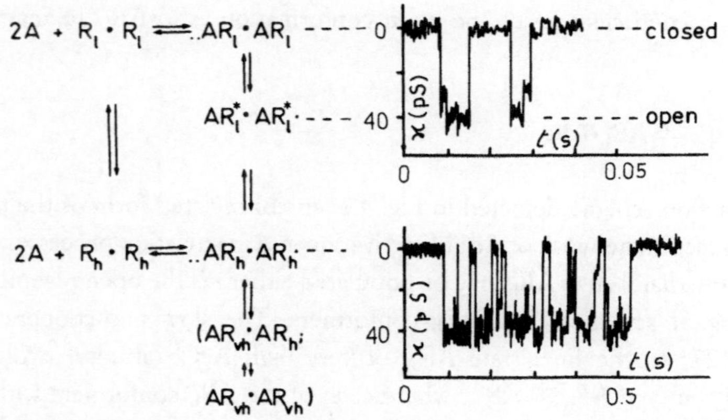


Fig. 4 - Reaction scheme (short form) for the AcChR dimer channel activation. Top: normal low affinity pathway of activation leading to *single-channel* events according to $AR_l \bullet AR_l$ (closed) ⇄ $AR^*_l\, AR^*_l$ (open). Bottom: burst of repetitive opening and closing events under desensitizing conditions, modelled by $AR_h \bullet AR_h$ (closed) ⇄ $AR^*_l \bullet AR^*_l$ (open).

Receptor inactivation procedes in at least two phases (see *e.g.* Ref. 9), a fast cooperative one [28, 29] and a slow non-cooperative one [27, 29]. Recent dialysis data show that at low AcCh concentrations ($\leqslant$ 50 n$M$) the slow transition to the final equilibrium state of very high affinity denoted by $R_{vh}$ ($K \approx 5$ n$M$) is *induced* by agonist binding [5]. The preexisting high affinity receptor state $R_h$ to which the agonists may directly bind is of smaller affinity ($K_h \approx 0.1\ \mu M$ for AcCh).

In order to facilitate a comparison with schemes developed previously, the following correspondences to Ref. 9 hold:

$R_l \triangleq$ R of Ref. 9; activatable, low AcCh affinity ($K_l \approx 10^{-4}\ M$).
$R_l^* \triangleq$ A of Ref. 9; activated, conducting ($K_l \approx 10^{-5}\ M$).
$R_h \triangleq$ I of Ref. 9; inactivated, high AcCh affinity ($K_h \approx 10^{-7}\ M$).
$R_{vh} \triangleq$ D of Ref. 9; fully desensitized, very high AcCh affinity ($K_{vh} = 5 \times 10^{-9}\ M$).

Channel activation from desensitized AcChR. — Channel acitivity is also observed under conditions where the AcChR protein is usually assumed to be inactivated or desensitized [20,11]. Whereas the response immediately after agonist addition are frequently occurring as short single channel events, longer exposure to agonists gives rise to occasional bursts of repetitive opening and closing events [20, 30, 31, 11]: *Nachschlag* or *flickering* [20] (see Fig. 4). The occasional dimer-channel activation under desensitizing conditions requires an extension of the receptor activation scheme by the cooperative coupling (HILL coefficient 2) of the open conformation $AR_l^* \bullet AR_l^*$ to inactivated states according to:

$$AR_l^* \bullet AR_l^* \rightleftarrows AR_h \bullet AR_h$$

The reaction scheme depicted in Fig. 4 is an abbreviated form of the possibly more complicated reaction network of AcChR activation and inactivation processes. In detail, the inactive (intermediate) state $AR_h$ may be populated either *via* the open receptor states or by direct binding of activator to the $R_h$ conformers. The slow non-cooperative phase of inactivation [27] to the final state $AR_{vh}$ of very high AcCh affinity involves structural transitions of the type $AR_h \rightleftarrows AR_{vh}$, where each of the $AR_h$ conformers within the dimer $AR_h \bullet AR_h$ can independently convert to the $AR_{vh}$ state (HILL-coefficient 1, Refs. 27, 29).

Monomer — The isolated AcChR monomer (fragment) in detergent solution [24] as well as reconstituted in planar lipid bilayers [11] appears to have a more loose structure compared to the dimer. The high affinity AcCh binding to the M-species is a factor of about 10 ($K \approx 5 \times 10^{-8}\ M$) less stable [7] than to the dimer ($K \approx 5 \times 10^{-9}\ M$). The hydrodynamic properties of the M-fragment suggest a more extended, perhaps partially unfolded structure [24]. As seen from Fig. 3, the occurrence of the substates of lower conductance are much more frequent for the monomers than for the dimers. The monomer channels are thus more *flexible* than the dimer channels. The more loose structure of the monomer may also be the reason why membrane fragments in the presence of detergents (0,1 % Triton X-100) bind two α-Bgt molecules per monomer compared to only one high affinity bound α-Bgt per monomer in the absence of detergents [4].

## 3. The acetylcholine binding hysteresis

*Torpedo californica* membrane fragments and the isolated receptors bind AcCh along a hysteresis loop [4]. Since AcCh is a cation [$(CH_3)_3N^+ - CH_2CH_2OCOCH_3$] and the receptor protein at pH 7 is net anionic, the hysteresis may be caused by mainly electrostatic contributions, similar to the case of the proton-binding hysteresis of certain polynucleotide complexes [32, 33]. Because hysteresis is a physical mechanism for memory imprint [34-36] it is very tempting to associate synaptic AcCh hysteresis with neural memory processes. AcCh binding hysteresis is obtained when *T. cal.* membrane fragments rich in the nicotinic acetylcholine receptor (nAcChR) are mixed with AcCh of increasing concentration (pulse mode addition of AcCh) up to [A] ≈ 1 *μM*, yielding the lower branch, followed by a gradual dilution of AcCh resulting in the upper branch of the hysteresis loop (see Fig. 5). If, however, the AcCh binding is studied under the open system conditions afforded by equilibrium dialysis, a concentration-dilution cycle always traces the upper curve. Although

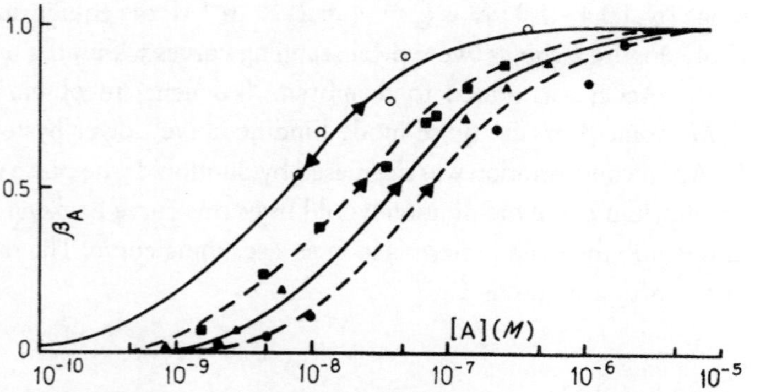


Fig. 5 - The hysteresis of the high affinity acetylcholine binding to *T. cal.* AcChR in membrane fragments at 4 °C. (○) equilibrium dialysis data of $^3$HAcCh binding, $[R_T] = 0.4\ \mu M$; the curve is calculated using $\beta_A = [A]/([A] + \bar{K})$ with $\bar{K} = (5 \pm 1) \times 10^{-9}\ M$; (■, ▲, ●) pulse mode addition (mixing) of $^3$HAcCh at $[R_T] = 0.1$, 0.4 and 1.0 $\mu M$ total receptor concentration, respectively, (α-Btx sites in the absence of detergent). The data are corrected for radioimpurity, acetylcholinesterase activity and non ($^3$HAcCh) – radioactivity (9).

the binding of the pulse mode addition is time-independent (≥ 17 h, 4 °C), it is only the upper curve that reflects true equilibrium binding [4].

Because the rapid, pulse-like addition of AcCh to the AcChR leads to less binding, receptor binding sites $R_1$ of lower affinity are preserved. Indeed, it is known that in the range of the acetylcholine concentration $[A] \leq 1\ \mu M$ where hysteresis is observed only the high affinity receptor conformers, and not the low affinity ones, of the conformational equilibria ($R_1 \rightleftarrows R_h$) are the targets of AcCh binding. Obviously, the extent to which the conformational equilibria are shifted to the side of the high affinity conformers is dependent on the way the acetylcholine concentration is increased (simple mixing *versus* dialysis).

In a recent study the existence of scanning curves [37, 38] was shown [5]. The kinetics of AcCh binding under dialysis conditions and the detergent-induced binding of a second α-bungarotoxin molecule to the AcChR in membrane fragments were analyzed. The main result is that the final very high affinity conformer $R_{vh}$ does not pre-exist in its unliganded form. Rather, it is induced by AcCh binding and is only stable as the acetylcholine binding complex $AR_{vh}$ [5].

Scanning curves — The hysteresis loops in Fig. 5 appear to increase in size with increasing receptor concentration. Since the scatter of the $^3$HAcCh data points is, however, rather large, the present accuracy of the method is not sufficient to argue in terms of a concentration dependence. The half-binding distribution constant $Q$ of the receptor concentration range $[R_T]$ 0.4 – 1.0 $\mu M$ is $Q = (4 \pm 2) \times 10^{-8}\,M$; the equilibrium constant is $\bar{K} = 5 \times 10^{-9}\,M$ [4]. Figure 6 shows two typical scanning curves within the main hysteresis loops. For instance, AcCh was added to membrane fragments to obtain $\beta_A = 0.3$ at $[A] = 4 \times 10^{-8}\,M$, point B on the pulse mode binding curve (lower hysteresis branch). Subsequently the AcCh concentration was decreased by dilution. If the pulse mode binding curve were an equilibrium curve the dilution would trace this curve backward. Instead the dilution curve enters into the main hysteresis loop as a scanning curve. The same is true for the dilution curve starting at point C.

## 4. Detergent effect on membrane bound AcChR

The binding of α-Btx is a well established method for quantitatively determining AcChR binding sites. Recently, it has been discovered that, when membrane fragments are exposed to 0.1 % (≈ 2 mM) Triton X-100, they bind twice as much α-Btx compared to the absence of the detergent. Figure 7 shows that increasing amounts of Triton X-100 cause a gradual increase in the binding of a second α-Btx molecule per monomer ($M_r$=290 000) of the receptor dimer R•R. Titration with Lubrol WX leads to a similar binding curve. The data in Fig. 7 suggest that detergent binding causes changes in the membrane bound AcChR that permit the binding of a second α-Btx molecule. The data can be described in terms of the steric hindrance model for the second α-subunit of the monomer ($M_r$ = 290 000). It appears that the increase in the concentration of the detergent removes the hindrance such that the second α-site is also occupied by α-Btx [5].

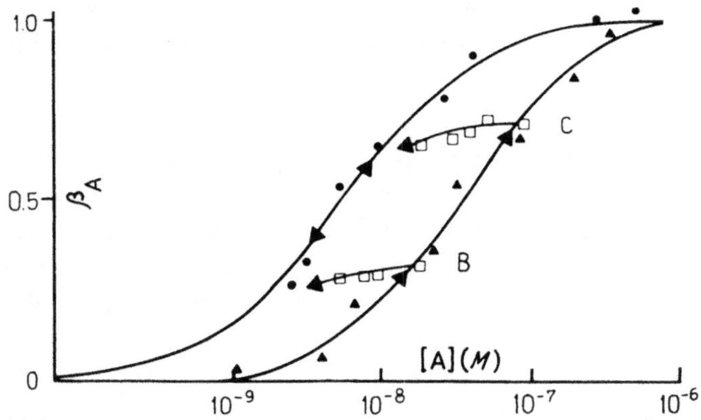


Fig. 6 - Two examples of scanning curves of the AcChR of *T. cal.* membrane fragments. Main hysteresis loop of pulse-like addition of AcCh (▲, →) and of dilution from $[A] > 1\,\mu M$ along the equilibrium (dialysis) curve (●, ⟷). □, dilution curves starting from the lower hysteresis branch. $[R_T] = 1\,\mu M$, $\bar{K} = 5 \times 10^{-9}\,M$ and $Q \approx 4 \times 10^{-8}\,M$. See Fig. 5.

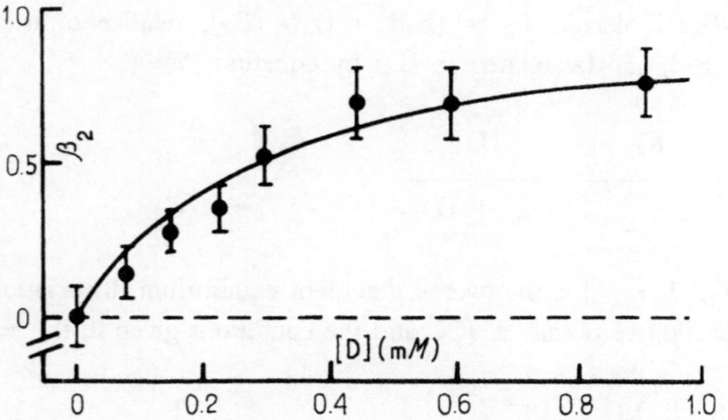


Fig. 7 - Detergent effect on α-Btx binding. $\beta_2$ is the degree of the binding of the second α-Btx molecule to AcChR of *T.cal.* membrane fragments, [D]: Triton X-100 concentration at 4 °C; 30 m*M* NaCl, 1 m*M* $CaCl_2$, 4 m*M* KCl, 4 m*M* Na-phosphate, 1 m*M* Pipes, 0.05 m*M* EDTA, 0.3 m*M* $NaN_3$, pH 7. Initial (at [D] = 0) concentration of AcChR (in terms of $M_r$ = 290 000): $[R_T] = [BR]_o = 0.03$ μ*M*, of α-Btx: $[B_T] = 0.24$ μ*M*; equilibrium constants of α-Btx binding: $K_1 \approx K_2 \approx 10^{-11}$ *M*.

Whereas in the absence of detergents (D) the binding of α-Btx (B) to the receptor monomer (R) is described globally by:

$$B + R \overset{(K_1)}{\rightleftarrows} BR, \tag{1}$$

the reaction sequence for the D titration in the presence of an excess of B is given by:

$$\begin{array}{l} nD + BR \overset{(K_D)}{\rightleftarrows} \ldots \rightleftarrows BR \cdot D_{n-1} + D \overset{(K_D)}{\rightleftarrows} BR \cdot D_n \\ \qquad\qquad\qquad\qquad\qquad\qquad\qquad\qquad \uparrow\downarrow \; (K_c) \\ \qquad\qquad\qquad\qquad\qquad\qquad\qquad\qquad BR' \cdot D_n \end{array} \tag{2}$$

$$B + BR' \cdot D_n \overset{(K_2)}{\rightleftarrows} B_2R' \cdot D_n \tag{3}$$

The structural transition R ⇄ R' converts the second α-site into a conformation that binds α-Btx with high affinity as soon as n detergent molecules have bound. Because of the high-affinity binding of the second B the smooth curve in Fig. 7 indicates the independent and allosteric nature of the detergent binding. Reaction (3) apparently does not shift the reaction (2) which is only dependent on the concentration of D. The reaction (3) may be used to measure the detergent effect on the receptor. Therefore the degree of the binding of the second α-Btx molecule, $\beta_2 = [B_2R' \bullet D_n]/[R_T]$, relative to the total receptor concentration $[R_T]$, can be related to [D] by equation [5]:

$$\beta_2 = \frac{K_C}{1 + K_c} \; \frac{[D]}{\bar{K}_D + [D]} \tag{4}$$

where $\bar{K}_D = K_D/(1 + K_c)$ is the overall detergent equilibrium dissociation constant. For D = Triton X-100, we obtain at 4 °C and the conditions given in the legend of Fig. 7:

$$\bar{K}_D = 2.8\ (\pm\ 0.3) \times 10^{-4}\ M.$$

In the case of the detergent Lubrol WX we obtain:

$$\bar{K}_D = 3.3\ (\pm\ 0.4) \times 10^{-4}\ M.$$

These constants are apparent quantities partially reflecting lipid-detergent exchange of the receptor protein.

## 5. Domain structure of AcChR

The AcCh binding hysteresis and the existence of scanning curves, see Fig. 6, qualify the AcChR as a physical memory molecule with co-operative non-equilibrium domain structure [37, 38]. The scanning curves are additional evidence for the previous conclusion that the pulse mode addition of AcCh to AcChR membrane fragments does *not* yield equilibrium values for $\beta_A$ and [A]. Rather, at the level of high affinity AcCh binding at [A] $\leqq$ 1 $\mu M$ simple (non-dialytic) mixing of AcChR with AcCh leads to non-equilibrium distributions of low and high affinity conformers.

The non-equilibrium distribution is extremely long-lived, it may be characterized by a half-binding constant *Q*. However, any equilibrium analysis of the pulse mode data in terms of equilibrium binding cooperativity (upward curvature of Scatchard plots) is forbidden. Already the fact that only one AcCh molecule binds to the high affinity receptor conformer in equilibrium excludes any ligand binding co-operativity. The non-equilibrium co-operativity indicated by hysteresis and scanning [36-38] can be successfully analyzed in terms of the receptor dimer concept [4, 39]. In particular the notion of a dimer hybrid $R_l \bullet R_h$ has proven to be very useful.

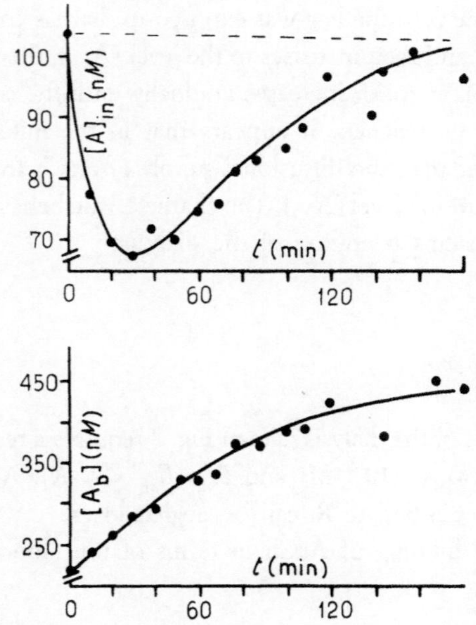


Fig. 8 - Top: Change in the $^3$HAcCh concentration of membrane fragments $[A_{in}]$ / [$^3$HAcCh] in solution prepared by the pulse mode, inside the dialysis bag as a function of dialysis time. At the start of the dialysis at $t$ = 0, $[A_{in}] = [A_o]$ and $\beta_A = \beta_o$. Bottom: Time course of additional $^3$HAcCh binding caused by the dialysis. Dashed line: $[A_{out}]$ in the absence of AcChR.

It is known that also in the absence of AcCh binding a conformational equilibrium of the type $R_l \rightleftarrows R_h$ exists between low affinity ($R_l$) and high affinity ($R_h$) conformers. The data suggest that the equilibrium constant $K_b$ of AcCh binding to $R_h$ has a value of about $10^{-7}$ $M$. Therefore $R_h$ is a conformer of intermediate affinity for AcCh; it is not the final conformer of highest affinity. The data in Fig. 8 suggest that the final receptor conformer, leading to the overall value $\bar{K} = 5 \times 10^{-9}$ $M$ at 4 °C, is only stable in the AcCh binding form $AR_{vh}$.

## 6. Kinetics of the dialysis mode of $^3$HAcCh binding

Figure 8 shows the time courses of the concentrations of free $^3$HAcCh, $[A_{in}]$, and of the bound $^3$HAcCh, $[A_b]$, both inside the dialysis bag. $^3$HAcCh was added to membrane fragments in the bag (pulse mode, yielding an initial value of $\beta_o$ and of $[A_{o,\,in}]$ on the lower hysteresis branch. At time zero the bag was exposed to dialysis conditions. It can be seen that $[A_{in}]$ first decreases and then increases to the level $[A_{out}]$, the $^3H$ AcCh concentration outside the bag. The bound AcCh increases gradually until the equilibrium value on the upper hysteresis branch is reached. It appears that in the initial phase the additional binding of AcCh is faster than the diffusional supply of AcCh from the outside solution such that $[A_{in}]$ sinks below the level $[A_{out}]$. The characteristic behaviour of $[A_{in}]$ and $[A_b]$ is intrinsic to the open-systems condition of the dialysis.

## 6.1. *Cyclic reaction scheme*

The kinetic analysis of the dialysis data in Fig. 8 requires a reaction scheme involving $R_l$ ($K_l \approx 10^{-4}$ *M*), $R_h$ ($K_h \approx 10^{-7}$ *M*) and $R_{vh}$ ($K_{vh} << \bar{K}$). At AcCh concentrations $[A] < 1\ \mu M$, the AcCh binding to $R_l$ can be neglected ($< 1$ %); hence the minimum reaction scheme for the binding of AcCh in terms of the monomer units R reads:

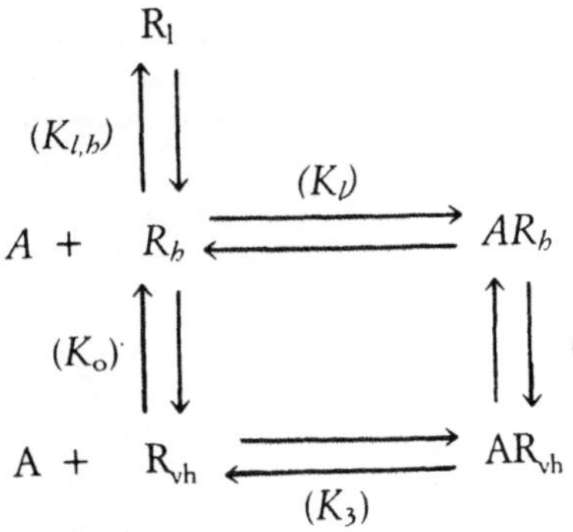


(5)

## 6.2. *The induced-fit model*

In scheme (5) the final high affinity complex $AR_{vh}$ can be obtained along two limiting pathways. The kinetic analysis will show that only the *induced fit-pathway*:

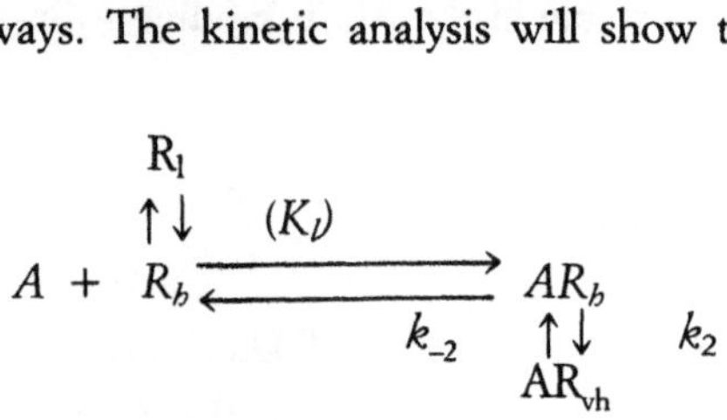


(6)

is consistent with the data. The pathway of direct, selective binding of A to the conformer $R_{vh}$ is negligible, *i.e.* practically excluded. The practical, apparent equilibrium constant for the high affinity AcCh binding for the general scheme (5) is given by equation (7):

$$\bar{K} = \frac{[A]\,([R_h] + [R_{vh}])}{[AR_h] + [AR_{vh}]} = \frac{K_l\,(1 + K_o^{-1})}{1 + K_2^{-1}} \tag{7}$$

where the individual apparent equilibrium and rate constants are given by:

$$\begin{aligned} K_1 &= [A]\,[R_h]/[AR_h] \\ K_2 &= [AR_h]/[AR_{vh}] = k_{-2}/k_2 \\ K_3 &= [A]\,[R_{vh}]/[AR_{vh}] \\ K_0 &= [R_h]/[R_{vh}] = k_{-0}/k_0 \\ K_{1,h} &= [R_1]/[R_h] \end{aligned} \tag{8}$$

For the induced-fit model in scheme (6), equation (7) is reduced to equation

$$\bar{K} = \frac{K_1}{1 + K_2^{-1}} = \frac{K_1 K_2}{1 + K_2} \tag{9}$$

6.3. *Rate equation*

The fundamental rate equation for the slow increase in the concentration of the complex $AR_{vh}$ (observed in Fig. 8) is:

$$\frac{d\,[AR_{vh}]}{dt} = k_2 [AR_h] - k_{-2} [AR_{vh}] \tag{10}$$

The degree of total high affinity binding is defined by $\beta = [A_b]/[R_T]$ or explicitly by equation (11):

$$\beta = \frac{[AR_{vh}] + [AR_h]}{[R_T]} = \frac{[AR_{vh}]\,(1 + K_2)}{[R_T]} \tag{11}$$

Mass conservation implies that the total receptor (binding site) concentration is:

$$[R_T] = [R_h] + [R_1] + [AR_h] + [AR_{vh}] = [R_h]\,(1 + K_{1,h}) + \beta \cdot [R_T] \tag{12}$$

Because of the min/h kinetics indicated in Fig. 8 it is reasonable to assume that the bimolecular binding step (associated with $K_l$) is fast compared to the slow structural transition ($k_2$, $k_{-2}$). Therefore, $[AR_h]$ is given by the pre-equilibrium:

$$[AR_h] = K_I^{-1}\,[A]\,[R_h] \tag{13}$$

Equation (12) yields $[R_h] = (1-\beta)\,[R_T]/(1 + K_{l,h})$, substitution into equation (13) and then into equation (10) finally leads to the practical rate equation:

$$\frac{d\beta}{dt} = a\,[A]\,(1-\beta) - k_{-2}\,\beta \tag{14}$$

where with equations (8) and (9)

$$a = \frac{k_2\ (1+K_2)}{K_1\ (1+K_{l,h})} = \frac{k_{-2}}{\bar{K}\ (1+K_{l,h})} \tag{15}$$

### 6.4. *Initial rate*

It is readily seen that

$$\frac{d\beta}{dt} = -\frac{d\,(1-\beta)}{dt} = -\frac{d\,[A]\,/\,dt}{[R_T]} \tag{16}$$

At the start of the dialysis at $t = 0$, $\beta = \beta_0$ and $[A_{in}] = [A_0]$, inside the dialysis bag. Because $\bar{K}^{-1} >> 1$, the initial rate of $^3$HAcCh binding at $t = 0$ is given by equation (17):

$$\left(\frac{d\beta}{dt}\right)_0 = a\,[A_0]\,(1-\beta_0) \tag{17}$$

From equation (16) we obtain:

$$\left(\frac{d[A]}{dt}\right)_0 = -\,a\,(1-\beta_0)\,[R_T]\,[A_0] \tag{18}$$

Lumping the experimental quantities together we may define a relative initial rate:

$$v_0 = -\frac{(d\,[A]\,/\,dt)_0}{(1-\beta_0)\,[R_T]} = a\,[A_o] \tag{19}$$

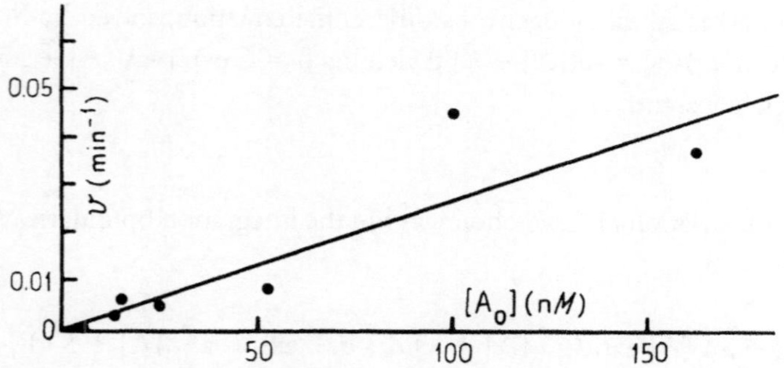


Fig. 9 - The initial relative rate $v_o$ as a function of initial AcCh concentration $[A_o]$ in the dialysis bag, see equation (19) of the text.

It can be seen from equations (18) and (19) that the induced-fit model in scheme (5) predicts that the initial rate of the decrease in $[A_{in}]$ is proportional to the initial AcCh concentration at the start of the dialysis. On the other hand the limiting case of rapid direct, selective binding $A + R_{vh} \rightleftarrows AR_{vh}$ coupled to the slow, rate-determining step $R_{vh} \rightleftarrows R_h$; $k_{-o}$, $k_o$) leads to an expression for the initial relative rate:

$$v_o = - \frac{(d[A] / dt)_o}{(1-\beta_o)\ [R_T]} = k_o \tag{20}$$

This means that $v_0$ is a constant, independent of $[A_o]$. Although analysis of the dialysis data according to equation (19) is rather inaccurate because of the inaccuracy of graphical slope determinations, the data of Fig. 5 show clearly that the $v_o$-values are dependent on $[A_o]$. The linear approximation yields $a \approx 3 \times 10^5\ M^{-1}\ \text{min}^{-1}$.

## 6.5. *Integral rate equation*

Because two sets of data, [A] and $\beta = \beta_A$, are available for every time point, the entire time courses, $[A]\,(t)$ and $\beta(t)$, may be treated analytically. For this purpose, equation (14) is rewritten as:

$$\frac{d\beta}{dt} = a\ [A] - a\ ([A] + \bar{K})\ \beta \tag{21}$$

Equation (21) is an inhomogeneous differential equation, solved by first solving the homogeneous form $d\beta/dt = -a([A] + \bar{K})\,\beta$ yielding $\beta = C \exp[-a\int([A] + \bar{K})\,dt]$, where $C$ is an integration constant.

By the method of variation of $C$ we obtain within the integration boundaries $t$ and $t_o = 0$, $\beta$ and $\beta\ (t = 0) = \beta_o$:

$$\beta = \left(\beta_o + a\int [A] \exp\{a\int [A] + \bar{K})\,dt\}\,dt\right) \exp\{-a\int ([A] + \bar{K})dt\} \tag{22}$$

The available database permits a graphical evaluation replacing the integrals by sums according to

$$\int_0^t [A(t)]\,dt = \Sigma_{i=1} [\bar{A}_i] \Delta t_i \tag{23}$$

where $[\bar{A}_i] = ([A_i] + [A_{i-1]})/2$ and $\Delta t_i = t_i - t_{i-1}$ with $t_{i-1} = t_o$ at $i = 1$.

Introducing the terms

$$y = \beta - \beta_o b \tag{24}$$

$$x = (\Sigma[\bar{A}_i] \exp[a\ \Sigma\ ([\bar{A}_j] + \bar{K})\ \Delta t_j]_i\ \Delta t_i)\ b \tag{25}$$

$$b = \exp[-a\Sigma([\bar{A}_i] + \bar{K})\ \Delta t_i] \tag{26}$$

the dialysis data can be evaluated according to $y = ax$. As shown in Fig. 10 the dependence is linear yielding $a = 6.5 \times 10^5\ M^{-1}\ \text{min}^{-1}$.

The fraction of $R_1$ conformers (at $[A] = 0$) was estimated $\approx 0.8$ [40]; thus, $K_{1,h} \approx 0.8/0.2 = 4$. From equation (15), we obtain $k_{-2} = a\,\bar{K}\,(1+K_{l,h}) \approx 10^{-4}\,(2.7 \pm 0.2)\ s^{-1}$. Since $[AR_{vh}]$ appears to be large compared with $[AR_h]$ it is obvious that $K_2^{-1} << 1$, and therefore $\bar{K} \approx K_1K_2 = K_1k_{-2}/k_2$. If we take the estimate $K_1 \approx 10^{-7}\ M$, we find $k_2 \approx K_1k_{-2}/\bar{K} \approx 5.4 \times 10^{-3}\ s^{-1}$ and $K_2 \approx 0.05$.

The numerical values of $k_2 \approx 5.4 \times 10^{-3}\ s^{-1}$ and $k_{-2} \approx 0.27 \times 10^{-3}\ s^{-1}$ correspond to the rather slow kinetics of the additional AcCh binding under dialysis conditions. The intrinsic relaxation rate of the $AR_h \rightleftarrows AR_{vh}$ equilibration is given by

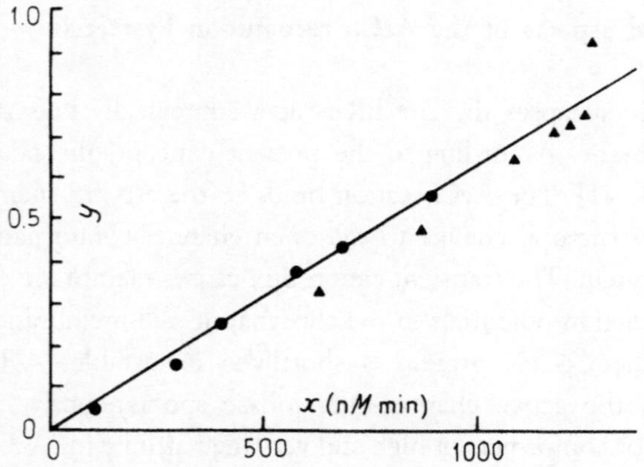


Fig. 10 - Dialysis data evaluation according to $y = ax$ derived from equation (22) of the text. (●, ▲) two different protein concentrations.

$$\frac{1}{\tau} = k_2 + k_{-2} \approx 5.67 \times 10^{-3}\ s^{-1}$$

or $\tau \approx 3$ min. This value may be compared to the time constant of AcCh equilibration across the dialysis bag in the absence of binding (no membrane fragments): $\tau_{equil.} \approx 30$ min [5]. Therefore the kinetics of [A] changes in the presence of membrane fragments must be rate-limited by the receptor structural changes. The rather dramatic decrease of $[A_{in}]$ within about 30 min after start of the dialysis is caused by the rather limited size of the surface of the dialysis bag limiting the net amount of AcCh which can enter the bag. In the initial dialysis phase, the AcCh rich membrane fragments bind AcCh more rapidly than the rate of supply by diffusion from the outside of the dialysis bag.

### 6.6. *Reaction scheme of the AcCh recognition hysteresis*

On the basis of the kinetic data, the reaction scheme previously developed to rationalize thermodynamic data [4] can be specified in terms of the induced-fit mechanism, excluding major amounts of direct binding of AcCh to the very high affinity conformer because its concentration is negligibly small (see Fig. 11).

## 7. Physiological aspects of the AcCh recognition hysteresis

In cholinergic synapses the AcChR is at a strategically important position of the postsynaptic membrane. According to the present concept, the nerve impulse leads to release of AcCh [1, 41]. The AcCh cation binds to the AcChR membrane receptor and causes a transient structural change to the open channel conformation of this $Na^+/K^+$ transport-gating protein. The transient cation flux causes membrane depolarization which finally may trigger action potentials in the subsynaptic cell membrane. The open channel conformation of the AcChR protein is shortlived metastable [42]. In the prolonged presence of AcCh the active channel-open phase spontaneously converts to inactive, desensitized receptor conformers of high and very high affinity for AcCh [43]. It is only the $R_l$ conformers that are involved in channel activity, most likely as highly co-operative dimers (double-channels), as shown recently [11]. However, due to the conformational coupling ($R_l \rightleftarrows R_h$) the low affinity receptor conformers are also affected by the hysteresis of the high affinity conformers.

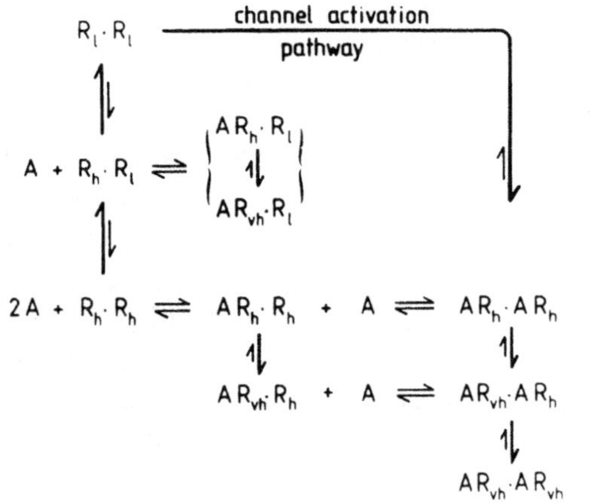


Fig. 11 - Reaction scheme of high affinity AcCh binding to the *T.cal.* AcChR. The simple bimolecular processes are represented by the horizontal sequences, whereas the vertical steps model the various slow structural isomerizations. $R_l$: low affinity conformer ($K_l \approx 10^{-4}\ M$); $R_h$: high affinity conformer ($K_h = 10^{-7}\ M$); $R_{vh}$: very high affinity conformer ($K_{vh} << \bar{K} = (5 \pm 1) \times 10^{-9}\ M$). The $R_h \bullet R_l$ *hybrid form* is the branching point for the dead end hybrid $AR_{vh} \bullet R_l$ of the pulse mode non-equilibrium states. The thick arrows indicate the preferential position of the isomerization equilibria. The large upper arrow indicates the channel activation pathway. At high AcCh concentration ($> 10^{-6}\ M$), the low affinity conformer, $R_l$, is directly involved in the binding which results in channel opening and, at prolonged exposure to AcCh, is subsequently transformed to the high affinity conformers. At low AcCh concentrations ($< 10^{-6}\ M$), the $R_h$ conformer is the dominant direct reaction partner for AcCh and thus constitutes the direct route of the high affinity state pathway to the final complex $AR_{vh}$. Hysteresis: *Single pulse mode addition* of AcCh (closed system) favors the hybrid $AR_{vh} \bullet R_l$ binding state, whereas *dialysis mode addition* (open system) ultimately leads to the $AR_{vh} \bullet AR_{vh}$ conformer.

The longevity of the metastable hybrid conformations $AR_h \cdot R_l$ and $AR_{vh} \cdot R_l$ may be caused by steric hindrance in the dimer preventing the $R_l \rightarrow R_h$ transition. Functionally, the partially occupied hybrids may be viewed as a saving device for the channel-active $R_l$ conformers. At low AcCh concentrations, $[A] < 1\ \mu M$, hysteresis comprises a mechanism to prevent the total conversion to the inactivated, desensitized conformers. It has recently been shown [6] that AcCh binding enhances the phosphorylation of phosphatidylinositol by the isolated nAcChR-lipid complex as well as the autophosphorylation of the receptor. If now the AcCh binding hysteresis is concomitant with a phosphorylation hysteresis, the (desensitized) AcChR may serve as a memory molecule in the transsynaptic information flux of nerve-nerve and nerve-muscle synapses. Thus, the AcCh binding hysteresis may serve as part of the molecular mechanism of synaptic memory imprint and recognition memory.

## Acknowledgements

We thank M. Pohlmann for the typing of the manuscript and the Deutsche Forschungsgemeinschaft for financial support, grant Cl, SFB 223, to E.N.

## References

[1] D. Nachmansohn, *Harvey Lect.,* **49,** 57 (1955).
[2] J.P. Changeux, J. Giraudat and M. Dennis, *TIPS,* **8,** 459 (1987).
[3] J. Cartaud, in *Ontogenesis and Functional Mechanisms of Peripheral Synapses,* J. Taxi (Editor) Elsevier, New York, (1980) pp. 199-210.
[4] H.W. Chang, E. Bock and E. Neumann, *Biochemistry,* **23,** 4546 (1984).
[5] E. Neumann, E. Boldt, B. Rauer, H. Wolf and H.W. Chang, *Bioelectrochem. Bioenerg.,* **20,** 45 (1988).
[6] R. Kiehl, M. Varsányi and E. Neumann, *Biochem. Biophys. Res. Commun.,* **147,** 1251 (1987).
[7] H.W. Chang and E. Bock, *Biochemistry,* **16,** 4513 (1977).
[8] S.L. Hamilton, M. McLaughlin and A. Karlin, *Biochem. Biophys. Res. Commun.,* **79,** 692 (1977).
[9] J.P. Changeux, A. Devillers-Thiéry and P. Chemouilli, *Science,* **225,** 1335 (1984).
[10] C. Toyoshima and N. Unwin, *Nature (London),* **336,** 247 (1988).
[11] H. Schindler, F. Spillecke and E. Neumann, *Proc. Natl. Acad. Sci. USA,* **81,** 6222 (1984).

[12] F. Spillecke, *Thesis,* University of Konstanz, Martinsried, G.F.R. (1984).
[13] V. Volz, *Diploma thesis,* University of Bielefeld, G.F.R. (1987).
[14] R. Rüchel, D. Watters and A. Maelicke, *Eur. J. Biochem.,* **119,** 215 (1981).
[15] H. Schindler and U. Quast, *Proc. Natl. Acad. Sci. USA,* **77,** 3052 (1980).
[16] A. Karlin and E. Bartels, *Biochim. Biophys. Acta,* **126,** 525 (1966).
[17] J.W. Walker, R.J. Lukas and M.G. McNamee, *Biochemistry,* **20,** 2191 (1981).
[18] T. Schürholz, J. Weber and E. Neumann, *Bioelectrochem. Bioenerg.,* **21,** 71 (1989).
[19] O.P. Hamil and B. Sakmann, *Nature (London),* **294,** 462 (1981).
[20] B. Sakmann, J. Patlak and E. Neher, *Nature (London),* **286,** 71 (1980).
[21] I. Silman and A. Karlin, *Science,* **164,** 1420 (1969).
[22] R.N. Cox, M. Kawai, A. Karlin and P.W. Brandt, *J. Membr. Biol.,* **51,** 145 (1979).
[23] H.A. Lester, M.E. Krouse, M.M. Nass, N.H. Wassermann and B.F. Erlanger, *J. Gen. Physiol.,* **75,** 207 (1980).
[24] A. Maelicke, *Angew. Chem.,* **96,** 193 (1984).
[25] J.M. Wolosin, A. Lyddiatt, J.O. Dolly and E.A. Barnard, *Eur. J. Biochem.,* **109,** 495 (1980).
[26] H.W. Chang, E. Bock and E. Neumann, in *Molecular Basis of Nerve Activity,* J.P. Changeux, F. Hucho, A. Maelicke and E. Neumann, (Editors), W. de Gruyter, Berlin, (1985), pp. 369-385.
[27] R.R. Neubig, N.D. Boyd and J.B. Cohen, *Biochemistry,* **21,** 3460 (1982).
[28] P. Pennefather and D.M. Quastel, *Br. J. Pharmacol.,* **77,** 395 (1982).
[29] T. Heidmann, J. Bernhardt, E. Neumann and J.P. Changeux, *Biochemistry,* **22,** 5452 (1983).
[30] G. Boheim, W. Hanke, F.J. Barrantes, H. Eibl, B. Sakmann, G. Fels and A. Maelicke, *Proc. Natl. Acad. Sci, USA,* **78,** 3586 (1981).
[31] M. Montal, P. Labarca, D.F. Fredkin, B.A. Suarez-Isla and J. Lindstrom, *Biophys. J.,* **45,** 165 (1984).
[32] M. Spondheim and E. Neumann, *Biophys. Chem.,* **3,** 109 (1975).
[33] D. Thiele and W. Guschlbauer, *Biopolymers,* **8,** 361 (1969).
[34] A. Katchalsky and R. Spangler, *Q. Rev. Biophys.,* **1,** 127 (1968).
[35] A. Katchalsky and E. Neumann, *Int. J. Neurosci.,* **3,** 175 (1972).
[36] E. Neumann, *Angew. Chem.,* **85,** 430 (1973); *Angew. Chem., Int. Ed. Engl.,* **12,** 356 (1973).
[37] D.H. Everett, *Trans. Faraday Soc.,* **51,** 1551 (1955).
[38] D.H. Everett, in *The Solid-Gas Interface,* E.A. Flood (Editor), Marcel Dekker, New York, (1967) Vol. **2,** p. 1055.
[39] E. Neumann, in *Nicotinic Acetylcholine Receptor,* A. Maelicke (Editor), NATO ASI Serien, (1986), Vol. **H3,** pp. 177-196.
[40] N.D. Boyd and J.B. Cohen, *Biochemistry,* **19,** 5344 (1980).

[41] B. Katz, *Release of Neural Transmitter Substances,* Liverpool University Press, Liverpool, (1969) p. 55.

[42] E. Neumann and J. Bernhardt, *J. Physiol. (Paris),* **77,** 1061 (1981).

[43] B. Sakmann, J. Patlack and E. Neher, *Nature (London),* **286,** 71 (1980).

# CYTOCHROME *c* OXIDASE: STRUCTURE

ROBERTO BISSON
*C.N.R. Centro per lo Studio della Fisiologia dei Mitocondri,*
*Laboratorio di Biologia e Patologia Molecolare,*
*Istituto di Patologia Generale, Università di Padova,*
*via Loredan 16, 35131 Padova, Italy*

CONTENTS



*Bioelectrochemistry III*
Edited by G. Milazzo and M. Blank
Plenum Press, New York, 1990

## 1. Introduction

The cytochrome oxidases are a wide and heterogeneous family of electron transfer hemoproteins which play a central role in the energy metabolism of aerobic organisms. They catalyze the terminal electron transfer steps in the complex branched respiratory systems of prokaryotes as well as in the mitochondrial respiratory chain [1]. The structural and functional variety of these enzymes is perhaps exemplified by *P. denitrificans* where, depending on the growth conditions, up to five different terminal oxidases termed $aa_3$, cytochrome *o*, cytochrome $a_1$, cytochrome *d* and cytochrome $cd_1$ can be synthesized [2]. The differences among the various oxidases are consistent. Copper and heme *a*, the redox centers present in the mitochondrial enzyme, may be replaced in the bacterial terminal oxidases by different prosthetic groups, while cytochrome *c* and oxygen may not always be the electron donor and acceptor. Also the integration in the membrane does not appear as a general feature. Depending on their spectroscopic properties and prosthetic groups, the cytochrome oxidases have been arranged in different classes [2, 3]. Nevertheless, the structure and the kinetic competence of most of them remain poorly understood. In this situation, changes of class and nomenclature are not surprising [4, 5]. Only recently the structural organization of these proteins in bacteria began to be elucidated at molecular level. In *E. coli* the two terminal oxidases, termed cytochrome *d* and cytochrome *o*, have been characterized. In spite of their similar function, namely the transfer of electrons from ubiquinol to oxygen, the two enzymes have different prosthetic groups (two protoheme IX groups and two hemes *d* in cytochrome *d*, two copper atoms and two protoheme IX in cytochrome *o*) and different subunit composition (two and four polypeptides, respectively). As a consequence, the two oxidases exhibit different $K_m$ for the substrate and different sensitivity to inhibitors (at last one order of magnitude); these properties may offer a selective advantage to the organism under different environmental conditions [3].

By far the best characterized class of oxidases remains the $aa_3$-type, whose members are generally termed cytochrome *c* oxidases. It is the only one to be represented both in eukaryotes (with the sole exception of some protozoa where the presence of cytochrome *o* has been demonstrated) and in most aerobic bacteria. The high efficiency in energy transduction of the $aa_3$ oxidases, which can couple proton translocation to electron transfer from cytochrome *c* to oxygen, may perhaps explain this wide distribution among

## 2. Purification

A basic step in the study of a protein is its purification from suitable sources. In the case of cytochrome *c* oxidase this is facilitated by the fact that the enzyme has a characteristic spectrum and an easily measurable activity. The fact that cytochrome *c* oxidase is present at a relatively high concentration in the inner mitochondrial membrane, can explain why the beef heart was for a long time the most frequent choice [14, 15]. The conventional procedure, later extended to some other eukaryotic sources [16-18], was based on the differential solubilization of the protein complexes with detergents and salts. In the most common approach, the so called *green membrane splitting*, the anionic detergents cholate and deoxycholate and KCl were used, taking advantage of the low solubility of the oxidase with respect to the other mitochondrial enzymes under these conditions. The oxidase-enriched membranes were then solubilized and the enzyme isolated by repeated fractionation with ammonium sulphate, in a procedure that could last from many hours to several days. The modifications to the general method, introduced by various laboratories, led to a series of preparations which frequently appear to be heterogeneous in some of their spectroscopic and functional properties [19-22] (see Section *8.1*). The idea of an inadequacy of the different isolation techniques was also emphasized by the possibility of protein contamination, suggested after the advent of the sodium polyacrylamide gel electrophoresis (SDS-p.a.g.e.) as additional tool for testing the enzyme purity (see Section **3**). To overcome these problems which, in part, could be the result of a prolonged exposure of the enzyme to the artificial environment created by detergents and salts, rapid purification methods based on column chromatography have been developed. In the most common procedure, an affinity column is used after a preliminary ion exchange or hydrophobic chromatography step on the solubilized mitochondria or oxidase–enriched membranes [23, 25–29]. More recently, methods based on h.p.l.c. and f.p.l.c. systems have also been proposed [30–32]. Nevertheless, although simpler and straightforward with respect to the classical procedure, the chromatographic techniques do not allow the isolation of *purer* enzymes (at least from eukaryotes). Rather they have made possible the isolation of cytochrome *c* oxidases from new sources where the previous methods would have been difficult to apply because of the limited amount of available material, its sensitivity to prolonged manipulations and the need for a laborious setting in order to find the most useful experimental conditions [5, 23, 33–35]. However, when there are no limitations on the biological sources and large amounts of enzyme are required, the development of *ad hoc* procedures may still be the best solution [36, 37].

As in the case of the purification of other membranous enzymes, a major problem is the choice of the detergent. Because of the loose interaction of certain subunits with the rest of the complex, even mild detergents such as lauryl maltoside, Triton X-100 and dodecyldimethylamine oxide (LDAO) can lead to loss of specific enzyme components

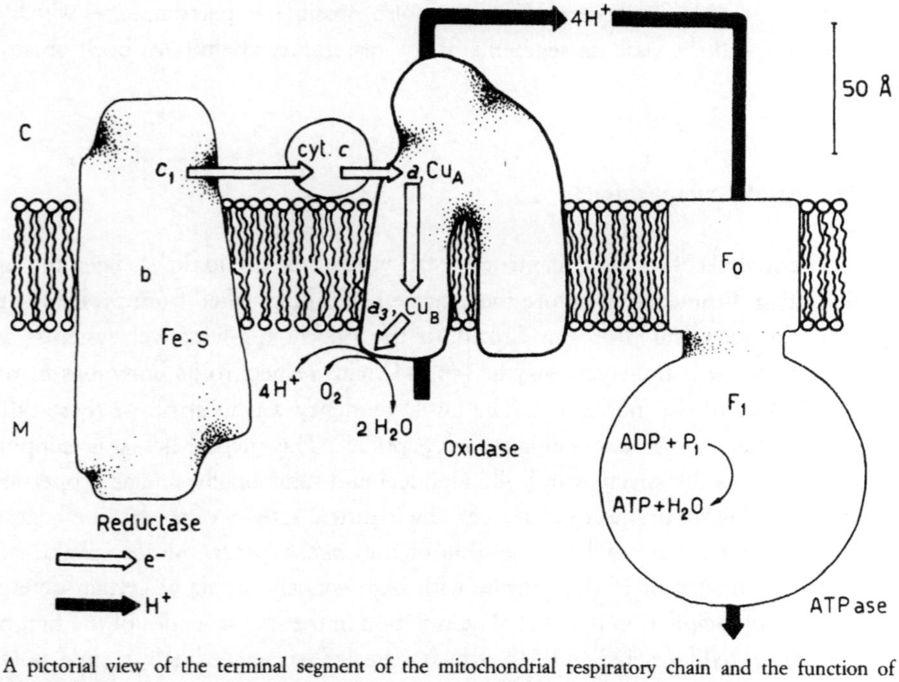


Fig. 1. – A pictorial view of the terminal segment of the mitochondrial respiratory chain and the function of cytochrome *c* oxidase.

so many different organisms favored by evolution. The observation that cytochrome *c* oxidase *is probably responsible for more than 90 % of the oxygen consumption by living organisms in the earth* [6] very effectively summarizes the important role of this group of hemoproteins.

Most of the available structural and functional data concern the mammalian form of the enzyme and in particular the bovine heart cytochrome *c* oxidase (Fig. 1). Only recently, information on the bacterial form of the protein has become available and is rapidly increasing, because of the interest in the much simpler polypeptide composition [2] (Section **3**). Many technical limitations, which have hampered their study, are now overcome by molecular genetics that, under certain conditions, may also offer the possibility of studying the enzyme directly *in vivo*.

This chapter is an overview of recent developments and present information on the structure of cytochrome *c* oxidase. For a more comprehensive view the reader may refer to the excellent reviews that have been published recently [2, 6-13].

[28, 29, 32, 38-40]. The reverse situation is also possible; supercomplexes which include both the $bc_1$ and the oxidase segments of the respiratory chain have been obtained [41, 42].

### 3. Polypeptide composition

The analysis of the subunit structure of cytochrome *c* oxidase has been the subject of a long-lasting dispute. Well before the enzyme could be purified from prokaryotes and in spite of the technical problems linked to the poorly resolving gel systems, a higher complexity of the mammalian enzyme [43, 44] with respect to its homologs in yeast and fungi [17, 45, 46] was noticeable. The initial tendency was to attribute these differences to a copurification of contaminant polypeptides. This hypothesis was supported by evidence such as the spectroscopically identical and functionally similar properties of the enzyme from higher to lower eukaryotes, the identical activity of the bovine heart oxidase before and after the proteolytic digestion of the *contaminants*, the loss of the *impurities* either by the treatment of the enzyme with high concentrations of certain detergents or after immunoprecipitation [47, 48]. The isolation in the late seventies of the first bacterial cytochrome *c* oxidases [23, 49-51], which revealed an exceedingly simple polypeptide composition (only two three subunits), appeared as an indirect, further support to the possibility of contamination in the isolated mammalian enzyme. About the same time, new data began to challenge this view. Newly developed, high-resolution gel systems were able to completely resolve the different polypeptides suggesting a subunit structure for the mammalian enzyme even more complex than previously thought: 13 subunits with a molecular weight ranging from 50 to 5 kDa [52]. The polypeptides appeared to be present in a 1:1 stoichiometry [53], a very unlikely event for copurified contaminants. Also the loose binding of some components to the rest of the complex was not maintained as a significant proof for contamination when it was realized that suitable detergents and pH conditions can have the same effect on subunit III, a polypeptide universally present in cytochrome *c* oxidase of eukaryotes and many bacteria [39, 54-56]. A complete agreement exists now on this conclusions supported by the isolation and sequencing of all thirteen polypeptides from bovine heart cytochrome *c* oxidase [57-59]. The different and confusing nomenclature that still is found in the literature is perhaps the only negative aftermath of the settled dispute (Fig. 2).

The improvement in the separation technique has also demonstrated a slight increase in the complexity of the yeast oxidase which, with its 9 polypeptides [60], remains well below the mammalian counterparts.

The composition of the bacterial cytochrome *c* oxidase is clearly very simple. Several prokaryotic enzyme forms appear to contain three subunits homologous to the

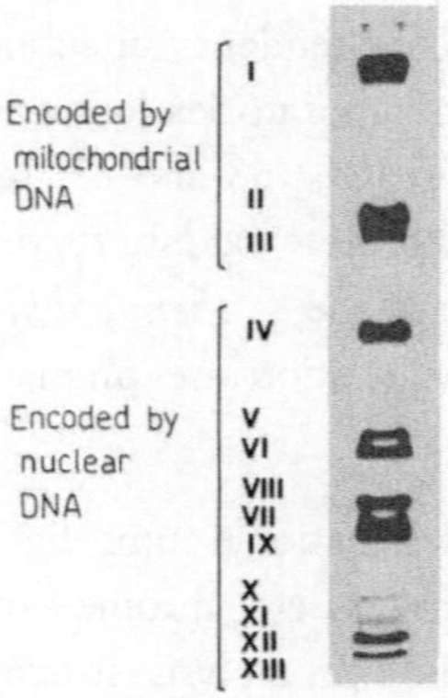


| Buse 1988 | Capaldi 1986 | Kadenbach 1980 | Capaldi 1976 | Buse 1976 | Mr | N-terminal sequence |
|---|---|---|---|---|---|---|
| I | $Mt_I$ | I | I | I | 56993 | Formyl-MFI |
| II | $Mt_{II}$ | II | II | II | 26049 | Formyl-MAY |
| III | $Mt_{III}$ | III | III | III | 29918 | (M)TH |
| IV | $C_{IV}$ | IV | IV | IV | 17153 | AHG |
| V | $C_V$ | Va | V | V | 12436 | SHG |
| VI | $C_{VI}$ | Vb | a | VIa | 10668 | ASG |
| VII | AED | VIb | c | VII | 10063 | Acetyl-AED |
| VIII | ASA | VIa | b | VIb | 9419 | ASA |
| IX | STA | VIc | VI | VIc | 8480 | STA |
| X | $C_{VII}$ | VIIa | | VIIIc | 6243 | FED |
| XI | IHQ | VIIb | VIIs | | 6000? | IHQ |
| XII | $C_{VIII}$ | VIIc | | VIIIa | 5441 | SHY |
| XIII | $C_{IX}$ | VIII | | VIIIb | 4962 | ITA |

Fig. 2 - Subunit composition and nomenclature of bovine cytochrome *c* oxidase. For historical reasons (see Section 3), five different nomenclatures are currently used in literature. In the present review the nomenclature recently proposed by Buse and colleagues [91], after the complete definition of the subunit structure of the enzyme, is taken. To avoid confusion, the results mentioned in the text have been *translated in this language* using the *dictionary* reported in the figure. The genetic origin of the enzyme subunits and their migration, in the SDS-p.a.g.e. system of Kadenbach *et al.* [92], is shown.

three largest mitochondrial polypeptides [5, 23, 33, 61, 62]. This is true also in the case of the *Paracoccus* complex in spite of the initial isolation of a two-subunits enzyme [50]. Only after the identification of the *Paracoccus* subunit III gene, was a method devised to isolate the three-subunits oxidase; again a major problem was the choice of the detergents [32, 63]. There is, however, evidence that does not support the three-subunits enzyme as a general model for the prokaryotic cytochrome *c* oxidase. The $aa_3$ oxidase purified from the thermophilic bacterium *T. thermophilus* contains only two subunits and the peculiar characteristic of an heme *c* covalently linked to the smaller polypeptide [51, 64]. The largest subunit, homologus to the mitochondrial subunit I [65], has been proposed as a *single-subunit* oxidase [66]. Very recently, however, it has been demonstrated that the smaller polypeptide has sequence homologies with both cytochrome *c* and the mitochondrial subunit II [65]; hence, it is probably the product of a fused gene. The same results have been found in the oxidase of *PS3* (N. Sone, personal communication), another thermophilic bacterium, but with an important difference: here a subunit apparently similar to subunit III is present [61]. Hence, in spite of some differences which can be present in the structure of the enzyme of organisms living under extreme conditions, the latter result adds further support to the three-subunits complex as the minimal structural unit *in vivo*. The isolation of two-subunits bacterial $aa_3$ oxidases, however, is not limited to the *Thermus* enzyme and it has been reported also in the case of *Thiobacillus novellus* [49, 67], *Nitrobacter agilis* [49, 68], *Rhodobacter sphaeroides* [24], *Nitrosomonas europaea*

[4], *Pseudomonas AMI* [69] and the alkalophilic bacterium *Bacillus firmus* [70] (here a third low-molecular-weight polypeptide is stoichiometrically present but it corresponds to a cytochrome *c*). However, considering the loose binding and the tendency of subunit III to aggregate, and some conflicting results obtained in different laboratories (compare, for example, the oxidase preparations from *N. agilis* according to Refs. 68 and 23; see also the polypeptide composition of *Nitrosomonas europaea* cytochrome *c* oxidase reported in Refs. 4 and 5, and the intriguing results obtained in the case of the *T. thermophilus* [64, 71]), it seems clear that conclusions based only on the purification are premature.

The important question about the enzyme minimal functional unit has been considered in very recent experiments. A subcomplex of *Paracoccus* cytochrome *c* oxidase containing only the larger subunit has been prepared by limited proteolysis; it exhibits a similar $K_m$ value for cytochrome *c* and a significant electron transfer activity [72, 73]. In this context, another intriguing result has been found. From the aerobic photoheterotroph *Erythrobacter longus*, an $aa_3$ oxidase which appears to be composed of two identical subunits has been purified [35]. The structural characterization of the enzyme, however, deserves further study. Analysis of the polypeptide composition by SDS p.a.g.e. cannot be considered completely reliable by itself. On this basis, for example, the *PS3* oxidase was originally described as a single subunit enzyme [74] and the *Nitrosomonas europoea* $aa_3$ oxidase as a two-subunits complex [4]; later it was shown that both were *classical* bacterial three-subunits cytochrome *c* oxidases [5, 61].

In conclusion, a general agreement exists on the increasing complexity of cytochrome *c* oxidase with the stage of evolution of the organism. Within the limits mentioned above, the bacterial enzyme contains at most three polypeptides which are homologous to the three largest subunits present in the eukaryotic enzyme. To the latter, a progressively increasing number of polypeptides is added from 4 in the slime mold *D. discoideum* [29] to 5 in *N. Crassa* [46], 6 in the yeast *S. Cerevisiae* [60], 7-9 in fishes and birds [75, 76], 10 in mammals [52]. On the basis of the sequences now available for the bovine and yeast enzymes, it would appear that all the smaller polypeptides found in the lower eukaryote have an homologous counterpart in the mammalian oxidase; four polypeptide are probably unique to higher eukaryotes [13, 57, 77, 78].

Pioneering work on the mitochondrial biosynthesis in *Neurospora crassa* and *Saccharomyces cerevisiae*, later extended to other organisms, has demonstrated that the three largest subunits of cytochrome *c* oxidase are synthesized in mitochondria while the remaining polypeptides are coded for by the nuclear genome and imported into the mitochondria [79-84]. This digenomic origin of the enzyme poses additional interesting problems concerning the coordinate expression and the correct assembly of the different subunits.

The organization and transcription of the oxidase genes, based on investigation of the yeast oxidase, appears to be complex. Though present in single copy, the nuclear structural genes are unlinked and scattered on different chromosomes; some of them can produce multiple transcripts and contain introns [85]. Preliminary experiments do not support coordination of the expression at transcription level [85]. An intricate network of regulatory, trans-acting nuclear genes appear to exert specific effects on the expression of oxidase genes encoded on mitochondrial DNA [86] while others have an essential influence on the assembly of a functional enzyme [87].

Because of the interest in the mitochondrial genome, the DNA sequences of the genes for the three largest subunits are now available from many different organisms (see legends to Figs, 7 and 9). Recently also the genes for the bacterial cytochrome *c* oxidase of *Paracoccus* (subunit I, II and III), *PS3* (subunit I and II) and *T. thermophilus* (subunit I) have been cloned and sequenced, after a partial characterization of the subunits by protein sequencing [63, 65, 88-90]. The evident homologies with the three largest eukaryotic subunits, which support the endosymbiotic hypothesis, could be very useful for the identification of the enzyme functional domains (see Section 7).

## 4. Shape

The low solubility of the enzyme membranous form is not only an advantage for the purification but it was probably an important factor also for the preparation of two-dimensional crystals of the protein, obtained by treating beef heart mitochondrial membranes with suitable detergents [93-95]. Electron microscopy of the *green membranes*, prepared in the presence of Triton X-114 and Triton X-100, occasionally showed large patches of ordered cytochrome *c* oxidase molecules [96, 97]. A second differently ordered array could also be obtained by deoxycholate extraction of mitochondria [98, 99]. In the first case, image reconstruction analysis demonstrates the presence of collapsed vesicles containing dimeric cytochrome *c* oxidase with all molecules having the same vectorial orientation relative to the inside of the vesicle. The enzyme was asymmetrically disposed across the membrane projecting most of its mass toward the inside of the flattened vesicles and very little beyond the external surface of the lipid bilayer.

It was previously shown, by subunit specific antibody-binding [97], that the smaller domain is exposed on the matrix (M) side of the inner mitochondrial membrane. Taken together, these observations suggest that the large hydrophilic domain, inside the flattened vesicles, is located on the cytoplasmic (C) side of the mitochondrial membrane, where it binds cytochrome *c*.

A different arrangement was found in the deoxycholate crystal; here the enzyme was monomeric, packed in rows with alternative orientation, in detergent–rich sheets with no membrane bilayer present. Nevertheless, the overall shape and mass distribution with respect to the membranes remained the same [98]. Studies performed at higher resolution (2.0 nm) have shown a Y–shaped molecule exposing one third of its mass to the cytosolic side and spanning the membrane with two domains (termed M1 and M2) that protrude 1.0–2.5 nm on the matrix side. Contact between monomers in the dimer appears to occur through the M2 arm, but favourable intermolecular contacts seem predominantly located in a more specialized area on the C-domain, up to about 3.0 nm above the membrane surface. Further away from the bilayer, the C-domains of the dimer split apart to form a large cleft between the enzyme monomers [98, 100, 101, 103]. From these studies, a volume between 0.150 and 0.220 $nm^3$ (150 and 220 $Å^3$) was calculated for the monomer; this corresponds to a molecular weight between 110 and 165 kDa [101]. These large oscillations and the fact that we now know that the exact molecular weight is 205 kDa, gives a clear idea of the poor degree of definition of the reconstructed image (Fig. 3). Nevertheless, this *out-of-focus* box has stimulated a large number of studies on the enzyme functional aggregation state and on the subunit folding and interactions. Attemps to improve the quality of the two-dimensional crystals (a resolution of 0.7 nm was sufficient to understand some details of the tertiary structure of bacteriorhodopsin [102]) or to localize the subunits through selective antibody binding to the enzyme molecules in the crystals have been so far unsuccessful [103, 104]. Reports have been published on the preparation of three–dimensional crystals, but only one of them provided some evidence about their possible use for X-ray diffraction studies [27, 105].

## 5. Aggregation state

The presence of dimers in two-dimensional vesicular crystals immediately raised great interest because of the possible functional implications in the control of the enzyme activities. As perhaps one would expect because of the existence of the alternative monomeric form in the deoxycholate rich sheet, it became rapidly clear that the dimeric state was reversible. Under high detergent concentrations, extreme pH conditions, suitable ionic strength, the enzyme can be monomerised. For example, monomerisation was reported to occur at physiological pH in the presence of 0.2 % Triton, 200 m*M* phosphate [38] or in 5 mg TX-100 / mg protein in 200 m*M* Tris-Cl [106] and the same result was obtained by alkaline treatment in 5 % Triton X-100 [48]. The monomeric form appears to be rather stable, but conditions that promote the formation of dimers can be found. A partial dimerisation seems to occur spontaneously in certain detergents such as cholate but it may be favoured by a membranous milieu [38]. The latter possibility has also been raised on the basis of kinetic evidence suggesting that reconstitution of

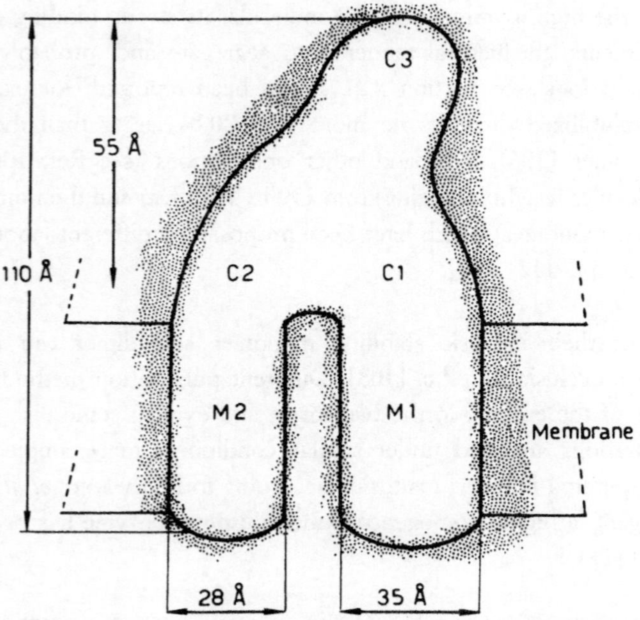


Fig. 3 - Shape and size of cytochrome *c* oxidase as deduced by image reconstruction analysis [7, 100, 101]. For technical reasons, in these studies up to 1/3 of the protein mass (schematically indicated by the dotted surface) was not detectable (see also legend to Fig. 4). Depending on the detergent used in the preparation of the two-dimensional crystals, the enzyme was found as monomer (in deoxycholate) or dimer (in Triton).

monomeric oxidase yields vesicles containing more than one monomer, possibly assembled as stable dimers [107]. Conditions for slow dimerisation, on a time scale of hours, under turnover conditions have been recently reported [108]. Certain subunits, specifically subunit III, may play a role in the process [48] though clear experimental evidence is still not available. Cross-linking experiments with cleavable reagents on the dimeric and monomeric forms of the enzyme suggest the involvement of subunit VII in the intermonomeric contacts [109]. Interestingly, this polypeptide is removed with subunit III in treatments that have been claimed to induce destabilization of the dimer [48, 110]. However, monomers and dimers with an identical complete subunit set have been recently described [106, 108]. A direct influence of the subunit composition on the aggregation state of the enzyme is questioned also by other experiments; after alkaline treatment at low Triton X-100 concentration, the enzyme loses subunits III, VII, VIII and X but still it appears to be dimeric when analyzed by discontinuous gradient polyacrylamide gel electrophoresis, under non-denaturing conditions [111].

Some of these contradictory results should be evaluated conscientiously. There are in fact several technical problems in the determination of the molecular weight of the protein, due to the high asymmetry of the molecule, its strong binding capacity toward lipids and detergents, the natural tendency to aggregate and, probably, unpredictable effects of pH and ions (see Section 9.2). It has been reported, for example, that the Triton X-100 solubilized dimer is no more than 20 % larger than the Triton X-100 solubilized monomer [106]. This and other observations (see Ref. 106) can perhaps explain the molecular weights (ranging from 130 to 400 kDa) and the number of subunits (from 6 to 12 per monomer) which have been proposed by different laboratories, even in recent years [48, 54, 112, 113].

Because of their intrinsic stability, monomer and dimer can coexist and be separated by size exclusion h.p.l.c. [108]. Different purification methods may influence the distribution of the enzyme in the two forms and explain some discrepancies among different preparations analyzed under similar conditions; for example, monomer and dimer can be separated in lauryl maltoside according to HAKVOORT *et al.* [108] while, in the same detergent, a highly homogeneous monodisperse enzyme has been described by SUAREZ *et al.* [113].

The functional role of the hypothetical monomer-dimer equilibrium, if it really exists *in vivo*, is not clear. In both conditions the enzyme is able to transfer electrons; occasional activation or inhibition cannot be discriminated from possible damages due to the procedure used to obtain the monomer or the dimer. Both forms exhibit biphasic steady-state kinetics with cytochrome *c* [108]. The discrepancy of the latter data with a previous result, suggesting linear kinetics for the monomer [112], has been tentatively interpreted as the consequence of a different phospholipid content of the enzyme preparations [108], since it was proposed that removal of the tightly bound cardiolipin can linearize the kinetics [114]. An additional observation, arguing against a physiological function of the dimer, is the isolation of monomeric enzymes from certain eukaryotic organisms [18, 115]. Cytochrome *c* oxidase is also monomeric as purified from *Paracoccus* [11, 116, 117]. In this case all the measurements refer to the so called subunit III-less enzyme since, until very recently, this was the only form available [32]. Unexplained results, such as the block of the electron transfer activity when one molecule of cytochrome *c* is covalently bound to a *half molecule* of oxidase [118, 119], and more recent data suggesting an influence of the aggregation state on proton pumping [38] leave the question open.

## 6. Subunit folding and interactions

### 6.1. *Arrangement of subunits*

A variety of labelling and cross-linking techniques have been used to investigate the arrangement and the interactions of the polypeptide components of cytochrome *c* oxidase. However, most of the original studies were hampered by the poor resolution of the gel electrophoretic systems used to analyze the data and some of them have been more recently reinvestigated [140, 259].

The subunit exposure to water and the sidedness in the membrane have been studied by using subunit specific antibodies and hydrophilic protein modifying reagents both on oriented mitochondrial membranes and reconstituted and detergent-solubilized preparations of the enzyme [47, 97, 120-124]. Technical problems, such as the possible steric hindrance of the large antibody molecules or the low reactivity of most amino acid residues toward chemical reagents, were partially overcome by the additional use of a probe, NAP-taurine, containing a photoreactive aryl azido group [124]. By this approach it became evident that most of the subunits were accessible from the water phase, in the detergent-solubilized enzyme.

The high and relatively broad reactivity of a hydrophilic photoreactive probe such as azido taurine was, however, a problem by itself because of the competition of water in the photochemical reaction. A more interesting application of the photolabelling technique was the design and use of hydrophobic probes able to investigate interactions within the lipid bilayer, a region normally restricted to a chemical reagent but accessible to the uncharged photoactivatable moiety [125, 126]. A further advantage was the possibility of anchoring the photoreactive group at different length in the aliphatic chain of phospholipids thus obtaining a *chemical ruler* able to monitor the depth of penetration of a protein within the membrane [125, 126]. The information obtained by this approach [124, 127-129, 153], complementary to those described above for the water-exposed protein domains, have shown that most of the enzyme subunits are in contact also with lipids. These results can be correlated to the sequence data now available for 12 of the 13 subunits of cytochrome *c* oxidase [57]. With the exception of subunit V, VI and VII, ten polypeptides of cytochrome *c* oxidase contain long uncharged stretches[1]; eight of them are labelled by azidophospholipid (as determined in a high resolving gel system; BISSON, unpublished observations) and hence they must be in contact with lipids. The location of the hydrophobic segments that belong to the two unlabeled subunits (subunit X and XIII) remains uncertain. They could be either buried in the enzyme hydrophobic sector but not in contact with lipid, or face the lipid, but be unreactive because of the particular amino acid composition and/or orientation [125, 126]. In more general terms, the possibility that long hydrophobic stretches might reside outside the membrane, buried in the interior of the protein, can not be excluded. The *symmetrical* situation, namely that relatively charged segments might be transmembrane but shielded from contact with

[1] Long stretches formed by 18-25 uncharged amino acid residues are common in membrane proteins. On the basis of theoretical considerations and experimental data, they have been shown to be involved in the formation of the protein membranous sector, spanning the bilayer in helical conformation [102, 126, 130].

lipids by other polypeptides, is also found in membrane proteins such as bacteriorhodopsin.

The search for putative transmembrane segments in protein sequences is facilitated by the use of hydropathic plots; it should be stressed, however, that this method cannot take into account the different possibilities mentioned above. Therefore, in the absence of a direct experimental support, it can offer only an approximate evaluation. For example, in the case of cytochrome *c* oxidase this method predicts a total of 28 transmembrane helices: subunit I, 12; subunit II, 2; subunit III, 6-7, subunit IV, VIII, IX, X, XI, XII, XIII, 1 [131-133]. It is evident that such number can not fit with the size of the M1 and M2 domains (the surface area in projection is 9.00 and 6.50 $nm^2$, respectively) determined by image reconstruction analysis (see legend to Fig. 4). In more recent studies, the labeling techniques described above have been combined with limited proteolysis experiments on oriented membrane preparations, using subunit specific antibodies and N-terminal sequencing analysis as tools for the identification of the cleavage products from the different subunits [134-136]. These investigations have demonstrated conclusively that subunits IV, IX and XI project their N-terminus in the mitochondrial M-side.

In the bovine enzyme, the subunit near-neighbor relationships have been analyzed with cross-linking reagents [137-139], using low resolution SDS p.a.g.e. systems. More recently, these experiments have been repeated in cytochrome *c* oxidase from rat liver [140]. The obtained data provide information for most of the enzyme subunits. Several nuclear-encoded polypeptides appear to contact subunit II, including those previously shown to contain the carboxyl groups that are protected by cytochrome *c* against carbodiimide modification (see Section *6.3*). A lower number of polypeptides interact with subunit I while the contact with subunit III seems limited to a single protein component. This latter information would fit very well with the known loose binding of subunit III to the rest of the complex (but could also be the result of the lack of reactive residues in the subunit contact sites). As expected from the hydrophobic photolabelling data and supported by these results, the nuclear encoded polypeptides appears to *coat* the two largest mitochondrial subunits limiting their contact with the membranes, possibly to influence the enzyme activity (see Section *8.2*).

A different approach to the investigation of lipid-protein and protein-protein interactions in the membrane was stimulated by the progressive accumulation of sequences for the three largest subunits of cytochrome *c* oxidase, obtained by DNA sequencing. The analysis was based on the assumption, suggested by experimental observations [141], that domains involved in protein-protein interaction should be more conserved than those in contact with lipids. Several different clusters of highly and poorly conserved residues could be identified on the surface of the putative transmembrane

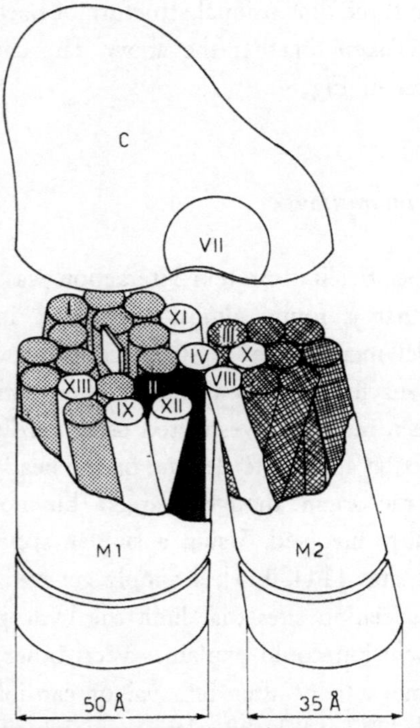


Fig. 4 - An hypothetical scheme for the arrangement of the subunits in the hydrophobic sector of cytochrome *c* oxidase. Because of the large number of predicted transmembrane segments (see Section *6.1* and Fig. 10), subunits I and III (shaded helices) are placed in the M1 and M2 domains, respectively; subunit II is shown in black (see also Fig. 7 B), and the nuclear coded subunits in white. The small rectangle in M1 represents heme *a* (see Fig. 11). The two hydrophobic regions are probably joined by subunit VII, a polypeptide located in the C-domain [136] which may contribute to the monomer-monomer interaction [109] and which contacts simultaneously subunits II and III [140]. The same polypeptide, with subunits III, VIII and X is lost by alkaline treatment of the enzyme [143]; for this reason, in this scheme, the latter two transmembrane subunits share the same hydrophobic domain of subunit III. As suggested by lack of labeling with photoreactive lipids and small lipophilic probes [124, 127 and R. Bisson, unpublished results], the hydrophobic stretch of subunit X may be shielded by other polypeptides from contact with the membrane. As indicated by cross-linking data [140], which however refers to the water exposed protein surfaces, several nuclear encoded subunits (IV, VIII, IX and XII) may interact simultaneously with subunit II, the polypeptide that contains part of the binding site for cytochrome *c* (see Figs. 5 and 7), and lipids [124, 127, 230]. So far the subunit orientation with respect to the membrane is known only for the subunits IV, IX and XI, which project their N-terminus in the M-side [136], and for subunit II (see Fig. 7). Several lines of evidence (see text) suggest that the transmembrane peptide segments span the bilayer in α-helical conformation. Assuming that the average diameter of an α-helix is 1.0 nm the calculated surface area in projection is approximately 19.00 $nm^2$ and 8.00 $nm^2$ for the M1 and M2 domain, respectively. The corresponding values suggested by image reconstruction analysis, 9.0 $nm^2$ and 6.50 $nm^2$ [7], appear largely underestimated (see Fig. 3). The alternative possibility that at least some of the uncharged stretches might not be transmembrane, in spite of theoretical predictions based on hydropathic plots, cannot be excluded (see Section *6.1*).

helices leading to a model where only the latter were placed at the boundary of the hydrophobic sector in contact with lipids [125]. This proposal is now firmly supported by comparative studies on the three-dimensional structure of bacterial reaction centers from different organism [142]. Taken together the above data can be used to propose the model schematically shown in Fig. 4.

### 6.2. *Lipid-protein specific interactions?*

The possibility of specific lipid-protein interaction was raised by the observation that cardiolipin, a lipid that is found almost exclusively in the inner mitochondrial membrane and in bacterial membranes, tends to copurify with cytochrome *c* oxidase affecting, apparently, the enzyme kinetics and interaction with cytochrome *c* [114, 144, 145]. The problem has been recently investigated using cardiolipin spin label analogues [146]. The selective interaction appears to depend on the head group structure but it may not be of purely electrostatic origin, though charged (but not specific) residues on the protein surface are probably involved. Using a similar approach, rather than a small number of highly specific sites [144, 145], a simple generalized increase in cardiolipin selectivity (of the lipid association sites that limit the hydrophobic sector surface) was proposed [147]. This conclusion could explain several other observations such as: the conservation of the enzyme activity after removal of cardiolipin when assayed in the presence of excess of lipids [147, 148], the relatively high activity (in the detergent lauryl maltoside) of rat liver oxidase containing a substoichiometrical amount of cardiolipin [28], the similar labeling patterns found after chemical modification with cardiolipin, phosphatidic acid and phosphatidylcholine benzaldehyde derivatives [149]. Moreover, the oxidase of certain organisms does not show any preference for cardiolipin [150, 151].

However, all the experiments mentioned above have been performed with the detergent solubilized or reconstituted purified enzyme; hence, the possibility that the generalized selectivity for cardiolipin might become higher and functionally very important in the native membrane cannot be excluded. In conclusion, the requirement for this phospholipid does not appear to be a general one, though other lipids could substitute for the possible role[2] played by cardiolipin. In fact, it has been reported that reconstitution of purified enzyme into acidic phospholipid vesicles partially restore the extended *low affinity* phase observed in the kinetics of the mitochondrial membrane bound oxidase [154].

[2] It has been implied that cardiolipin binds to cytochrome *c* and influences the oxidase activity [114]. This suggestion was extrapolated from photoaffinity labelling data with azido cytochrome *c* [119]; however, it should be pointed out that there is no clear experimental evidence in support of this speculation.

### *6.3. Interaction with cytochrome c*

Azido phospholipids and azido-taurine were not the first photoreactive derivatives to be used in labelling experiments on cytochrome *c* oxidase. The location of the cytochrome *c* binding site was in fact determined earlier by using photoactivatable analogs of cytochrome *c* [118]. The design of these probes took advantage of the well known three-dimensional structure of the substrate and, particularly, of the characteristics of its binding site [152]. The azido moiety was covalently linked to a residue in the center of this site and a derivative containing the same group but out-of-site (presumably on the back face of the molecule) was also prepared to be used as control [118, 119]. The specific photocross-linking of subunit II was the first evidence for a possible role of one of the thirteen components of the mitochondrial enzyme [118].

The identification of the peptide containing the residue(s) cross-linked by the azidocytochrome *c*, performed by protein fragmentation and sequencing, established its location on the C-side and the orientation of subunit II with respect to the inner mitochrondrial membrane [155]. Interestingly, this peptide contained three invariant residues (Asp.-158, Val-159 and His-161) out of the eight that are strictly conserved in the subunit II sequences of more than twenty different organism (see Fig. 7). On these bases, His-161 was proposed as a possible ligand to a prosthetic group and Asp-158 as one of the acidic residues responsible for the binding of cytochrome *c* [155].

The presence of the cytochrome *c* binding site on subunit II was confirmed and defined in more detail by a different approach based on chemical modification experiments with water soluble carbodiimides [156-159]. The reactive acidic residues were present predominantly on subunit II. Some of them (in the sequence of bovine subunit II: Asp-112, Glu-114 and Glu-198), identified by peptide mapping, were selectively shielded by the presence of cytochrome *c* [158]. Noteworthy, Asp-112 and Glu-114 were located in a segment containing a cluster of amino acid residues with conserved aromatic character, an observation that led to speculations about their involvement in the electron transfer mechanism [7, 158]. In one of these chemical modification studies, the possible involvement of some cytoplasmic subunits in the binding of the substrate was also suggested [159] and more recently confirmed, in cytochrome *c* oxidase from different organisms and tissues, using a highly resolving gel system [37, 160].

Another interesting derivative of cytochrome *c*, used in labelling studies of oxidase, was obtained by chemical activation of the sulfhydryl group present in the single cysteine (Cys-107) of the yeast molecule with thionitrobenzoate [161]. Covalent cross-linking involved in this case subunit III [161], and the same result was found using bovine oxidase [162] or different yeast Cys-107 modified cytochrome *c* derivatives [163]. The fact that the same result is obtained with the enzyme from different organism, regardless of the different subunit distribution of the reactive sulfhydryl groups among the various oxidase subunits[3], indicates that random labeling is not involved in the reaction.

In the bovine enzyme the site of cross-linking was tentatively assigned to Cys-115 of subunit III [164]. The different results obtained with photoreactive and chemical cytochrome *c* derivatives, are explained by the different location of the two reactive moieties. The azido group was attached to Lys 13 in the center of the cytochrome *c* binding site on the so called *front face* of the molecule, while cys-107 is located on the *back face* of the protein. This observation was used to suggest that the binding site for cytochrome *c* is a cleft of the dimer formed by subunit II of one monomer and subunit III of the second opposing monomer [7] (Fig. 5). A substantial support for this view came from studies on the kinetics of the enzyme cross-linked to the thiol-activated yeast cytochrome *c* in its monomeric and dimeric forms [165]. The labeling inhibits the enzyme activity, measured spectrophotometrically[4], only in the dimer but not in the monomer. Taken together, these results would suggest that subunit III is topologically *at the border* of the cytochrome *c* binding site but that the residues controlling the electrostatic interaction are located on subunit II and, possibly, on other smaller polypeptides. Very recent data add further support to this model; cytochrome *c* quenches the fluorescence of a probe covalently linked to the reactive sulfhydryl group of subunit III in the dimer but not in the monomer [166][5].

The possibility of more than one cytochrome *c* binding site per oxidase molecule has been considered on the basis of the enzyme steady state kinetics and equilibrium binding experiments [167, 168]; these observations have led to the proposal of an high and a low affinity site for the substrate with apparent $K_m$ of $10^{-8}$ $M$ and $10^{-6}$ $M$,

[3] In the yeast enzyme, for example, subunit II is predominantly labelled by thiol modifying reagents [161].

[4] An intriguing result concerns the polarographic activity of the labeled dimer which, with exogenous cytochrome *c*, ascorbate and TMPD is essentially maximal [162]. In a tentative explanation the monomerisation of the dimer under the conditions of the polarographic assay was assumed [165]. The possibility that the complex might not be stable under the reducing conditions used for the polarographic assay was suggested also by others [170].

[5] The general conclusions described above hardly fit with recent findings [7, 73] which indicate no substantial changes in the $K_M$ for cytochrome *c* in *Paracoccus* oxidase after proteolysis of subunit II. In the light of this result, the identification of the carboxyl groups involved in cytochrome *c* binding by shielding experiments [158, 159] have been questioned. However, the suggestion that the modified acidic residues are not universally conserved [73] is not supported by sequence homologies (see Fig. 7 A). On the other hand, functionally significant changes in evolutionary distant organims are expected, even in the binding site region as result of the coevolution of the enzyme-substrate interaction [174]. Hence, some amino acid replacements may simply reflect different structural requirement for better protein-protein interaction and substrate binding [89].

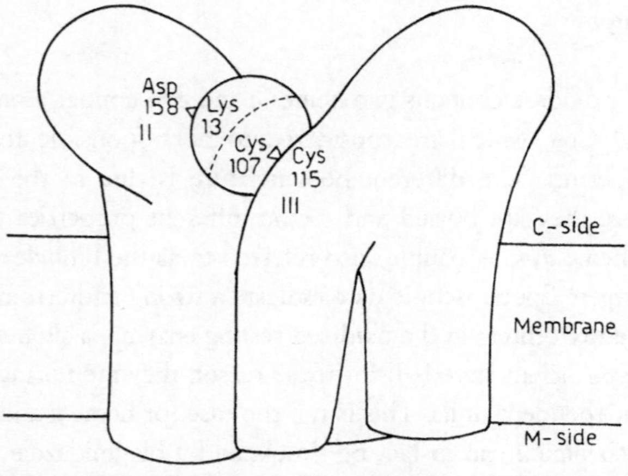


Fig. 5 - The cytochrome *c* binding site on cytochrome *c* oxidase. Most of the structural and kinetic data concerning the enzyme-substrate interaction (see text) can be correlated by the scheme shown in this figure, modified from an original model proposed by CAPALDI and colleagues [7]. Cytochrome *c* would bind to dimeric cytochrome *c* oxidase in a cleft delimited by subunit III of one monomer and subunit II of the other one. Only the latter polypeptide, however, is important for the strenght of the interaction. In this process, a possible involvement of some smaller subunits has also been suggested [159, 160]. The figure schematically shows the results of the two key cross-linking experiments that led to this model; the dashed line on cytochrome *c* is to recall that two different derivatives of the substrate were used in the experiments [118, 161, 162]. Recent data add further support to this model [166].

respectively. Although both sites were originally suggested to be kinetically competent, it now seems clear that only the high-affinity site is functional in electron transfer [169-173]. This conclusion is strongly supported by the results of the enzyme-substrate cross-linking experiments discussed above. One molecule of cytochrome *c* covalently linked to cytochrome *c* oxidase *via* disulfide crosslinking is in fact sufficient, in the spectrophotometric assay, to inhibit electron transfer from added ferrocytochrome *c*. In photoaffinity labeling experiments with arylazido cytochrome *c*, the complete block of the oxidase activity is obtained when a *half molecule* of cytochrome *c* is covalently bound per oxidase monomer, a result that supports the idea of the dimer as *physiological aggregation state* of the enzyme [119]. The low affinity site(s) could be associated with lipids and have a regulatory role [154, 170) (see also Section *6.2*), although clear evidence is still lacking[6].

[6] Recently, a different view was proposed to explain the multiphasic kinetics, previously interpreted in terms of multiple (catalytic and non-catalytic) interacting binding sites. Conformational forms of the enzyme influenced by the concentration of cytochrome *c* has been suggested [175].

## 7. Prosthetic groups

Cytochrome *c* oxidase contains two hemes *a* and two copper atoms, termed heme *a*, heme $a_3$, $Cu_A$ and $Cu_B$, which are considered to be responsible for its catalytic and spectroscopic properties. The different nomenclature is due to the different chemical environments where they are buried and to the different properties that, consequently, they exhibit. The heme $a_3$-$Cu_B$ couple, also referred to as the binuclear center, is the site of reduction of oxygen. Spectroscopic data indicate a strong antiferromagnetic interaction between the two redox centers in the oxidized resting enzyme, a situation that determines the unusual $Cu_B$-e.p.r.-silent state [6]; for these reason they are normally considered as a pair rather than independent units. This is not the case for heme *a* and $Cu_A$. The spectral properties of cytochrome *a* can in fact be simulated by bis-imidazole model compounds (see Ref. 176 and references therein) and the two centers are magnetically and probably physically isolated[7]. Nevertheless, because of their rapid redox equilibrium [169], that makes it difficult to discriminate the primary electron acceptor from cytochrome *c*, they also are collectively identified as the heme *a*-$Cu_A$ site [8, 12].

None of the prosthetic groups is covalently linked to the protein. Hence, in the absence of the three-dimensional structure of the enzyme, their location in the oxidase structure and particularly the identification of the amino acid ligands are difficult. Most of the present information on the coordination environment of the prosthetic groups comes from spectroscopic data. As mentioned above, e.p.r., Raman and studies with model compounds suggest a bis-imidazole coordination in cytochrome *a* [177-181]. The involvement of histidine is further strongly supported by the comparison of the data obtained using electron nuclear double resonance (e.n.d.o.r.) spectroscopy on the native and $^{15}N$ histidines substituted cytochrome *c* oxidase [176]. The observation of only one new resonance in the e.n.d.o.r. spectrum of the isotopically substituted enzyme demonstrates the coordination of at least one histidine ligand to cytochrome *a*; this result does not exclude the presence of the second histidine ligand, with a coincident resonance as consequence of the symmetry of the coordination environment, as suggested by parallel e.n.d.o.r. studies performed with two bisimidazole porphyrin model compounds. Hence the low potential, low spin iron of cytochrome *a* is most probably coordinated by the four in-plane nitrogen ligands of the heme and by two axial histidines (Fig. 6).

[7] It is however clear that some form of *redox communication*, possibly through conformational changes, exists since a change in the redox state of one center can perturb the electronic structure of the others as reflected, for example, by modifications in midpoint potentials and subtle alterations in spectral properties [188, 211-213].

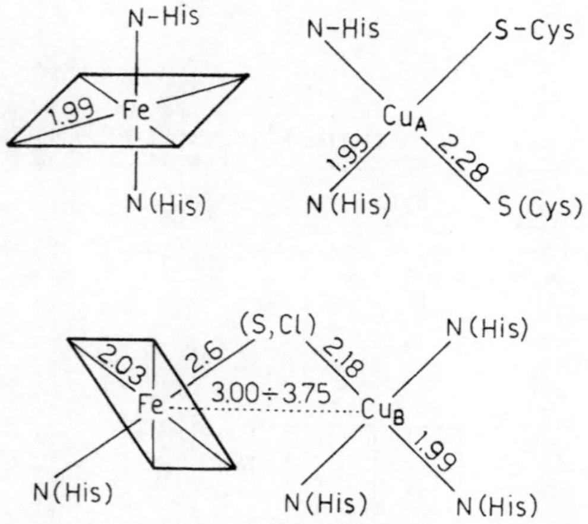


Fig. 6 - Proposed coordination environment for the active sites in resting cytochrome *c* oxidase. The average interatomic distances (in mm$^2$) obtained by e.x.a.f.s. studies are also reported. Residues whose involvement as ligands to the redox centers is highly probable but not conclusively demonstrated, are shown in brackets. References are given in text.

Isotopically substituted cytochrome *c* oxidase, obtained by incorporation of $^{15}$N histidine, $^2$H cysteine or $^{13}$C cysteine and investigated by e.p.r. and e.n.d.o.r. spectroscopy, also provided useful information for the definition of the inner coordination sphere of $Cu_A$ [182, 183]. These studies indicate at least one histidine and one cysteine as ligands to $Cu_A$. Copper extended X-ray absorption fine structure (e.x.a.f.s.) measurements performed on a $Cu_A$-depleted, chemically-modified oxidase [184] and the comparison with the native enzyme, suggest the involvement of a second cysteine in addition to another nitrogenous ligand [185]. The latter approach has also allowed the deconvolution of the e.x.a.f.s. data for the two copper sites, normally hampered by the overlapping absorption; three (N, O) ligands and one (S, Cl) ligand may be assigned to the inner coordination sphere of $Cu_B$ [185]. These conclusions are supported by spectroscopic studies on e.p.r. visible $Cu_B$ species trapped during turnover conditions [186] indicating three nitrogenous ligands at the $Cu_B$ site. Previous suggestions, based on e.x.a.f.s. studies [187], that $Cu_B$ resembles the *blue* copper protein stellacyanin are not supported by more recent investigations [185, 188]; rather than cysteine, methionine has been proposed as candidate for the sulfur ligand of $Cu_B$, but the alternative possibility of a chloride ligand was also considered [185, 189]. Always on the basis of e.x.a.f.s. data, histidine is one of the axial ligands of iron in heme $a_3$ of the resting enzyme [187, 188], a conclusion that is supported also by other spectroscopic studies [181]. As shown in Fig. 6, a sulfur-

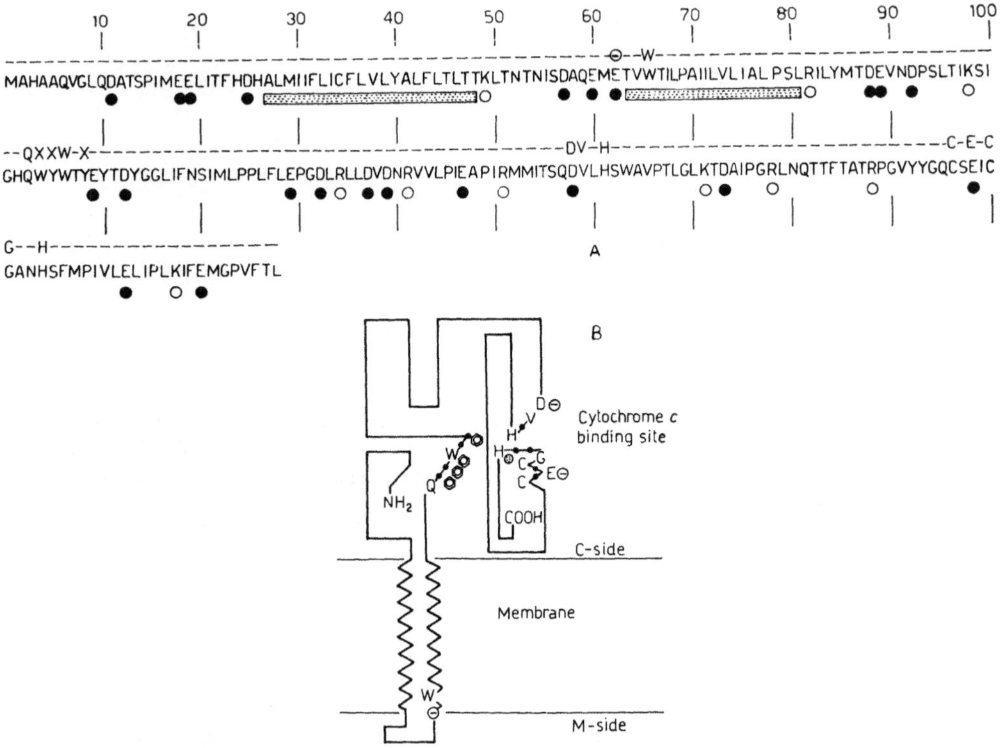


Fig. 7 - Invariant residues and proposed folding for cytochrome *c* oxidase subunit II.

**A**: Numbering and position of the conserved amino acids (shown in one–letter notation) refer to the subunit II sequence of human cytochrome *c* oxidase. Charged residues (○ positive and ● negative) and predicted transmembrane segments ( ▒ ) are indicated. The letter X identifies conserved aromatic character. The 19 organisms considered for sequence comparison are: man [214], beef [215], mouse [216], rat [217], *Neurospora crassa* [218], *Saccharomyces cerevisiae* [219], *Drosophila melanogaster* [220], *Drosophila yakuba* [221], *Trypanosoma brucei* [222, 223], *Xenopus laevis* [224], *Leishamania tarentolae* [225], *Paracoccus denitrificans* [89, 63], the thermophilic bacteria *PS3* (Sone, 1988 personal communication) and (partial) *Thermus thermophilus* [65], *Zea mays* [226], wheat [227], *Oenothera berteriana* [228] and *Crithidia fasciculata* [229]. Cys-200 does not appear to be conserved in wheat (but see footnote 8). Contact with lipids have been probed by hydrophobic photolabelling, protein fragmentation and partial sequencing of the labelled peptide [230].

**B**: Schematic diagram of subunit II folding. The two invariant acidic residues, Asp-158 and Glu-198, are clustered to the highly conserved histidines (His-161, His-204) and cysteines (Cys-196, Cys-200) which different lines of evidence indicate as the most probable ligand to $Cu_A$ ( ◍ ). The two highly conserved regions identified by Asp-158 and Glu-198 must be spatially related because both residues contact the cytochrome *c* binding site in the enzyme–substrate complex [155, 158]. Another important region appears to be represented by the stretch of aromatic residues indicated by ⏣ . Worthy of mention is the very strict homology maintained in this segment from higher eukaryotes to bacteria. The homology, but not the character, is lost only when the sequence comparisons are extended to thermophilic organism; hence, a more complex role than a simple structural function may be associated with these residues.

containing ligand may bridge the iron to $Cu_B$ which is found about 0.3–0.38 nm apart from the heme iron [187, 188]. These observations offer a structural basis for the tight magnetic coupling between the two redox centers mentioned above.

In a complex multisubunit protein such as cytochrome *c* oxidase, the localization of the binding sites for the prosthetic groups would appear hopeless, even knowing the ligand structure. The problem, however, is simplified by two facts. First, the isolation of the enzyme from bacterial sources demonstrates that the redox centers are present in the two largest subunits. Second, because of the interest in the mitochondrial genome, the sequence of these polypeptides is now (1988) known from approximately twenty organisms. Hence the identification of the highly conserved residues in subunit I and II, knowing the nature of the ligands to the redox centers, may offer a simple way to try to identify heme and copper sites. Since conserved cysteines are not found in subunit I but are present in subunit II (corresponding to position 196 and 200[8], see Fig. 7 A), the latter polypeptide appears as an obvious candidate for binding $Cu_A$. On the other hand only two strictly invariant histidines are present on subunit II (His-161 and His-204, in Fig. 7 A), sufficient for the coordination of $Cu_A$ only; the remaining copper and the hemes must therefore be placed in subunit I where seven conserved histidines (Fig. 8) are present (histidines 61, 233, 240, 290, 291, 368 and 376).

In addition to the spectroscopic studies mentioned above, several independent lines of evidence based on chemical labeling and protein modification experiments, support the involvement of the conserved histidines and cysteins of subunit II as ligands to $Cu_A$. As shown by the subunit II amino acid sequence (Fig. 7 A), one of the two conserved histidines (His-204) is located close to the cysteines and, even more interesting, one of the two strictly invariant acidic residues, Glu-198, is between the two cysteines. The second invariant histidine (His-161) is in a segment apparently far from the sequence just mentioned, but again very close to the remaining strictly conserved acidic residue Asp-158. That the two peptide are spatially contiguous is suggested by the fact that both are part of the cytochrome *c* binding site. His-161 was in fact identified earlier as a possible target of the Lys-13-arylazido cytochrome *c* derivative used in the photoaffinity labelling of the enzyme [155] and Glu-198 is protected against chemical modification by the presence of cytochrome *c* [158] (see Section *6.3*). The fact that this residue is exposed to water soluble reagents indicates that the two cysteines are very close to the protein surface though protected by the protein environment and, possibly, by the coordination

[8] In more than twenty sequences described so far, there is one exception. From the sequence of the subunit II gene of wheat, it appears that Cys-200 is not conserved [227]. This result, which would support only Cys-196 as ligand to $Cu_A$, is in contrast with e.n.d.o.r. data that suggest two cysteine ligands. On this basis, significant differences in the spectroscopic and functional properties of the enzyme have been predicted [183]. Hence the purification of the wheat oxidase is now a crucial step to clarify this point, especially in the light of more recent data favouring subunit I as the catalytic core of the enzyme [66, 72, 73].

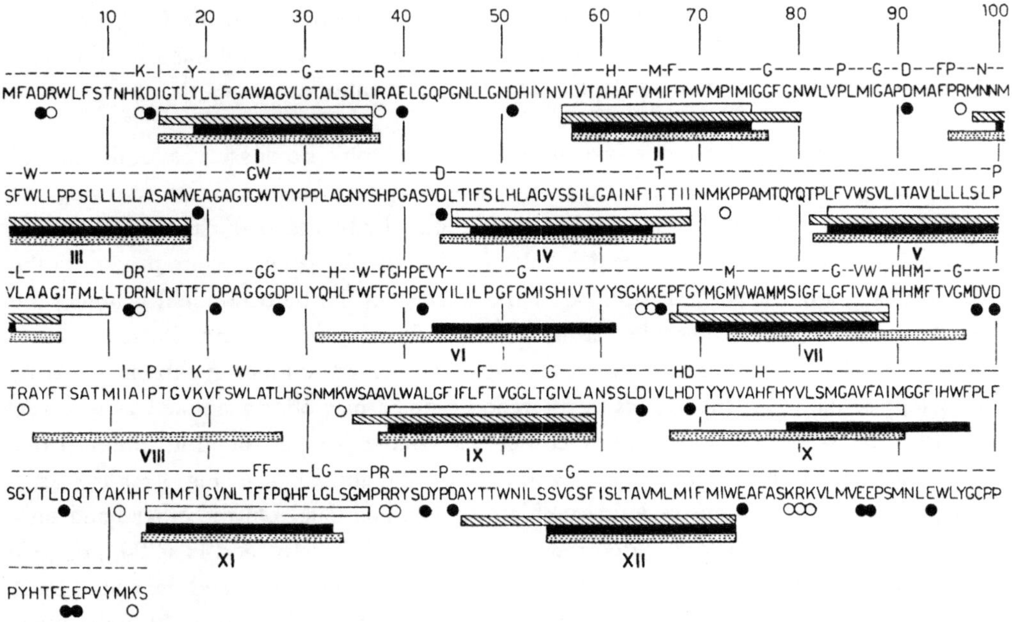


Fig. 8 - Invariant residues in cytochrome *c* oxidase subunit I. As for subunit II in Fig. 7, human subunit I is taken as reference for numbering and position of the conserved amino acids and to show the distribution of the charged residues. Horizontal bars indicate the predicted transmembrane segments according to WELINDER and MIKKELSEN [201] ( ☐ ), BISSON and MONTECUCCO [202] ( ▨ ), LUNDEEN [131] ( ■ ) and HOLM *et al.* [132] ( ▩ ). There is a considerable agreement, among different laboratories, about the location of the hydrophobic stretches. In the scheme of HOLM *et al.*, however, two major assumptions seem to be present: in spite of the positive charge (K-329), segment VIII is assumed to be in the membrane, while segments VI and X are moved toward the amino terminus to include four invariant histidines and also an additional charged residue (Glu-242). Although this proposal may appear opposed to predictions based on hydropathic plots, the subunit I folding obtained on this basis, leads to a model which is in agreement with most of the present experimental data concerning location and spatial relationships among the enzyme redox centers (see Fig. 9, 10 and 11). The invariant residues have been obtained from sequence comparison in 20 organisms: man [214], beef [215], mouse [216], rat [217], *Neurospora crassa* [231], *Saccharomyces cerevisiae* [232], *Drosophila melanogaster* [220], *Drosophila yakuba* [233], *Trypanosoma brucei* [222], *Xenopus laevis* [224], *Leishmania tarentolae* [225], *Paracoccus denitrificans* [63], the thermophilic *Bacterium PSE* [90], *Paramecium* [234], *Aspergillus nidulans* [235], *Chlamydomonas reinhardtii* [236], *Zea mays* [237], wheat [238], *Oenothera berteriana* [239] and, partially, *Podospora anserina* [240].

to copper. That this can be the case is also suggested by experiments were $Cu_A$ is specifically removed [166, 190] or its interaction with the protein perturbed, by thiol-reactive compounds, without dramatic changes in the protein structure [184, 191]. Under these conditions sulfhydryl groups present in subunit II become accessible and can be labeled [166]. These conclusions are summarized in Fig. 7.B.

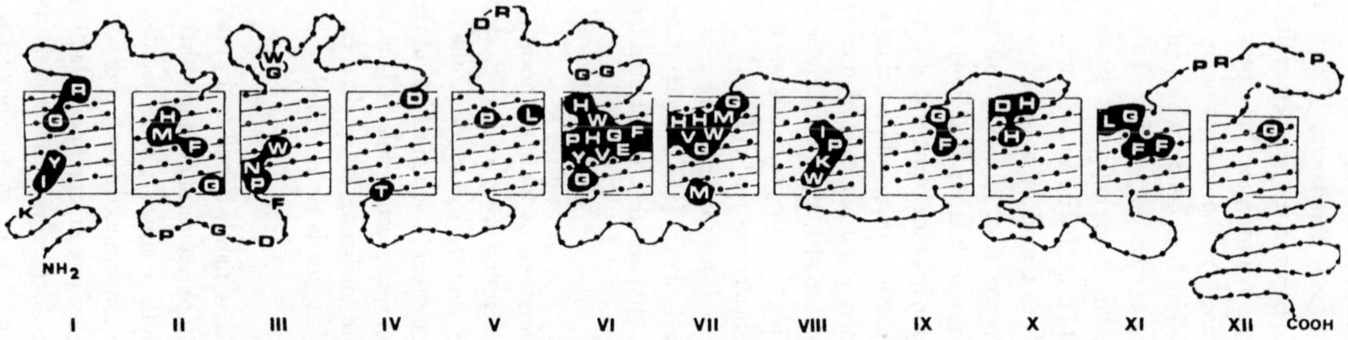


Fig. 9 - A proposed arrangement for cytochrome *c* oxidase subunit I in the membrane. The scheme has been modified from Holm *et al.*, [132] to take into account additional sequence information. Most of the invariant residues are clustered on the surface of the putative transmembrane helices. It may be noticed that six out of seven conserved histidines, are present on the helices identified as VI, VII and X. In the contiguous helices VI and VII, the invariant residues are distributed on the whole surface; this observation suggests that the two helices are buried in the protein, out of contact with lipids [125, 142]. Very likely the histidines present in these segments are involved in the coordination of the hemes and $Cu_B$. The two histidines present in helix X complete, according to model building experiments [132], the proposed coordination environment for the redox centers (see Fig. 6). It may be noticed that the carboxyl residue of helix VI is in the back of the histidines; in that position, it could pair with the lysine of helix VIII, increasing the structural stability of the hydrophobic domain.

A possible arrangement of the remaining three prosthetic groups in subunit I can be tentatively suggested only on the basis of sequence homologies (Fig. 8) and the available information on the geometrical relationship between the two hemes (Fig. 10). In a recent model [132] (see Figs. 8 and 9) all the conserved histidines have been proposed to reside in the transmembranous segment determined by the analysis of the subunit I hydropathic profiles. Six of them are found clustered in three helices which represent the most conserved regions in the primary structure of the polypeptide. This situation suggests that the three peptide segments are involved in important protein-protein contacts in the interior of the hydrophobic sector [125, 142]. The co-ordination of the six histidines to heme *a* and the binuclear center leads to a model that seems to satisfy the expected spatial relationships among the redox centers (Fig. 10). As predicted on the basis of spectrocscopic studies on oriented membrane multilayers [192-194], the hemes, sandwiched by the helices, are perpendicular to the membrane plane and at a distance of 1.2–1.6 nm, a value in reasonable agreement with previous proposals based on e.p.r. and resonance X-ray diffraction data [195-200], considering the relatively high experimental error of these types of measurements. In the same model, the Fe-Cu distance in the binuclear centre is 0.4 nm, a number sufficiently close to those previously suggested from e.x.a.f.s. [187, 188]. These observations are schematically reported in Fig. 11. At variance with the case of subunit II, there are no chemical proofs for the arrangement of subunit I. For this reason, the subunit I folding mentioned above [132], which represents a significant evolution of earlier proposals [131, 201-203] remains to be tested (see legend to Fig. 8).

A general agreement on this model seems prevented by findings concerning the *T. thermophilus* enzyme [66] and a proteolitically digested form of the *P. denitrificans* cytochrome *c* oxidase [72, 73] (see Sections 3 and 6.3). However, the suggested possibility of a *single subunit* enzyme (four redox centers in subunit I) is, in the case of the *thermophilic* enzyme, questionable in the light of very recent results [65] mentioned in Section 3. The case of *Paracoccus* is perhaps more intriguing, since clear experimental evidence for the presence of $Cu_A$ in subunit I has not been provided.

Recently, the presence of additional metal atoms, tightly bound to the enzyme (they are not removed by dialysis against EDTA, 1,10-phenanthroline and other chelating agents media) has been suggested using inductively coupled plasma atomic emission spectrometry (i.c.p. a.e.s.) [31, 180, 204] and proton induced X-ray emission (p.i.x.e.) [205]. There is a general consensus on the presence of one Zn and one Mg atom localized somewhere on the nuclear-encoded subunits. The disagreement concerns the possibility that an additional copper atom might be bound, either in stoichiometric or sub-stoichiometric ratio to the oxidase molecule. However, since it is found also in bacterial

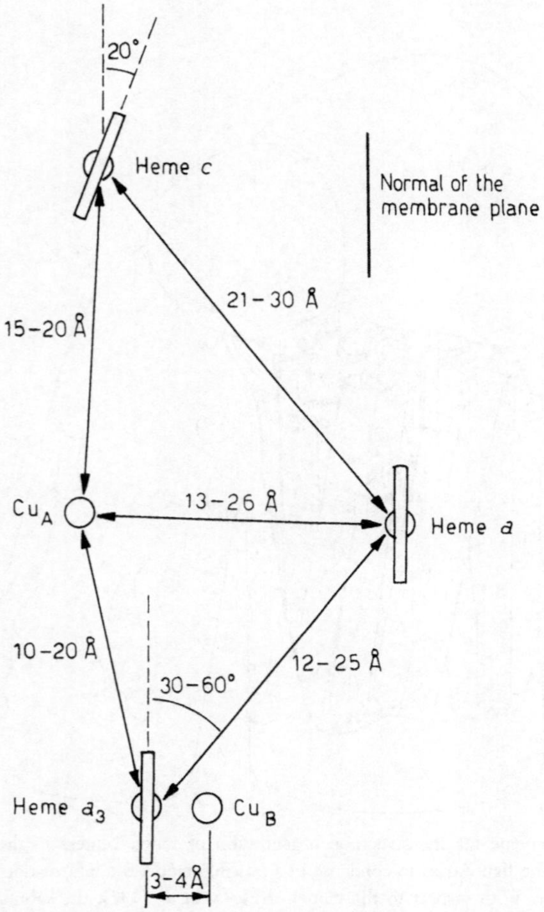


Fig. 10 - Spatial relationships among redox centers in cytochrome *c* oxidase. For references see text. The distance between $Cu_A$ and heme *c* is proposed by the author, based on the observation that the invariant acidic residue Glu-198 present in subunit II, is close to the cysteines ligating copper and, at the same time, faces the cytochrome *c* binding site that includes the heme *c* exposed edge (see also Fig. 6 in Ref. 183). The distance of heme *c* from heme *a* and its relative orientation with respect to the membrane are from references [241-244]. The absolute orientations of the hemes relative to each other are not considered in this figure.

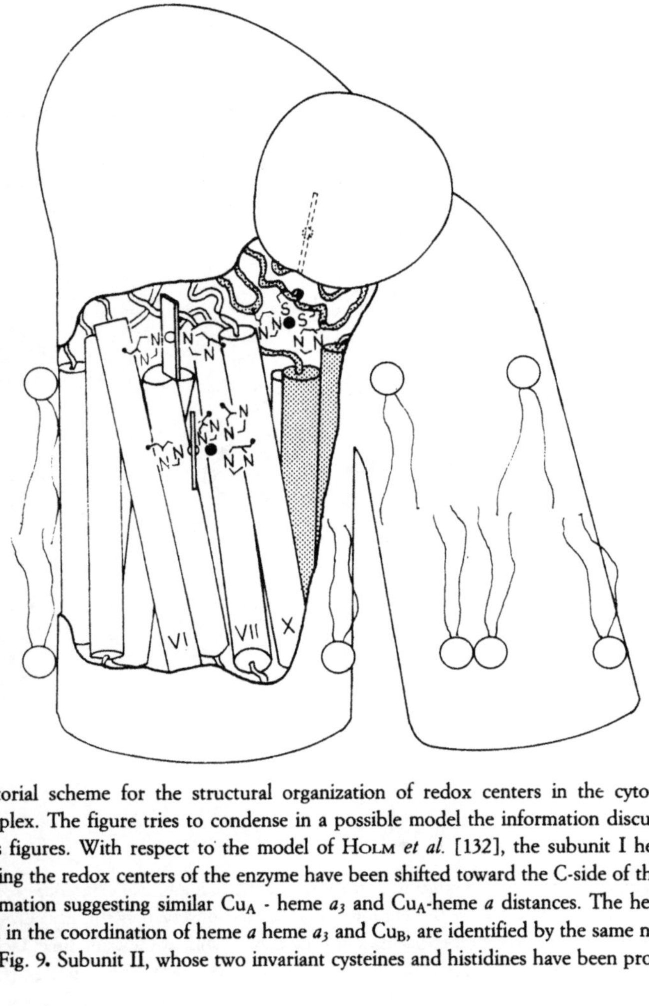


Fig. 11 - A pictorial scheme for the structural organization of redox centers in the cytochrome *c* oxidase - cytochrome *c* complex. The figure tries to condense in a possible model the information discussed in the text and shown in previous figures. With respect to the model of HOLM *et al.* [132], the subunit I helices containing the residues coordinating the redox centers of the enzyme have been shifted toward the C-side of the membrane to take into account information suggesting similar $Cu_A$ - heme $a_3$ and $Cu_A$-heme $a$ distances. The helices, containing the histidines involved in the coordination of heme $a$ heme $a_3$ and $Cu_B$, are identified by the same nomenclature (roman numbers) used in Fig. 9. Subunit II, whose two invariant cysteines and histidines have been proposed to coordinate $Cu_A$, is shaded.

oxidases, it must reside in the two largest subunits [31]. i.c.p. a.e.s. investigation on alkaline-treated oxidase, where subunits III, VII, VIII and X are removed, shows a metal composition identical to the native enzyme [206][9]; this observation reduces to seven the number of nuclear encoded subunits possibily involved in metal binding. A very recent e.x.a.f.s. study suggests that Zn is most probably ligated to three sulfur and one nitrogen (or oxygen) atom. On the basis of the amino acid composition and secondary structure prediction, subunit VI was proposed as a possible polypeptide involved in the formation of the binding site for the metal [207]. In the same study a partial depletion of Zn (up 50 %) was obtained either by trypsin digestion or by treatment with the chelating agent dipicolinic acid but, at variance with a previous suggestion based on Zn removal by protein chemical modification [208], no effect on the enzyme activity was found. Hence it would appear that Zn is either tightly bound to the protein in an adventitious site, without any specific function, or it plays simply a structural role. This may not be the case for the additional copper. Preliminary results on anaerobic titrations of bovine heart cytochrome *c* oxidase seems to indicate its possible involvement in the redox activity of the enzyme [209]. In an extreme attempt, to reconcile these observations with the *single subunit hypothesis* (namely, the suggestion that the active core of the enzyme is in subunit I), the third copper has been proposed to be in subunit II in an e.p.r. invisible state while $Cu_A$, $Cu_B$ and the hemes would be in subunit I [31]. Previous measurements, which have shown only two copper atoms in the *Thermus* enzyme [210] in spite of the concomitant presence of subunit II [65] (see Section **3**), argue against this possibility.

## 8. Heterogeneity

### *8.1. Active-site conformers*

Active-site heterogeneity is frequently observed in different oxidase preparations [19, 21, 22, 245, 246]. Originally it was suggested by the variable position of the Soret band and the multiple kinetic phases of the enzyme reaction with cyanide, a specific ligand of the binuclear center [247-249]. In addition to CN two other ligands, NO and fluoride, have been used in a combination of optical and e.p.r. studies to characterize three active-site conformers in oxidized cytochrome *c* oxidase, as isolated from beef heart (also referred to as *resting*). The conformers were not in rapid equilibrium and their relative amount was found to vary in different preparations, an observation that

[9] In their work, Yewey and colleagues [206] report the removal of subunit VI (termed a in the first Capaldi's nomenclature, see Fig. 2). This result is not in agreement with other reports that, in high resolution gels, have shown loss of subunits III, VII, VIII and X but not of subunit VI after alkaline treatment [111, 143]. The low resolution of the gel used by Yewey *et al.* could be at the origin of this discrepancy. On the other hand, subunit VI is just the polypeptide that Naqui *et al.* [207] have proposed to bind Zn; our observation can perhaps explain the apparent contradiction between the two papers.

emphasizes the importance of the different *environments* that the enzyme may experience during isolation [19, 21]. The e.p.r. signal associated with one of these oxidase forms in fact, was not present in submitochondrial particles but it appeared during purification and it could be correlated to the magnitude of the slow phase of the reaction of cytochrome *c* oxidase with cyanide [21]. On the basis of these observations, a *slow* and a *fast* reacting form of the enzyme was defined. The fast enzyme reacted rapidly not only with cyanide but also with nitric oxide, fluoride and hydrogen peroxide indicating an increased accessibility of heme $a_3$ to small ligands. Conditions that could limit the conversion from the fast to the slow form during purification were described; the stimulation of the reverse transition was not possible reflecting, presumably, a permanent structural modification at the binuclear center[10] [21]. Resonance Raman spectra of the two forms have shown various differences that have been interpreted as a global conformational change involving not only the axial histidine ligand of heme $a_3$ but also the heme *a* environment [250]. The complexity of the problem is stressed by the additional observation that a spectroscopically detectable heterogeneity persists even in the two forms of the enzyme [21, 22]. A method to quantitatively determine the distribution of active-site conformers generally applicable to different enzyme preparations was proposed recently [22].

Another interesting conformer is represented by the oxidase form termed *pulsed*, which is generated upon a cycle of reduction and reoxidation of the *resting* oxidase [251-252]. The resting-into-pulsed transition enhances the catalytic activity probably (although alternative explanations are possibile) as a consequence of a change in the coordination environment at the binuclear center [187, 253]. As shown by kinetic and spectroscopic evidence, a similar change appears to be generated by the removal of subunit III; the observation has led to the suggestion that the bridging ligand, supposed to be present in the resting state, may reside in this polypeptide [39][11]. In the decay back of the pulsed form to the resting state at least two distinct conformations, which appear to form sequentially and transiently in the presence of oxygen, have been partially characterized ([12] and references therein, [253, 254]) and defined as *pulsed* (also referred to as *peroxidatic*) and *pulsed-peroxide* forms respectively [10][12]. Following these

[10] The enzyme can still be pulsed but then it converts back to the slow form, thus preserving a memory of its original conformation; in this sense the slow form is a permanent state of the enzyme. The same happens to the fast form [21].

[11] This conclusion is not supported by recent experiments [255]. The subunit III depleted enzyme, obtained by heat treatment of reduced oxidase, displays spectroscopic signatures and CN binding kinetics similar to that of the resting enzyme.

[12] Interestingly, the spectroscopic signals associated to the peroxidatic form are not found in the pulsed oxidase from the thermophilic bacterium *PS3* suggesting some differences between the bacterial and the mammalian active site; an effect on the heme environment exerted by the cytoplasmic subunits has been speculated [256].

observations, a role for the pulsed form in a side reaction where it would act as peroxide scavenger has been proposed [10, 253].

A physiological function for the other active-site conformers described above remains doubtful; more properly, they may be considered as artifacts of protein manipulation, useful to better characterize the chemical environment surrounding the prosthetic groups and ultimately to gain insights into the oxygen reduction and proton translocation mechanisms[13]. Any conclusion, however, may be premature at this stage. For example, a general importance in metabolic control was proposed recently for two major molecular forms, designated as α and β which were observed in rat heart myocytes analyzing the relaxation kinetics of photodissociated CO by FOURIER transform IR spectroscopy at low temperature [258]. The difference between the two forms was interpreted as a difference in order and/or polarity at the $a_3Cu_B$ pocket, possibly at the level of the coordination of the copper ion. Again a preparation-dependent variability in the relative amount of the α and β forms was found [258].

### 8.2. *Isoenzymes*

Perhaps a more important role *in vivo* could have a quite different type of *structural heterogeneity* of the enzyme that depends on the nuclear-encoded subunits. Using high resolving gel systems and, more recently, molecular genetics, it has been possibile to demonstrate the existence of tissue-specific isologs to several cytoplasmic subunits ([113] and references therein). Striking observations have been made by protein and DNA sequence comparisons [260]. Only three nucleotides makes the difference between the subunit IX gene coding region in rat liver and heart. In contrast, in the same two tissues, the amino acid sequences of subunit VIII show only 50 % homology [261]. Surprisingly enough, subunit VIII exhibits a consistently higher homology (74 %) in the same tissue from different organisms (rat and bovine heart) (Fig. 12). These results demonstrate conclusisvely the hypothesis, formulated originally on the basis of immunological evidence, that tissue-specificity overrides species specificity [263]. Taken together, the above data suggest that the oxidase nuclear genes started to diverge very early (perhaps even before tissue development of higher organisms; see below, as an example, the case of two lower eukaryotes) but also that the process was strictly controlled in tissues, possibly to obtain an enzyme with a similar behaviour (regulation?) in a similar cellular

[13] In this context an elegant example has been provided recently, taking advantage of a previous observation showing loss of proton pumping (but not electron transfer) activity in oxidase exposed to a mild heat treatment [257]. The effect was interpreted as a local perturbation at the $Cu_A$ site as a result of the formation of three different active site conformers [255]. Since heme *a* and $a_3$ are apparently unperturbed, these observations have been taken as a support for a major fole of $Cu_A$ in the mechanism of proton pumping.

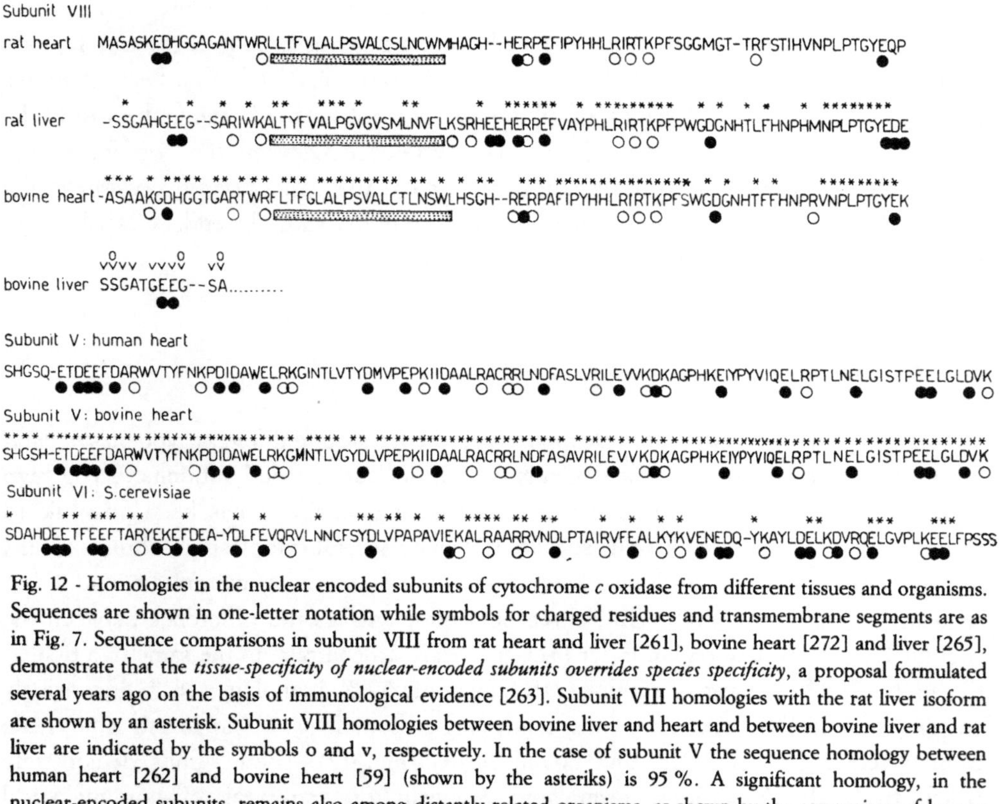


Fig. 12 - Homologies in the nuclear encoded subunits of cytochrome *c* oxidase from different tissues and organisms. Sequences are shown in one-letter notation while symbols for charged residues and transmembrane segments are as in Fig. 7. Sequence comparisons in subunit VIII from rat heart and liver [261], bovine heart [272] and liver [265], demonstrate that the *tissue-specificity of nuclear-encoded subunits overrides species specificity*, a proposal formulated several years ago on the basis of immunological evidence [263]. Subunit VIII homologies with the rat liver isoform are shown by an asterisk. Subunit VIII homologies between bovine liver and heart and between bovine liver and rat liver are indicated by the symbols o and v, respectively. In the case of subunit V the sequence homology between human heart [262] and bovine heart [59] (shown by the asteriks) is 95 %. A significant homology, in the nuclear-encoded subunits, remains also among distantly related organisms, as shown by the comparison of human subunit V and yeast subunit VI [273]. See text for a discussion of the possible functional implications of these observations.

environment. In this context it is also interesting to note that, regardless of the organism and the tissue where they are present, different forms of the same subunit conserve the same general features [78].

Evidence for multiple processed genes and pseudogenes have also been found. Some of them possess the characteristics of retroposons, since they do not have introns, the flanking regions are limited by direct repeats and a poly(A) tail is present after the polyadenilation signal [260]. The apparently large structural heterogeneity that the above observations seem to imply is however limited, so far, by the fact that for only three (VIII, X, XIII) of the ten cytoplasmic subunits, the presence of isoforms has been conclusively demonstrated [261, 264, 265]. Recent molecular genetics experiments suggest that this may be the case also for subunit IX in rat [260, 266].

The molecular mechanisms that promote the differential expression of the nuclear subunits and the role that they may play in the enzyme remain unknown. In this respect, the accumulation of structural details may offer in the future one of the keys for understanding the subunit function. For example, the comparison of the aminoacid sequences of subunit VIII in rat heart and liver shows, in the latter, the presence of five additional negative charges (Fig. 12). It has been proposed that this might alter differentially the electrostatic interaction with cytochrome *c* and be at the origin of the kinetic differences which have been claimed to exist in the two tissues [37, 160, 261].

The unexpected recent discovery of cytochrome *c* oxidase isoforms in two lower eukaryotes such as the yeast *S. cerevisiae* [267] and the slime mold *D. discoideum* [268] may offer new approaches to the study of oxidase gene expression, a problem that is still difficult in higher organisms. Taking advantage of the well known genetics of yeast, all the genes for the nuclear encoded subunits have been isolated and sequenced ([77] and references therein). This has allowed the characterization of an additional form of subunit V, termed Vb (recall that the yeast subunit V is homologous to the bovine IV), which has 67 % primary sequence homology to its counterpart termed Va [269]. The gene is expressed but at low levels and is not found assembled in aerobically grown cells. Experiments on mutated strains show that, when efficiently expressed, subunit Vb can assemble giving functional holocytochrome *c* oxidase molecules [267]. Major differences in the expression of the two genes seem due to the 5'-regions. The expression of COX5a, the structural gene for subunit Va, is largely predominant in all the conditions tested including heme deprivation and anaerobiosis, though the latter appear to increase COX5b and decrease COX5a transcription [85].

In *D. discoideum* the almost complete replacement of the smallest nuclear-encoded enzyme subunit by a larger polypeptide, occurs during the transition from the exponential to the stationary phase of growth [268]. Very recently it has been shown that the effect is promoted by modification in the environmental oxygen concentration, further supporting the idea of a regulative role for some of the cytoplasmic polypeptides [270, 271].

In summary, the enzyme appears to contain an evolutionary highly conserved, catalytic core formed by the mitochondrial subunits. The core, in eukaryotes, interacts with several smaller nuclear encoded subunits; some of them have become interchangeable with other polypeptides which can have a very different primary structure and yet maintain a similar spatial organization.

## 9. Conformations

### *9.1. Redox-linked conformational changes*

The localized structural changes which appear to be responsibile for the structural heterogeneity at the active sites, represent only part of the possible conformers of cytochrome *c* oxidase. Redox-linked, large-scale protein conformational changes have been suggested by early observations, such as an increased susceptibility to proteolysis of the reduced form of the enzyme [274], the change in the sedimentation coefficient on reduction of cytochrome *c* oxidase [275] and the dependence of optical rotatory dispersion and circular dichroism upon the oxidation state of the complex [276]. A redox-dependent antibody binding to subunit V has also been shown [277]. The latter result is interesting because it demonstrates that structural rearrangements in the coordination environment of the prosthetic groups can propagate to the nuclear-encoded subunits. On the other hand, the binding had only a limited influence on the enzyme activity suggesting that at least some conformational changes may be *loosely coupled* [277].

A relevant structural rearrangement is triggered upon reduction of heme *a* and/or $Cu_A$, as probed by energy transfer experiments with fluorescent derivatives of cytochrome *c*. The cytochrome *c* – heme *a* distance, approximately 2.3–2.4 nm in the resting fully oxidized complex, was found to increase up to 3.0–3.1 nm in the half–reduced enzyme, a value that does not change significantly upon complete reduction [244]. No such a change occurs in the transition from resting to pulsed oxidase but a modification on the charge distribution at the interaction site was reported [278]. That a major structural reorganisation of the complex accompanies the reduction of the heme *a*/$Cu_A$ site is also suggested by several other lines of evidence. The binding of cyanide at the binuclear center, which is a very slow process in the oxidized enzyme, is accelerated by the transfer of two electrons to cytochrome *a* and $Cu_A$ [248, 279]. This observation was interpreted as a transition from a *closed* to an *open* conformation where the cytochrome $a_3/Cu_B$ site is accessible for ligand binding. In the process, a volume change equivalent to about 80 $cm^3$ $mol^{-1}$ may be involved, as suggested by the analysis of the spectral perturbation induced by pressure on the aerobic steady state of cytochrome oxidase [280]. The apparent *contraction* and *expansion* of the molecule during the redox cycle could result from shifts in the polypeptide backbone possibly mediated by solvent movement into and out of protein pockets[14]; in the latter case five to ten solvent molecules could be sufficient to induce the observed effects [280]. That the environment around certain residues changes during the conformational transition is demonstrated by tryptophan fluorescence spectroscopy. Upon reduction of $Cu_A$, tryptophans appear to be in a more polar environment as compared to the oxidized enzyme [281-282].

These observations and previous data suggesting the need for two reducing equivalents for rapid intramolecular electron transfer to the binuclear center [283, 284], have led to the suggestion of a basic role played by the redox-linked conformational change in the mechanism for electron gating coupled to proton translocation [285-287]. A different model based on localized conformational transitions in the $Cu_A$ site (*i.e.*, metal ligand rearrangements) has also been proposed [288]. It should be pointed out, however, that evidence associating conformational changes to proton pumping are still rather indirect and other mechanisms have also been proposed [289].

### 9.2. *Conformational changes upon binding*

Recently it has become clear that significant conformational changes can also be triggered by some enzyme interactions. Spectroscopic analyses of the cytochrome *c* oxidase-cytochrome *c* complex show perturbations in the heme *a* and heme *c* regions that can not be explained by a direct contact between the two prosthetic groups since the heme-heme distance in the enzyme-substrate complex appears to be higher than 2.5 nm [241, 242, 244]. Rather, these effects were attributed to subtle conformational rearrangements in the environment of the redox centers induced upon binding of the substrate [173, 290]. The possibility that cytochrome *c* might be able to communicate conformational signals to heme $a_3$ in addition to heme *a*, in the absence of electron transfer, has also been proposed [278, 291].

Because of its possible physiological importance, the effect of the interactions of ATP and other ions has raised a considerable interest [292]. Early studies had shown that

[14] The possibility that the additional metal atoms present in eukaryotic oxidase may play an important role in this process has also been suggested [280]. In principle, this hypothesis could be tested by using a similar approach with the prokaryotic enzyme.

ATP, at physiological concentrations, can modify the kinetics of the cytochrome *c* interaction with oxidase [167, 293]. This behaviour has now been rationalized at the molecular level, since it appears that the nucleotide alters the distribution of negative charges on the enzyme surface responsible for binding the substrate [294]. This observation has been interpreted as an allosteric effect of ATP on cytochrome *c* oxidase [294]. From the different labelling studies reported, however, it is not clear if a specific or multiple sites are involved [294-296]. Their localization is complicated by the low affinity of ATP (which is in the *physiological* millimolar range) and the low specificity of the binding[15]. These features of the hypothetical binding site(s) make rather complicated the design of a reliable probe for photoaffinity or chemical labelling, while, under these conditions, other characteristics linked to a particular preparation of the enzyme (lipid content, buffer and detergents) could became important and influence the result. These latter observations could perhaps explain some conflicting data obtained by photoaffinity-labelling techniques [294, 295].

The above data have been used to support the hypothesis of a regulation of the enzyme activity by the nuclear-encoded subunits *via* interaction with allosteric effectors like substrates, ions, nucleotides, or hormones which would affect the protein conformation [292]. However, because of the limitations mentioned above, the labelling data can not offer a solid basis for this view. In this context the analysis of differential effects on the kinetics of bacterial and eukaryotic oxidases and their interpretation at the structural level, may be a more productive approach. In fact, the observation that intraliposomal ATP increases and ADP decreases the *Km* for cytochrome *c* of the bovine, but not of the *Paracoccus* reconstituted oxidase [297], appears as a more direct support to the possibility of a regulatory role exerted by the smaller subunits.

In summary, the physiological function of the conformational changes described above remains substantially unclear; though the large structural flexibility of the enzyme is evident. Both intramolecular electron transfer and chemical environment are able to induce rearrangements in the cytochrome *c* oxidase structure. As suggested by some very preliminary evidence, the coupling of the two processes could be a mechanism for controlling the enzyme activities. The picture is probably even more complex since there is indication of membrane potential controlled conformational transitions that could have

[15] Although at higher concentrations, ADP and phosphate can to some extent mimic ATP and almost identical effects are obtained by substituting ATP with UTP [294]. Moreover it has been reported that tripolyphosphate and inositolhexaphosphate influences the enzyme kinetics even more effectively than ATP [(296, 298]. These observations suggest that the anion charge density is more important than the adenosine moiety for triggering the conformational change. Each anion, however, seems able to promote specific structural rearrangements in the enzyme, as suggested by the different perturbations on the visible spectra [295] and the different patterns of intramolecular cross-linking in the presence of water-soluble carbodiimide [294].

an important function in modulating proton pumping [299] and, possibly, the accessibility of allosteric effectors on the protein surface [298].

## 10. Evolution

The possibility that $aa_3$-type oxidases might have been present on the hearth before oxygen has been raised by the analysis of the phylogenetic relationships among purple bacteria and mitochondria [65]. Support for the ancient origin of cytochrome *c* oxidase comes also from some observations suggesting the presence of the enzyme in archaebacteria such as *Halobacterium halobium* [300] and *Solfolobus acidocaldaricus* [301]. On the basis of this latter evidence, it would appear that the $aa_3$ oxidase genes can be traced at least to the divergence of the three cell lineages described as eubacteria, urkaryote and archaebacteria [302, 303]. In the *anaerobic* progenote (the universal ancestor in WOESE'S terminology) an $aa_3$ oxidase precursor may have been developed to act as a scavenger to remove traces of the *poisonous* oxygen released in the atmosphere (Fig. 13)[16]. Although in a much different context, this possibility recalls the hypothesis that has been proposed to explain the synthesis of a globin in legumes; here, leghemoglobin would trap any residual oxygen to prevent it from interfering with the strictly anaerobic nitrogen fixation process carried on by symbiontic bacteria [304]. Later, the catalytic core of cytochrome *c* oxidase may have evolved its function from oxygen trapping to oxygen utilization in the energy transduction process (Fig. 13). During this process alternative oxidases have also been developed and some of them, as supported by very recent evidence, may be phylogenetically linked to the $aa_3$ enzyme. This seems the case for *E. coli* cytochrome *o* since its primary structure shows striking sequence homologies with the three largest subunits of cytochrome *c* oxidase [305]. The observation however cannot be generalized. Most of the bacterial terminal oxidases are clearly unrelated to cytochrome *c* oxidase and this may be the case also for the recently discovered $ba_3$ oxidase, a single polypeptide enzyme isolated from the bacterium *T. thermophilus* [65, 306].

The choice of cytochrome *c* oxidase by eukaryotic evolution is not surprising considering the high efficiency in energy transduction, because of proton pumping, of the $aa_3$ enzyme with respect to other bacterial oxidases. The presence of additional protein components, whose number increased concomitantly with the development of progressively sophisticated cellular organizations, is a characteristic feature of the eukaryotic enzyme. The function of these subunits is currently under intense investigation but, so far, their possible involvement in enzyme regulation is not clearly demonstrated.

[16] Obviously, at present, there are no reasons to exclude alternative possibilities for the presence of $aa_3$-type oxidases in so many different organisms (for example: lateral transfer of genetic information).

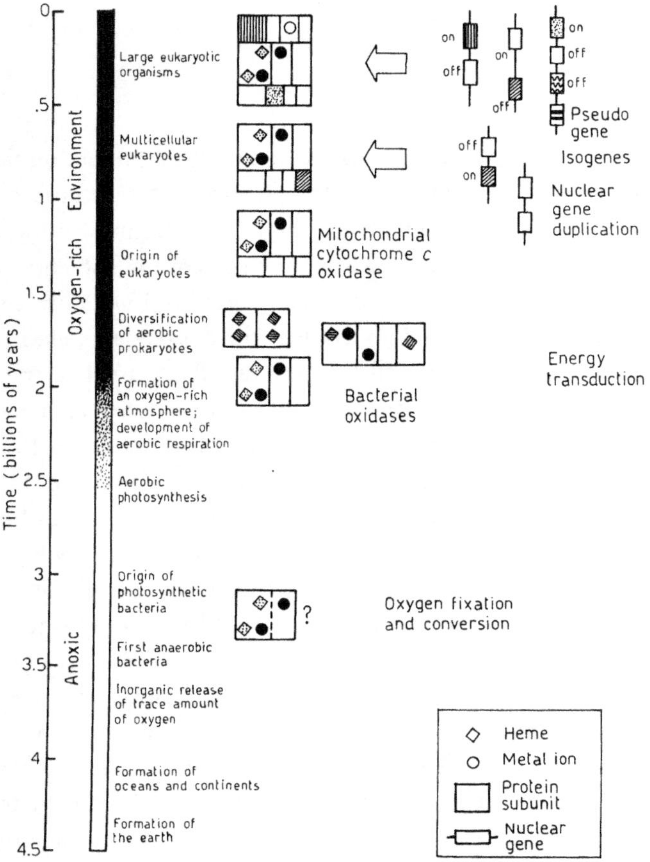


Fig. 13 - Cytochrome *c* oxidase: an hypothetical scheme for the evolution of an ancient enzyme. If, as suggested by the conserved secondary structure of rRNA, the three different cell lineages identified as archaebacteria, urkaryote and eubacteria have emerged from a common *progenote* [302, 303] then the structural similarity among cytochrome *c* oxidases of different organisms suggests that a prototype of the $aa_3$ oxidase was developed in this universal ancestor. This may have occurred more than 3 billions of years ago when, for organisms growing in a predominantly anaerobic atmosphere, traces of oxygen were probably poisonous. So the initial function of oxidase may have been the cell defense against oxygen. The situation changed radically about 2 billions of years ago when, in the oxygen-rich atmosphere created by photosynthesis, the protein became a key enzyme in energy transduction. Among the different terminal oxidases, developed in eubacteria to survive various hostile environments, cytochrome *c* oxidase was eventually chosen for eukaryote evolution and *improved* by the addition of small subunits to the catalytic core. Some of these polypeptides became interchangeable with isoforms, probably produced as result of gene duplication, giving rise to the complex modular enzyme present in higher organisms. At present, the reasons of this increasing complexity are still unknown.

However, the modular assembly of cytochrome *c* oxidase, suggested by the presence of its interchangeable subunits, remains an interesting feature that could become a fascinating example of how enzyme functions can be *finely tuned* by limited structural changes around an identical catalytic core. The solution adopted by bacteria is clearly different; in prokaryotes the synthesis of totally new enzymes (under different growth conditions) is preferred and perhaps justified by the more drastic modification in the external environment that they may have to cope with survival.

## Acknowledgments

I would like to thank Drs G. Buse, F. Millett, M. Muller, B. Kadenbach, R. Poyton, M. Saraste, N. Sone and R. Rizzuto for making their manuscripts available prior to publication.

## References

[1] R.K. Poole, *Biochim. Biophys. Acta,* **76,** 205 (1983).
[2] B. Ludwig, *FEMS Microbiol. Rev.,* **46,** 41 (1987).
[3] Y. Anraku and R.B. Gennis, *TIBS,* **12,** 262 (1987).
[4] T. Yamazaki, Y. Fukumori and T. Yamanaka, *Biochim. Biophys. Acta,* **810,** 174 (1985).
[5] A.A. Dispirito, J.D. Lipscomb and A.B. Hooper, *J. Biol. Chem.,* **261,** 17048 (1986).
[6] M. Wikstrom, K. Krab and M. Saraste, *Cytochrome c Oxidase. A Synthesis,* Academic Press, London, 1981.
[7] R.A. Capaldi, F. Malatesta and V.M. Darley-Usmar, *Biochim. Biophys. Acta,* **726,** 135 (1983).
[8] J.A. Freedman and S. Chan, *J. Bioenerg. Biomembr.,* **16,** 75 (1984).
[9] A. Azzi, K. Bill, R. Bolli, R.P. Casey, K.A. Nalecz and P. O'Shea, in *Structure and Properties of Cell Membranes,* G. Bengha (Editor), CRC Press Inc., Boca Raton, FL, (1988), pp. 105-138.
[10] A. Naqui, B. Chance and E. Cadenas, *Ann. Rev. Biochem.,* **55,** 137 (1986).
[11] M. Denis, *Biochimie,* **68,** 459 (1986).
[12] M. Brunori, G. Antonini, F. Malatesta, P. Sarti and M.T Wilson, *Adv. Inorg. Biochem.,* **7,** 93 (1987).
[13] B. Kadenbach, L. Kuhn-Nentwig and U. Buge, *Curr. Top. Bionerg.,* **15,** 113 (1987).

[14] T. YONETANI, *J. Biol. Chem.,* **236,** 1680 (1961).
[15] B. ERREDE, M.D. KAMEN and Y. HATEFI, *Methods Enzymol.,* **53,** 40 (1978).
[16] P.G. SHAKESPEARE and H.R. MAHLER, *J. Biol. Chem.,* **246,** 7649 (1971).
[17] R.O. POYTON and G. SCHATZ, *J. Biol. Chem.,* **250,** 752 (1975).
[18] V.M. DARLEY-USMAR, N. ALIZAI, A.I. AYASHI, G.D. JONES, A. SHARPE and M.T. WILSON, *Comp. Biochem. Physiol.,* **68B,** 445 (1981).
[19] G.W. BRUDVIG, T.H. STEVENS, R.H. MORSE and S.I. CHAN, *Biochemistry,* **20,** 3912 (1981).
[20] L. POWERS, B. CHANCE, Y. CHING and C.P. LEE, *J. Biol. Chem.,* **262,** 3160 (1987).
[21] G.M. BAKER, M. NOGUCHI and G. PALMER, *J. Biol. Chem.,* **262,** 595 (1987).
[22] L.J. YOUNG, *Biochemistry,* **27,** 5115 (1988).
[23] G.R. CHAUDHRY, I. SUZUKI and H. LEES, *Can, J. Microbiol.,* **26,** 1270 (1980).
[24] R.B. GENNIS, R.P. CASEY, A. AZZI and B. LUDWIG, *Eur. J. Biochem.,* **125,** 189 (1982).
[25] A. AZZI, K. BILL and C. BROGER, *Proc. Natl. Acad. Sci. USA,* **79,** 2447 (1982).
[26] Y. GOTO, N. AMURO and R. SHUKUYA, *Biochim. Biophys. Acta,* **719,** 102 (1982).
[27] T. OZAWA, M. TANAKA and T. WAKABAYASHI, *Proc. Natl. Acad. Sci. USA,* **79,** 7175 (1982).
[28] D.A. THOMPSON and S. FERGUSON-MILLER, *Biochemistry,* **22,** 3178 (1983).
[29] R. BISSON, G. SCHIAVO and E. PAPINI, *Biochemistry,* **24,** 7845 (1985).
[30] K.M.C. SINJORGO, T.B.M. HAKVOORT, I. DURAK, J.W. DRAIJER, J.K.P POST and A.P. MUIJSERS, *Biochim. Biophys. Acta,* **850,** 144 (1987).
[31] G.C.M. STEFFENS, R. BIEWALD and G. BUSE, *Eur. J. Biochem.,* **164,** 295 (1987).
[32] T. HALTIA, A. PUUSTINEN and M. FINEL, *Eur. J. Biochem.,* **172,** 543 (1988).
[33] W. DE VRIJ, A. AZZI and N. KONINGS, *Eur. J. Biochem.,* **131,** 97 (1983).
[34] B.S. BAINES, J.A.M. HUBBARD and R.K. POOLE, *Biochim. Biophys. Acta,* **766,** 438 (1984).
[35] Y. FUKUMORI, K. WATANABE and T. YAMANAKA, *J. Biochem.,* **102,** 777 (1987).
[36] Y. LI, A. NAQUI, T.G. FREY and B. CHANCE, *Biochem. J.,* **242,** 417 (1987).
[37] B. KADENBACH, A. STROH, M. UNGIBAUER, L. KUHN-NENTWIG, U. BUGE and J. JARAUSCH, *Methods Enzymol.,* **126,** 32 (1986).
[38] M. FINEL and M. WIKSTROM, *Biochim. Biophys. Acta,* **851,** 99 (1986).
[39] B.C. HILL and N.C. ROBINSON, *J. Biol. Chem.,* **261,** 15356 (1986).
[40] M. HEINRICHS and H. SCHONERT, *FEBS Lett.,* **223,** 255 (1987).
[41] E.A. BERRY and B.M. TRUMPOWER, *J. Biol. Chem.,* **260,** 2458 (1985).
[42] N. SONE, M. SEKIMACHI and E. KUTOH, *J. Biol. Chem.,* **262,** 15386 (1987).
[43] N.W. DOWNER, N.C. ROBINSON and R.A. CAPALDI, *Biochemistry,* **15,** 2930 (1976).
[44] G.J. STEFFENS and G. BUSE, *Hoppe-Seyler's Z. Physiol. Chem.,* **357,** 1125 (1976).
[45] W. SEBALD, W. MACHLEIDT and J. OTTO, *Eur. J. Biochem.,* **38,** 311 (1973).
[46] S. WERNER, *Eur. J. Biochem.,* **79** 103 (1977).

[47] B. Ludwig, N.W. Downer and R.A. Capaldi, *Biochemistry,* **18,** 1401 (1979).
[48] G. Georgevich, V.M. Darley-Usmar, F. Malatesta and R.A. Capaldi, *Biochemistry,* **22,** 1317 (1983).
[49] T. Yamanaka, K. Fujii and Y. Kamita, *J. Biochem.,* **86,** 821 (1979).
[50] B. Ludwig and G. Schatz, *Proc. Natl. Acad. Sci. USA,* **77,** 196 (1980).
[51] J.A. Fee, M.G. Choc, K.L. Findling, R. Lorence and T. Yoshida, *Proc. Natl. Acad. Sci. USA,* **77,** 147 (1980).
[52] P. Merle and B. Kadenbach, *Eur. J. Biochem.,* **125,** 239 (1982).
[53] P. Merle and B. Kadenbach, *Eur. J. Biochem.,* **105,** 499 (1980).
[54] M. Saraste, T. Penttila and M. Wikstrom, *Eur. J. Biochem.,* **115,** 261 (1981).
[55] L.J. Prochaska and K.A. Reynolds, *Biochemistry,* **25,** 781 (1985).
[56] L.J. Prochaska and P.S. Fink, *J. Bioenerg. Biomembr.,* **19,** 143 (1987).
[57] G. Buse, G.C.M. Steffens, R. Biewald, B. Bruch and S. Hensel, in *Cytochrome Systems. Molecular Biology and Bioenergetics,* S. Papa, B. Chance and L. Ernster (Editors), Plenum Publishing Corp., New York, (1987) pp. 261-270.
[58] S. Takamiha, M.A. Lindorfer and R.A. Capaldi, *FEBS Lett.,* **218,** 277 (1987).
[59] M. Tanaka, M. Hamin, K.T. Yasonobu, C.A. Yu, L. Yu, Y.H. We and T.E. King, *J. Biol. Chem.,* **254,** 3879 (1979).
[60] S.D. Power, M.A. Lochrie, K.A. Sevarino, T.E. Patterson and R.O. Poyton, *J. Biol. Chem.,* **259,** 6564 (1984).
[61] N. Sone and Y. Yanagita, *Biochim. Biophys. Acta,* **682,** 216 (1982).
[62] V. Artzatbanov, M. Muller and A. Azzi, *Arch. Biochem. Biophys.,* **257,** 476 (1987).
[63] M. Raitio, T. Jalli and M. Saraste, *EMBO J.,* **6,** 2825 (1987).
[64] K. Hon-nami and T. Oshima, *Biochemistry,* **23,** 454 (1984).
[65] G. Buse, S. Hensel and G.C.M. Steffens, *Ann. N.Y. Acad. Sci.,* **550,** 8 (1988).
[66] Y. Yoshida, R.M. Lorence, M.G. Choc, G.E. Tarr, K.L. Findling and J.A. Fee, *J. Biol. Chem.,* **259,** 112 (1984).
[67] T. Yamanaka, K. Fujii, *Biochim. Biophys. Acta,* **591,** 53 (1980).
[68] T. Yamanaka, Y. Kamita and Y. Fukumori, *J. Biochem.,* **89,** 273 (1981).
[69] Y. Fukumori, K. Nakayama and T. Yamanaka, *J. Biochem.,* **98,** 493 (1985).
[70] M. Kitada and T.A. Krulwich, *J. Bacteriol.,* **158,** 963 (1984).
[71] K. Hon-nami and T. Oshima, *Biochim. Biophys. Res. Comm.,* **92,** 1023 (1980).
[72] M. Muller, B. Schlapfer and A. Azzi, *Proc. Natl. Acad. Sci. USA,* **85,** 6647 (1988).
[73] M. Muller, B. Schlapfer and A. Azzi, *Biochemistry,* **27,** 7546 (1988).
[74] N. Sone, T. Ohyama and Y. Kagawa, *FEBS Lett.,* **106,** 39 (1979).
[75] P. Merle and B. Kadenbach, *Hoppe-Seyler's Z. Physiol. Chem.,* **361,** 1257 (1980).
[76] C. Montecucco, G. Schiavo, B. Bacci and R. Bisson, *Comp. Biochem. Physiol.,* **87B,** 851 (1987).

[77] T.E. Patterson, C.E. Trueblood, R.M. Wright and R.O. Poyton, in *Cytochrome Systems. Molecular Biology and Bioenergetics*, S. Papa, B. Chance and L. Ernster (Editors), Plenum Publishing Corp., New York (1987), pp. 253-260
[78] R.A. Capaldi, D. Gonzalez-Halphen and S. Takamiya, *FEBS Lett.,* **207,** 11 (1986).
[79] W. Sebald, H. Weiss and G. Jackl, *Eur. J. Biochem.,* **30,** 413 (1972).
[80] T.L. Mason and G. Schatz, *J. Biol. Chem.,* **248,** 1355 (1973).
[81] M.S. Rubin and A. Tzagoloff, *J. Biol. Chem.,* **248,** 4275 (1973).
[82] G. Schatz and T.L. Mason, *Annu. Rev. Biochem.,* **43,** 51 (1974).
[83] A. Tzagoloff, *Mitochondria,* Plenum Press, New York (1982).
[84] G. Schatz and R.A. Butow, *Cell,* **32,** 316 (1983).
[85] R.M. Wright, J.D. Trawick, C.E. Trueblood, T.E. Patterson and R.P. Poyton, *in Cytochrome Systems. Molecular Biology and Bioenergetics,* S. Papa, B. Chance and L. Ernster (Editors), Plenum Publishing Corp., New York (1987), pp. 49-55.
[86] B. Kloeckener-Gruissem, J.E. Mc Ewen and R.O.Poyton, *Curr. Genet.,* **12,** 311 (1987).
[87] J.E. Mc Ewen, C. Ko, B. Kloeckener-Gruissem and R.O. Poyton, *J. Biol. Chem.,* **261,** 11872 (1986).
[88] G.C.M. Steffens, G. Buse, W. Opplinger and B. Ludwig, *Biochem. Biophys. Res. Comm.,* **116,** 335 (1983).
[89] P. Steinrucke, G.C.M. Steffens, G. Panskus, G. Buse and B. Ludwig, *Eur. J. Biochem.,* **167,** 431 (1987).
[90] N. Sone, F. Yokoi, T. Fu, S. Ohta, T. Metso, M. Raitio and M. Saraste, *J. Biochem.,* **103,** 606 (1988).
[91] G. Buse, G.C.M. Steffens, R. Biewald, B. Bruch and S. Hensel in *Cytochrome Systems. Molecular Biology and Bioenergetics,* S. Papa, B. Chance and L. Ernster (Editors), Plenum Publishing Corp., New York, (1987) pp. 791-793.
[92] B. Kadenbach, J. Jarausch, R. Hartmann and P. Merle, *Anal. Biochem.,* **129,** 517 (1983).
[93] S. Seki, H. Hayashi and T. Oda, *Arch. Biochem. Biophys.,* **138,** 110 (1970).
[94] G. Vanderkooi, A.E. Senior, R.A. Capaldi and H. Hayashi, *Biochim. Biophys. Acta,* **274,** 38 (1972).
[95] R.A. Capaldi and Y.Z. Zhang, *Methods Enzymol.,* **126,** 22 (1986).
[96] R. Henderson, R.A. Capaldi and J.S. Leigh, *J. Mol. Biol.,* **112,** 631 (1977).
[97] T.G. Frey, S.H.P. Chan and G. Schatz, *J. Biol. Chem.,* **253,** 4389 (1978).
[98] S.D. Fuller, R.A. Capaldi and R. Henderson, *J. Mol. Biol.,* **134,** 305 (1979).
[99] S.D. Fuller, R.A. Capaldi and R. Henderson, *Biochemistry,* **21,** 2525 (1982).
[100] J.F. Deatherage, R. Henderson and R.A. Capaldi, *J. Mol. Biol.,* **158,** 487 (1982).

[101] J.F. DEATHERAGE, R. HENDERSON and R.A. CAPALDI, *J. Mol. Biol.,* **158,** 500 (1982).

[102] R. HENDERSON and P.N.T. UNWIN, *Nature (London),* **257,** 28 (1975).

[103] T.G. FREY, L.A. KUHN, J.S. LEIGH JR, M.J. COSTELLO and S.H.P. CHAN, *J. Inorg. Biochem.,* **23,** 155 (1985).

[104] U. JAYARAMAN, T. CHANG, T.G. FREY and J.K. BLASIE, *Biophys. J.,* **51,** 475 (1987).

[105] S. YOSHIKAWA, T. TERA, Y. TAKAHASHI, T. TSUKIHARA and S. CAUGHEY, *Proc. Natl. Acad. Sci. USA,* **85,** 1354 (1988).

[106] N.C. ROBINSON and L. TALBERT, *Biochemistry,* **25,** 2328 (1986).

[107] G. ANTONINI, M. BRUNORI, F. MALATESTA, P. SARTI and M.T. WILSON, *J. Biol. Chem.,* **262,** 10077 (1987).

[108] T.B.M. HAKVOORT, K. MOOLENAAR, A.H.M. LANKVELT, K.M.C. SINJORGO, H.L. DEKKER and A.O. MUIJSERS, *Biochim., Biophys. Acta,* **894,** 347 (1987).

[109] M. FINEL, *Biochim. Biophys. Acta,* **894,** 174 (1987).

[110] F. MALATESTA, G. GEORGEVICH and R.A. CAPALDI, in *Structure and Function of Membrane Proteins,* E. QUAGLIARELLO and F.PALMIERI (Editors), Elsevier, Amsterdam (1983), pp. 223-235

[111] M. HEINRICHS and G. BUSE, *Biochem. Biophys. Res. Comm.,* **151,** 623 (1988).

[112] R. BOLLI, K.A. NALECZ and A. AZZI, *Arch. Biochem. Biophys.,* **240,** 102 (1985).

[113] M.D. SUAREZ, A. REVZIN, R. NARLOCK, E.S. KEMPNER, D.A. THOMPSON and S. FERGUSON-MILLER, *J. Biol. Chem.,* **259,** 13791 (1984).

[114] S.B. VIK, G. GEORGEVICH and R.A. CAPALDI, *Proc. Natl. Acad. Sci. USA,* **78,** 1456 (1981).

[115] M.T. WILSON, W. LALLA-MAHARAJH, V. DARLEY-USMAR, J. BONAVENTURA, C. BONAVENTURA and M. BRUNORI, *J. Biol. Chem.,* **255,** 2722 (1980).

[116] B. LUDWIG, M. GRABO, I. GREGOR, A. LUSTIG, M. REGENASS and J.P. ROSENBUSCH, *J. Biol. Chem.,* **257,** 5576 (1982).

[117] R. BOLLI, K.A. NALECZ and A. AZZI, *J. Bioenerg. Biomembr.,* **18,** 277 (1986).

[118] R. BISSON, A. AZZI, H. GUTWENIGER, R. COLONNA, C. MONTECUCCO and A. ZANOTTI, *J. Biol. Chem.,* **253,** 1874 (1978).

[119] R. BISSON, B. JACOBS and R.A. CAPALDI, *Biochemistry,* **19,** 4173 (1980).

[120] D.L. SCHNEIDER, Y. KAGAWA and E. RACKER, *J. Biol. Chem.,* **247,** 4047 (1971).

[121] G.D. EYTAN, R.C. CARROL, G. SCHATZ and E. RACKER, *J. Biol. Chem.,* **250,** 8598 (1975).

[122] S.H.P. CHAN and R.P. TRACY, *Eur. J. Biochem.,* **89,** 595 (1978).

[123] G.D. EYTAN and R. BROZA, *J. Biol. Chem.,* **253,** 3196 (1978).

[124] L. PROCHASKA, R. BISSON and R.A. CAPALDI, *Biochemistry,* **19,** 3174 (1980).

[125] R. BISSON and C. MONTECUCCO, *in Progress in Protein-Lipid Interactions,* A. WATTS and J.J.H.H.M. DE PONT (Editors), Elsevier Scientific Publishing Co., Inc., New York, (1985), pp. 259-287

[126] R. Bisson and C. Montecucco, in *Techniques for the Analysis of Membrane Proteins*, C.I. Ragan and R. Cherry (Editors), Chapman & Hall, London, (1986), pp. 154-184

[127] R. Bisson, C. Montecucco, H. Gutweniger and A. Azzi, *J. Biol. Chem.*, **254,** 9962 (1979).

[128] N. Cerletti and G. Schatz, *J. Biol. Chem.*, **254,** 7746 (1979).

[129] G. Georgevich and R.A. Capaldi, *Biophys. J.*, **37,** 66 (1982).

[130] J. Deisenhofer, O. Epp, K. Miki, R. Huber and H. Michel, *Nature (London),* **318,** 618 (1985).

[131] M. Lundeen, B. Chance and L. Powers, *Biophys. J.*, **51,** 693 (1987).

[132] L. Holm, M. Saraste and M. Wikstrom, *EMBO J.*, **6,** 2819 (1987).

[133] M. Finel, T. Haltia, L. Holm, T. Jalli, T. Metso, A. Puustinen, M. Raitio, M. Saraste and M. Wikstrom, in *Cytochrome Systems. Molecular Biology and Bioenergetics*, S. Papa, B. Chance and L. Ernster (Editors), Plenum Publishing Corp., New York, (1987), pp. 247-252

[134] F. Malatesta, V.M. Darley-Usmar, C. de Jong, L. Prochaska, R. Bisson, R.A. Capaldi, G.C.M. Steffens and G. Buse, *Biochemistry,* **22,** 4405 (1983).

[135] Y.Z. Zhang, G. Georgevich and R.A. Capaldi, *Biochemistry,* **23,** 5616 (1984).

[136] Y.Z. Zhang, Lindorfer M.A. and R.A. Capaldi, *Biochemistry,* **27,** 1389 (1988).

[137] M.M. Briggs and R.A. Capaldi, *Biochemistry,* **16,** 73 (1977).

[138] M.M. Briggs and R.A. Capaldi, *Biochem. Biophys. Res. Comm.,* **80,** 533 (1978).

[139] J.A. Kornblatt and D. Lake, *Can. J. Biochem.*, **58,** 219 (1980).

[140] J. Jarausch and B. Kadenbach, *Eur. J. Biochem.*, **146,** 211 (1985).

[141] Von Heine, *Eur. J. Biochem.*, **120,** 275 (1981).

[142] T.O. Yeates, H. Komiya, D.C. Rees, J.P. Allen and G. Feher, *Proc. Natl. Acad. Sci. USA,* **84,** 6438 (1987).

[143] B. Kadenbach and P. Merle, *FEBS Lett.*, **135,** 1 (1981).

[144] N.C. Robinson, F. Strey and L. Talbert, *Biochemistry,* **19,** 3656 (1980).

[145] N.C. Robinson, *Biochemistry,* **21,** 184 (1982).

[146] G.L. Powell, P.F. Knowles and D. Marsh, *Biochemistry,* **26,** 8138 (1987).

[147] G.L. Powell, P.F. Knowles and D. Marsh, *Biochim. Biophys. Acta,* **816,** 191 (1985).

[148] A. Watts, D. Marsh and P.F. Knowles, *Biochem. Biophys. Res. Comm.,* **81,** 403 (1978).

[149] A. Kuppe, R.J. Mrsny, M. Shimizu, S.J. Firsan, J.F.W. Keana and O.H. Griffith, *Biochemistry,* **26,** 7693 (1987).

[150] W.F. Al-tai, M.G. Jones, K. Rashid and M.T. Wilson, *Biochem. J.*, **209,** 901 (1983).

[151] W.F. Al-tai, M.G. Jones and M.T. Wilson, *Comp. Biochem. Physiol.*, **77B,** 609 (1984).

[152] E. Margogliash and H.R. Bosshard, *Trends Biochem Sci.,* **8,** 316 (1983).
[153] H. Gutweniger, R. Bisson and C. Montecucco, *Biochim. Biophys. Acta,* **635,** 187 (1981).
[154] S.H. Speck, C.A. Neu, M.S. Swanson and E. Margogliash, *FEBS Lett.,* **164,** 379 (1983).
[155] R. Bisson, G.C.M. Steffens, R.A. Capaldi and G. Buse, *FEBS Lett.,* **144,** 359 (1982).
[156] C.H.A. Seiter, R. Margalit and R.A. Berreault, *Biochem. Biophys. Res. Comm.,* **86,** 473 (1979).
[157] F. Millet, V. Darley-Usmar and R.A. Capaldi, *Biochemistry,* **21,** 3857 (1982).
[158] F. Millet, C. de Jong, L. Paulson and R.A. Capaldi, *Biochemistry,* **22,** 546 (1983).
[159] R. Bisson and C. Montecucco, *FEBS Lett.,* **150,** 49 (1982).
[160] B. Kadenbach and A. Stroh, *FEBS Lett.,* **173,** 374 (1984).
[161] W. Birchmeier, C.E. Kohler and G. Schatz, *Proc. Natl. Acad. Sci. USA,* **73,** 4334 (1976).
[162] S.D. Fuller, V.M. Darley-Usmar and R.A. Capaldi, *Biochemistry,* **20,** 7046 (1981).
[163] R.B. Moreland and M.E. Dockter, *Biochem. Biophys. Res. Comm.,* **99,** 339 (1981).
[164] F. Malatesta and R.A. Capaldi, *Biochem. Biophys. Res. Comm.,* **109,** 1180 (1982).
[165] V.M. Darley-Usmar, G. Georgevich and R.A. Capaldi, *FEBS Lett.,* **166,** 131 (1984).
[166] J. Hall, A. Moubarak, P. O'Brie, L. Pin Pan, I. Cho and F. Millett, *J. Biol. Chem.,* **263,** 8142 (1988).
[167] S. Ferguson-Miller, D.L. Brautigan and E. Margoliash, *J. Biol. Chem.,* **251,** 1104 (1976).
[168] S. Ferguson-Miller, D.L. Brautigan and E. Margoliash, *J. Biol. Chem.,* **253,** 149 (1978).
[169] T.M. Antalis and G. Palmer, *J. Biol. Chem.,* **257,** 6194 (1982).
[170] S.H. Speck, D. Dye and E. Margogliash, *Proc. Natl. Acad. Sci. USA,* **81,** 347 (1984).
[171] K.M.C. Sinjorgo, J.H. Meijling and A.O.Muijsers, *Biochim. Biophys. Acta,* **767,** 48 (1984).
[172] K.M.C. Sinjorgo, O.M. Steinebach, H.L. Dekker and A.O. Muijsers, *Biochim. Biophys. Acta,* **850,** 108 (1986).
[173] B. Michel and R. Bosshard, *J. Biol. Chem.,* **259,** 10085 (1984).
[174] T. Luntz and E. Margogliash, in *Cytochrome Systems. Molecular Biology and Bioenergetics*, S. Papa, B. Chance and L. Ernster (Editors), Plenum Publishing Corp., New York, (1987), pp. 271-279.

[175] B. Bolgiano, L. Smith and H.C. Davies, *Biochim. Biophys. Acta,* **933,** 341 (1988).
[176] C.T. Martin, C.P. Scholes and S.I. Chan, *J. Biol. Chem.,* **260,** 2857 (1985).
[177] G.T. Babcock, L.E. Vickery and G. Palmer, *J. Biol. Chem.,* **251,** 7907 (1976).
[178] G.T. Babcock, P.M. Callahan, M.R. Ondrias and I. Salmeen, *Biochemistry,* **20,** 959 (1981).
[179] D.G. Eglinton, M.K. Johnson, A.J. Thomson, P.E. Gooding and C. Greenwood, *Biochem. J.,* **191,** 319 (1980).
[180] O. Einarsdottir and W.S. Caughey, *Biochem. Biophys, Res. Comm.,* **124,** 836 (1984).
[181] K. Carter and G. Palmer, *J. Biol. Chem.,* **257,** 13507 (1982).
[182] T.H. Stevens, C.T. Martin, H. Wang, G.W. Brudvig, C.P. Scholes and S.I. Chan, *J. Biol. Chem.,* **257,** 12106 (1982).
[183] C.T. Martin, C.P. Scholes and S.I. Chan, *J. Biol. Chem.,* **263,** 8420 (1988).
[184] J. Gelles and S.I. Chan, *Biochemistry,* **24,** 3963 (1985).
[185] P.M. Li, J. Gelles, S.I. Chan, R.J. Sullivan and R.A. Scott, *Biochemistry,* **26,** 2091 (1987).
[186] J. Cline, B. Reinhammar, P. Jensen, R. Venters and B.M. Hoffman, *J. Biol. Chem.,* **258,** 5124 (1983).
[187] L. Powers, B. Chance, Y. Ching and P. Angiolillo, *Biophys. J.,* **34,** 465 (1981).
[188] R.A. Scott and J.R. Schwartz, *Biochemistry,* **25,** 5546 (1986).
[189] D.F. Blair, W.R. Ellis, H. Wang, H.B. Gray and S.I. Chan, *J. Biol. Chem.,* **261,** 11524 (1986).
[190] S.T. Weintraub and D.C. Wharton, *J. Biol. Chem.,* **56,** 1669 (1981).
[191] T. Nilsson, J. Gelles, P.M. Li and S.I. Chan, *Biochemistry,* **27,** 296 (1988).
[192] H. Blum, H.J. Harmon, J.S. Leigh, J.C. Salerno and B. Chance, *Biochim. Biophys. Acta,* **502,** 1 (1978).
[193] M. Erecinska, J.K. Blasie and D.F. Wilson, *FEBS Lett.,* **76,** 235 (1977).
[194] M. Erecinska, D.F. Wilson and J.K. Blasie, *Biochim. Biophys. Acta,* **545,** 352 (1979).
[195] T. Ohnishi, R. Lo Brutto, J.C. Salerno, R.C. Bruckner and T.G. Frey, *J. Biol. Chem.,* **257,** 14821 (1982).
[196] J.K. Blasie, J.M. Pachence, A. Tavormina, M. Erecinska, P.L. Dutton, J. Stamatoff, P. Eisenberger and G. Brown, *Biochim. Biophys. Acta,* **679,** 188 (1982).
[197] R. Mascarenhas, Y.H. Wei, C.P. Scholes and T.E. King, *J. Biol. Chem.,* **258,** 5348 (1983).
[198] C.P. Scholes, R. Janakiraman, H. Taylor and T.E. King, *Biophys. J.,* **45,** 1027 (1984).
[199] G.W. Brudvig, D.F. Blair and S.I. Chan, *J. Biol. Chem.,* **259,** 11001 (1984).
[200] G. Goodman and J.S. Leigh Jr., *Biochim. Biophys. Acta,* **890,** 360 (1987).

[201] K.G. Welinder and L. Mikkelsen, *FEBS Lett.*, **157,** 233 (1983).
[202] R. Bisson and C. Montecucco, *J. Inorg. Biochem.*, **23,** 177 (1985).
[203] M. Lundeen, *Inorg. Chem.*, **25,** 4852 (1986).
[204] O. Einarsdottir and W.S. Caughey, *Biochem. Biophys. Res. Comm.*, **125,** 840 (1985).
[205] E. Bombelka, F.W. Richter, A. Stroh and B. Kadenbach, *Biochem. Biophys. Res. Comm.*, **140,** 1007 (1986).
[206] G.L. Yewey and W.S. Caughey, *Biochem. Biophys. Res. Comm.*, **148,** 1520 (1987).
[207] A. Naqui, L. Powers, M. Lundeen, A. Costantinescu and B. Chance, *J. Biol. Chem.*, **263,** 12342 (1988).
[208] A. Moubarak, L. Ping Pang and F. Millet, *Biochem. Biophys. Res. Comm.*, **143,** 1030 (1987).
[209] G.C.M. Steffens and G. Buse, *Eur. Bioenerg. Conf. Rep.*, **5,** 100 (1988).
[210] G. Buse. G.C.M. Steffens and L. Meinecke in *Structure and Function of Membrane Proteins*, E. Quagliarello and F. Palmieri (Editors), Elsevier, New York, (1983), pp. 131-138.
[211] W.R. Ellis, H. Wang, D.F. Blair, H.B. Gray and S.I. Chan, *Biochemistry,* **25,** 161 (1986).
[212] H. Wang, D.F. Blair, W.R. Ellis, H.B. Gray and S.I. Chan, *Biochemistry,* **25,** 167 (1986).
[213] C. Fan, J.F. Bank, R.G. Dorr and C.P. Scholes, *J. Biol. Chem.*, **263,** 3588 (1988).
[214] S. Anderson *et al.*, *Nature (London)*, **290,** 457 (1981).
[215] S. Anderson *et al.*, *J. Mol. Biol.*, **56,** 683 (1982).
[216] M.J. Bibb, R.A.V. Etten, C.T. Wright, M.W. Walberg and D.A. Clayton, *Cell,* **26,** 167 (1981).
[217] G. Pepe *et al.*, *Biochemistry Int.*, **6,** 553 (1983).
[218] G. Macino and G. Morelli, *J. Biol. Chem.*, **258,** 13230 (1983).
[219] G. Coruzzi and A. Tzagoloff, *J. Biol. Chem.*, **254,** 9324 (1979).
[220] M.H.L. Bruijn, *Nature (London),* **304,** 234 (1983).
[221] D.O. Clary and D.R. Wolstenholme, *Nucl. Acid Res.*, **11,** 4211 (1983).
[222] L.A.M. Hensgens, J. Brakenhoff, B.F. de Vries, P. Sloof, M.C. Tromp, J.H. Van Boom and R. Benne, *Nucl. Acid Res.*, **12,** 7327 (1984).
[223] R. Benne, J. van Den Burg, J.P.J. Brakenhoff, P. Sloof, J.H. Van Boom and M.C. Tromp, *Cell,* **46,** 819 (1986).
[224] B.A. Roe, D.P. Ma, R.K. Wilson and J.F.H. Wong, *J. Biol. Chem.,* **260,** 9759 (1985).
[225] V.F. de la Cruz, N. Neckelmann and L. Simpson, *J. Biol. Chem.*, **259,** 15136 (1984).

[226] D.T. Fox and C.J. Leaver, *Cell,* **26,** 315 (1981).

[227] L. Bonen, P.H. Boer and M.W. Gray, *EMBO J.,* **3,** 2531 (1984).

[228] R. Hiesel and A. Brennicke, *EMBO J.,* **2,** 2173 (1983).

[229] R. Benne, J. Van den Burg, J. Brakenhoff, B.F. de Vries, P. Nederlof, P. Sloof and A. Voogd, in *Achievements and Perspectives of Mitochondrial Research,* **2** E. Quagliariello, E.C. Slater, F. Palmieri, C. Saccone and A.M. Kroon (Editors), Elsevier, Amsterdam, (1985) pp. 325-336.

[230] R. Bisson, G.C.M. Steffens and G. Buse, *J. Biol. Chem.,* **257,** 6716 (1982).

[231] G. Burger, C. Scriven, W. Machleidt and S. Werner, *EMBO J.,* **1,** 1385 (1982).

[232] S.G. Bonitz, G. Coruzzi, B.E. Thalenfeld, A. Tzagoloff and G. Macino, *J. Biol. Chem.,* **255,** 11927 (1980).

[233] D.O. Clary and D.R. Wolstenholme, *Nucl. Acid Res.,* **11,** 6859 (1983).

[234] A.E. Pritchard, J.J. Seilhamer and D.J. Cummings, *Gene,* **44,** 243 (1986).

[235] R.B. Waring, T.A. Brown, J.A. Ray, C. Scazzocchio and R.W. Davies, *EMBO J.,* **3,** 2121 (1984).

[236] C. Vahrenholz, E. Pratje, G. Michaelis and B. Dujon, *Mol. Gen. Genet.,* **20,** 201 (1985).

[237] P.G. Isaac *et al., EMBO J.,* **4,** 1617 (1985).

[238] L. Bonen, P.H. Boer, J.E. McIntosh and M.W. Gray, *Nucl. Acid Res.,* **15,** 6734 (1987).

[239] R. Hiesel, W. Schobel, W. Schuster and A. Brennicke, *EMBO J.,* **6,** 29 (1987).

[240] T. Karsch, U. Kuck and K. Esser, *Nucl. Acid Res.,* **15,** 6743 (1987).

[241] J.M. Vanderkooi, R. Landerberg, G.W. Hayden and C.S. Owen, *Eur. J. Biochem.,* **81,** 339 (1977).

[242] M.E. Dockter, A. Steinemann and G. Schatz, *J. Biol. Chem.,* **253,** 311 (1978).

[243] J.A. Kornblatt, G. Hui Bon Hoa and A.M. English, *Biochemistry,* **23,** 5906 (1984).

[244] T.A. Alleyne and M.T. Wilson, *Biochem. J.,* **247,** 475 (1987).

[245] A. Naqui, C. Kumar, Y.C. Ching, L. Power and B. Chance, *Biochemistry,* **23,** 6222 (1984).

[246] F.M. Rusnak, E. Munck, C.I. Nitsche, B.H. Zimmerann and J. Fee, *J. Biol. Chem.,* **262,** 16328 (1987).

[247] A.O. Muijsers, R.H. Tiesjema and B.F. Van Gelder, *Biochim. Biophys. Acta,* **234,** 481 (1971).

[248] K.J.H. Van Buuren, P. Zuurendonk, B.F. Van Gelder and A.P. Muijsers, *Biochim. Biophys. Acta,* **256,** 243 (1972).

[249] K.J.H. Van Buuren, P. Nicholls and B.F. Van Gelder, *Biochim. Biophys. Acta,* **256,** 258 (1972).

[250] J.R. Schoonover, R.B. Dyer, W.H. Woodruff, G.M. Baker, M. Noguchi and G. Palmer, *Biochemistry,* **27,** 5433 (1988).

[251] E. Antonini, M. Brunori, A. Colosimo, G. Greenwood and M.T. Wilson, *Proc. Natl. Acd. Sci. USA,* **74,** 3128 (1977).
[252] M. Brunori, A. Colosimo, G. Rainoni, M.T. Wilson and E. Antonini, *J. Biol. Chem.,* **254,** 10769 (1979).
[253] B. Chance, C. Kumar, L. Powers and Y.C. Ching, *Biophys. J.,* **44,** 353 (1983).
[254] R.W. Shaw, R.E. Hansen and H. Beinert, *J. Biol. Chem.,* **253,** 6637 (1978).
[255] P.M. Li, J.E. Morgan, T. Nilsson, M. Ma and S.I. Chan, *Biochemistry,* **27,** 7538 (1988).
[256] N. Sone, A. Naqui, C. Kumar and B. Chance, *Biochem. J.,* **223,** 809 (1984).
[257] N. Sone and P. Nicholls, *Biochemistry,* **23,** 6550 (1984).
[258] F.G. Fiamingo, R.A. Altschuld and J.O. Alben, *J. Biol. Chem.,* **261,** 12976 (1986).
[259] J. Jarausch and B. Kadenbach, *Eur. J. Biochem.,* **146,** 219 (1985).
[260] X. Cao, L. Hengst, A. Schlerf, M. Droste, T. Mengel and B. Kadenbach, *Ann. N.Y. Acad. Sci.,* **550,** 337 (1988).
[261] A. Schlerf, M. Droste, M. Winter and B. Kadenbach, *EMBO J.,* **7,** 2387 (1988).
[262] R. Rizzuto, H. Nakase, M. Zeviani, S. Di Mauro and E. Schon, *Gene,* **69,** 245 (1988).
[263] J. Jarausch and B. Kadenbach, *Hoppe-Seyler's Z. Physiol. Chem.,* **363,** 1133 (1982).
[264] W. Yanamura, Y. Zhong-Zhang, S. Takamiya and R.A. Capaldi, *Biochemistry,* **27,** 4909 (1988).
[265] B. Kadenbach, R. Hartmann, R. Glanville and G. Buse, *FEBS Lett.,* **138,** 236 (1982).
[266] G. Suske, C. Enders, A. Schlerf and B. Kadenbach, *DNA,* **7,** 163 (1988).
[267] M.G. Cumsky, C. Ko, C.E. Trueblood and R. Poyton, *Proc. Natl. Acad. Sci. USA,* **85,** 2235 (1985).
[268] R. Bisson and G. Schiavo, *J. Biol. Chem.,* **261,** 4373 (1986).
[269] M.G. Cumsky, C.E. Trueblood, C. Ko and R.O. Poyton, *Mol. Cell Biol.,* **7,** 3511 (1987).
[270] R. Bisson and B. Bacci, in *Cytochrome Systems. Molecular Biology and Bioenergetics,* S. Papa, B. Chance and L. Ernster (Editors), Plenum Publishing Corp., New York, (1987), pp. 281-288.
[271] R. Bisson and G. Schiavo, *Ann. N.Y. Acad. Sci.,* **550,** 325 (1988).
[272] L. Meinecké and G. Buse, *Hoppe-Seyler's Z. Physiol. Chem.,* **367,** 67 (1985).
[273] R.M. Wright, C. Ko, M.C. Cumsky and R.O. Poyton, *J. Biol. Chem.,* **259,** 15401 (1984).
[274] T. Yamamoto and K. Okukuki, *J. Biochem., Tokyo,* **67,** 505 (1970).
[275] F. Cabral and B. Love, *Biochim. Biophys. Acta,* **283,** 181 (1972).
[276] T.E. King, P.M.B. Bayley and F.C. Yong, *Eur. J. Biochem.,* **20,** 103 (1971).

[277] J.A. Freedman and S. Chan, *J. Biol. Chem.,* **258,** 5885 (1983).
[278] J.A. Kornblatt and H.A. Luu, *Eur. J. Biochem.,* **159,** 407 (1986).
[279] P. Jensen, M.T. Wilson, R. Aasa and B. Malmstrom, *Biochem. J.,* **224,** 829 (1984).
[280] J.A. Kornblatt, G. Hui Bon Hoa and K. Heremans, *Biochemistry,* **27,** 5122 (1988).
[281] B.C. Hill, P.M. Horowitz and N.C. Robinson, *Biochemistry,* **25,** 2287 (1986).
[282] R.A. Copeland, P.A. Smith and S.I. Chan, *Biochemistry,* **26,** 7311 (1988).
[283] E. Antonini, M. Brunori, C. Greenwood and B.G. Malmstrom, *Nature (London),* **228,** 936 (1970).
[284] M. Fabian, P.E. Thornstrom, P. Brzezinski and B.G. Malmstrom, *FEBS Lett.,* **213,** 396 (1987).
[285] C.P. Scholes and B.G. Malmstrom, *FEBS. Lett.,* **198,** 125 (1986).
[286] P. Brzezinski and B.G. Malmstrom, *Biochim. Biophys. Acta,* **894,** 29 (1987).
[287] P.E. Thornstrom, P. Brzezinski, P.O. Fredriksson and B.G. Malmstrom, *Biochemistry,* **27,** 5441 (1988).
[288] J. Gelles, D.F. Blair and S.I. Chan, *Biochim. Biophys. Acta,* **853,** 205 (1986).
[289] K. Krab and M. Wikstrom, *Biochim. Biophys. Acta,* **895,** 25 (1987).
[290] C. Weber, B. Michel and H.R. Bosshard, *Proc. Natl. Acad. Sci. USA,* **84,** 6687 (1987).
[291] D. Bickar, J.F. Turrens and A.L. Lenhinger, *J. Biol. Chem.,* **261,** 14461 (1986).
[292] B. Kadenbach, *J. Bioenerg. Biomembr.,* **18,** 39 (1986).
[293] H. Roberts and B. Hess, *Biochim. Biophys. Acta,* **462,** 215 (1977).
[294] R. Bisson, G. Schiavo and C. Montecucco, *J. Biol. Chem.,* **262,** 5992 (1987).
[295] A. Reimann, F.J. Huter, J.A. Berden and B. Kadenbach, *Biochem. J.,* **254,** 723 (1988).
[296] F.J. Huter, J. Berden and B. Kadenbach, *J. Bioenerg. Biomembr.,* **20,** 503 (1988).
[297] F.J. Huter and B. Kadenbach, *Biochem. Biophys. Res. Comm.,* **153,** 525 (1988).
[298] F. Malatesta, G. Antonini, P. Sarti and M. Brunori, *Biochem. J.,* **248,** 161 (1987).
[299] M. Brunori, P. Sarti, A. Colosimo, G. Antonini, F. Malatesta, M.G. Jones and M.T. Wilson, *EMBO J.,* **4,** 2365 (1985).
[300] T. Fujiwara, Y. Fukumori and T. Yamanaka, *Plant. Cell Physiol.,* **28,** 29 (1987).
[301] S. Anemuller and G. Schafer, *Fifth Eur. Bioenerg. Conf.,* 92 (1988).
[302] C.R. Woese, *Sci. Am.,* **244,** 98 (1981).
[303] N.R. Pace, G.J. Olsen and C.R. Woese, *Cell,* **45,** 325 (1986).
[304] R.E. Dickerson and I. Geis, *Hemoglobin: Structure, Function, Evolution and Pathology*, Benjamin/Cummings, Inc. London (1983).
[305] M. Saraste, M.R. Raitio, T. Jalli, V. Chepuri, L. Lemieux and B.B. Gennis, *Ann. N.Y. Acad. Sci.,* **550,** 314 (1988).

[306] B.H. Zimmermann, C.I. Nitsche, J.A. Fee, F. Rusnak and E. Munk, *Proc. Natl. Acad. Sci. USA,* **85,** 5779 (1988).

# BACTERIORHODOPSIN – STRUCTURE AND FUNCTION

L. KESZTHELYI

*Institute of Biophysics, Biological Research Center, Szeged, Hungary, H-6701*

Contents



*Bioelectrochemistry III*
Edited by G. Milazzo and M. Blank
Plenum Press, New York, 1990

## 1. Introduction

The basic principle of energy transduction in living systems is to create a transmembrane electric potential difference and pH gradient, the so called electrochemical energy, by proton translocation [1]. Bacteriorhodopsin (BR), the protein in the plasma membrane of *Halobacterium halobium* turned out to be a very important molecule in understanding the mechanism of proton translocation [2]. Its main features are:
— it is a protein-pigment complex with retinal chromophore, similar to eye-rhodopsin,
— undergoes a cyclic reaction pathway after absorbing a photon, and
— this photocycle is connected with proton pumping from the inside of the cell to the outside.

There are also practical characteristics: BR is easy to separate from the cells, it is stable and performs the proton pumping function for long time. It is therefore not surprising that since the year of the discovery of the proton pumping activity of BR [3] a great number of studies were performed to elucidate the mechanism of proton translocation.

The amino acid sequence of BR was determined long ago [4, 5]. Recently, the gene of BR has been isolated [6], modified genes have been expressed, and the functioning of genetically modified BR has been studied [7].

The three-dimensional structure of BR has been known since 1975 to approx. 6 Å resolution [8] and since 1986 in projection to 3.5 Å resolution [9]. The data show that BR spans the membrane in seven helices, has a long tail (17 amino acids) at the cytoplasmic side and a shorter part (3 amino acids) at the external side. The retinal chromophore is linked by a protonated Schiff-base to the Lys 216 residue. Its location in the molecule has been studied by neutron diffraction [10, 11]. These data on the structure of BR are not sufficient to understand the proton translocation. Therefore, different physical, physico-chemical and biochemical methods were applied in order to gain a deeper insight into the mechanism of the BR proton pump. Description of results from light absorption (ultraviolet, visible, infrared), resonance Raman (RR), Fourier transformed infrared spectroscopy (FTIR), solid-state nuclear magnetic resonance (n.m.r.) studies and especially the endeavour of our laboratory to explore the charge motion inside BR is included in this chapter.

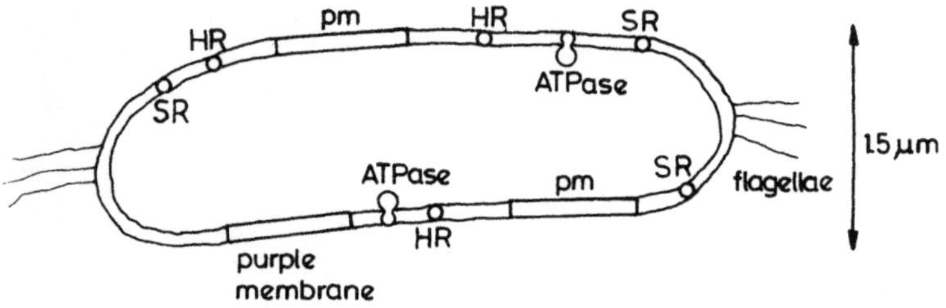


Fig. 1. – Scheme of *Halobacterium halobium*. Only selected membrane proteins are shown.

## 2. The bacterium

*Halobacterium halobium* belongs to the archaebacteriae. It lives in waters with high salt concentration (4 *M* NaCl). Salt ponds, highly salted meats are the places of its natural occurrence. Figure 1 shows the scheme of the bacterium indicating its approximate size and details of important proteins in its plasma membrane. The well separated membrane patches (which may be distinguished by electron microscopy) contain the protein BR and lipids only. Their average diameter is ~ 500 nm. The patches can be isolated [12], have a beautiful purple colour and are named purple membranes (pm). It is very important that the pm-s are so rigid that do not close to form vesicles as other isolated membrane fragments generally do. The purple membrane has a thickness of a usual lipid bilayer of 5 nm.

It is known that the plasma membrane of *H. halobium* contains at least three other retinal proteins: halorhodopsin (HR) which pumps chloride ions after light absorption [13], and two sensory rhodopsins (SRI, SRII) which, by sensing the intensity and colour of light induce the phototactic motion of the bacteria [14]. These pigments do not aggregate in the membrane. The membrane naturally contains many other proteins too. The scheme shows only those which are important in the energetics of the bacteria. It is known that hydrocarbons are used for energy source as well by the bacteria.

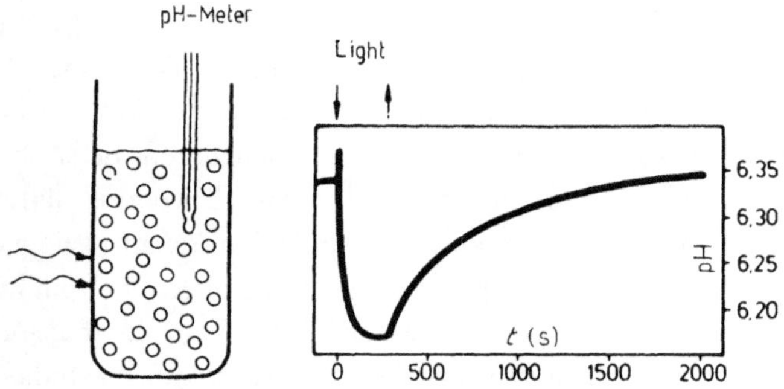


Fig. 2 – pH decreases (*i.e.* proton concentration increases) if *Halobacteria* in suspension are illuminated.

The decisive experiment demonstrating the effect of light in proton pumping is sketched in Fig. 2. Cells in solution produce a definite pH decrease when illuminated [3]. In another important experiment when ATP-synthase and pm were inserted into the same liposomes light induced ATP synthesis has been observed [15].

## 3. The bacteriorhodopsin molecule

### 3.1. *Structure*

The amino acid sequence of BR has been determined by two groups [4, 5]. Figure 3 shows the amino acids arranged into 7 helices; the carboxyl end is in the cytoplasmic, the amino end in the external side; the connecting parts are mainly outside the lipid phase; the retinal is connected to Lys 216 via a protonated SCHIFF-base. This arrangement is not quite certain. It may be that the connecting parts are longer or the helices contain more amino acids. The existence of 7 helices is supported by electron diffraction data taken from purple membranes [8, 9]. The electron density distribution — calculated from the diffraction pattern — shows a regular two dimensional hexagonal lattice the lattice points being in a distance of 6.3 nm. Every point is characterized by 3×7 helices (Fig. 4) in a symmetric arrangement, *i.e.* three BR molecules are at a lattice point. The regular arrangement and the high protein content of the pm (~ 75 % *w/w*) are responsible for its rigidity. The physiological importance of the pm structure is not yet understood: BR molecules pump protons in monomeric state as well as in the natural trimer state [16].

The assignment of the helices (1-7) in Fig. 5 to the helices determined from the amino acid sequence (A-G in Fig. 4) turned out to be very complicated. The best and most decisive approach comes from neutron diffraction data [10, 11]. Because $^2$H and $^1$H atoms have very different neutron scattering cross section, small parts of a large molecule may be visualized by $^1$H-$^2$H exchange. SCHIFF *et al.* extracted the natural retinal from BR and reconstituted it with two differently deuterated retinals:

1) chain deuterated when ten hydrogens were exchanged by deuterons in the polyene chain [10], and
2) ring deuterated when eleven $^1$H were exchanged in the cyclohexane ring [11].

The neutron diffraction patterns taken on natural BR and reconstituted with the two synthetic deuterated retinals located in the polyene chain and the ring among the 7 helices as shown in Fig. 5. The length of the polyene chain and the side chain of the lysine residue allow to locate helix G that contains Lys 216 at helix 2 or 6. Both cases — together with the possible assignment of the other helices — are shown in Fig. 5.

Attempts to grow three dimensional crystals from BR suitable for high resolution X-ray diffraction experiment failed until now [17].

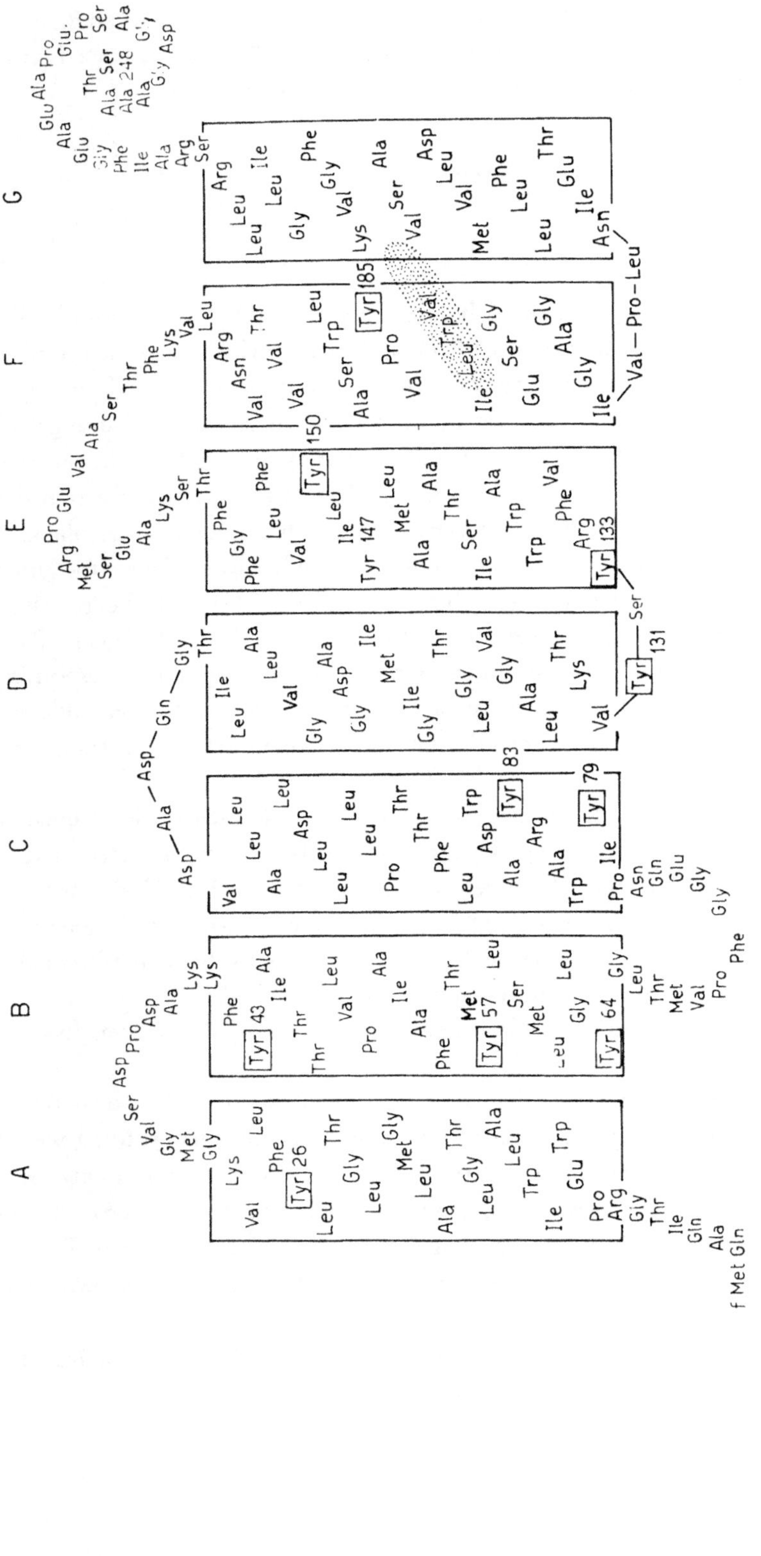


Fig. 3 – The amino acid sequence of BR (from Ref. 7).

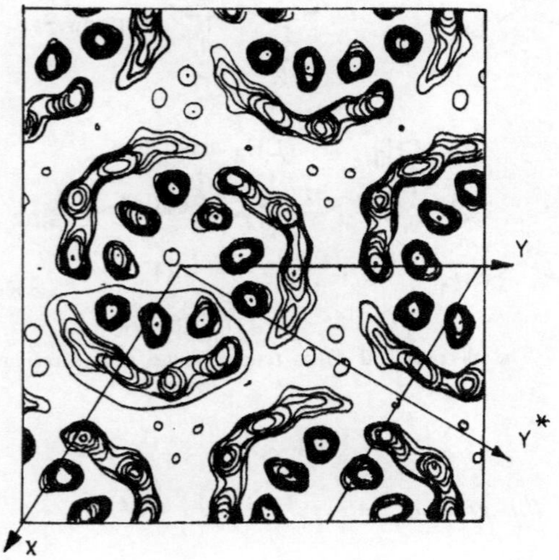


Fig. 4 – Distribution of electron density of BR-s arranged in triplets in pm (from Ref. 17).

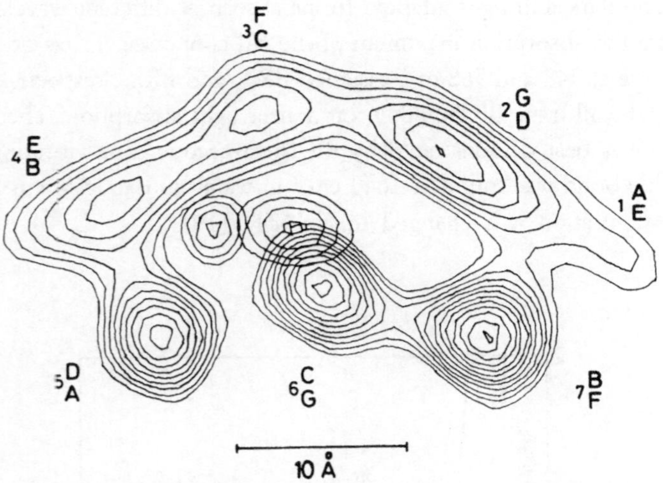


Fig. 5 – Possible spatial arrangements of the helices known from the amino acid sequence of BR (A-F) and the retinal (after Ref. 11).

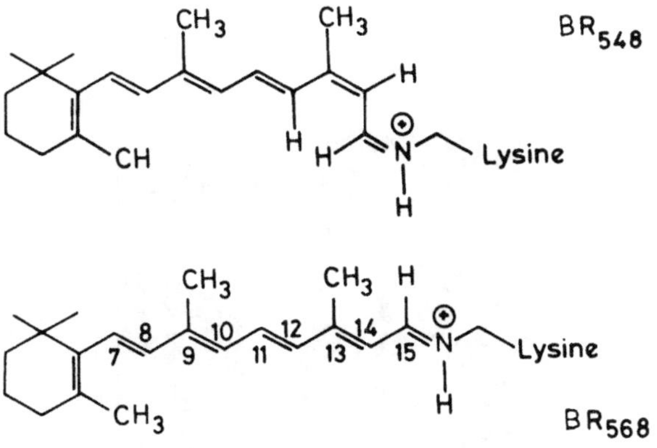


Fig. 6 – The all-trans and 13-cis retinal bound to Lys residue of BR.

### 3.2. *Spectroscopy of the ground state*

Dark adapted BR at 20 °C contains approximately 40 % all-trans and 60 % 13-cis retinal (Fig. 6) [18]. Intensive illumination, called light adaptation, converts all the retinals into all-trans form which reisomerises in the dark within 30 min at room temperature. The dark and light adapted forms absorb at different wavelengths (Fig. 7). From these data the absorption maximum of the BR containing 13 cis or all-trans-retinal is calculated to be at 548 and 568 nm, assigned $BR_{548}$ and $BR_{568}$, respectively. It is easy to show that only the all-trans BR pumps proton after light absorption. The protonation of the SCHIFF base is best demonstrated by RR spectroscopy. The vibration frequency of the C=N double bond was found at 1640 $cm^{-1}$ in water and at 1624 $cm^{-1}$ in $D_2O$ when the $C={}^1N$ configuration was changed to $C=N^2H$ [19].

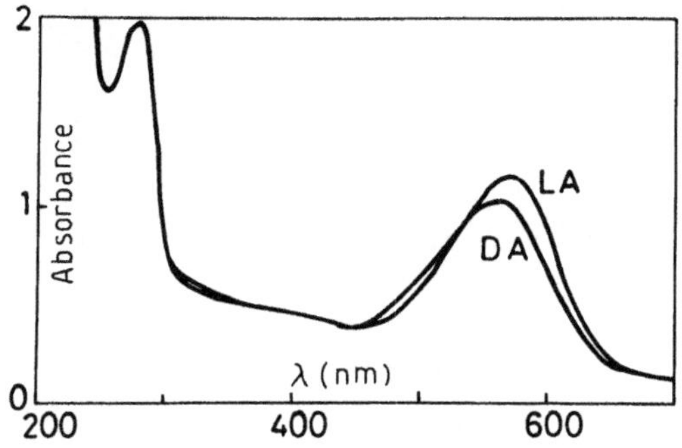


Fig. 7 – Absorption spectra of dark (DA) and light (LA) adapted BR in the U.V. and visible band.

The protonated SCHIFF-base presumably forms a weakened H-bond with a neighbouring $O^-$ of the carboxylic end of an amino acid residue [20]. n.m.r. measurements also provide information showing that a positive and a negative charge is located near to the β-ionone ring of the retinal [21] (Fig. 6). The data were used to interpret the absorption maximum ($\lambda = 568$ nm) of all-trans-BR theoretically [22]. According to this interpretation the difference in the absorption maxima between protonated SCHIFF-base in solution and in BR (the so-called opsin shift $\Delta\lambda = 568 - 440 = 128$ nm) is due to:
(1) isomerization of the $C_6$–$C_7$ bond,
(2) the negative and positive charges near $C_6$ and $C_7$, and
(3) the weakened hydrogen bond to a negative counterion.

It is interesting to note that intensive dehydration of BR lowers the absorption maximum to $\lambda = 530$ nm [23].

### 3.3. *The photocycle of BR*

The main function of BR is to pump protons from the cell to the external medium after photon absorption. During this process cyclic changes occur in the protein. The changes may be classified into four cycles:

*a*) Light absorption cycle. Characteristic changes occur in the light absorption at different times after photon absorption. The first description of the photocycle has been based on observation of this kind in the visible spectral region [24].

*b*) The retinal cycle. RR [25] and FTIR spectroscopy [26] point out changes of the isomerization state of the retinal during the phtocycle.

*c*) The opsin cycle. Changes occur in the U.V. absorption bands corresponding to alteration in protonation of tyrosines and changes in the environment of tryptophan [27]. FTIR spectroscopy proved suitable to follow protonation changes of tyrosine and aspartic acid side chains [28, 29].

*d*) Charge motion cycle. The pumped proton gets through the BR molecule during the photocycle. The process has at least two steps: uptake at the intracellular and release at the extracellular side of the BR. This process is called the *charge motion cycle*. The study of the charge motion cycle started with pm-s joined to a sort of lipid bilayer separating two conductive phases with electrodes [30-33] and continued by our method with pm-s oriented in a suspension [34].

### 3.4. *The light absorption cycle of BR*

The method of flash photolysis, *i.e.* the observation of the time course of light absorption changes at different wavelengths after a laser flash, offered the first insight into the cyclic nature of the photoreactions of BR. Figure 8 shows the principle of the measurement. Apparatuses with more and more sophistication provided better and better results but the photocycle is still in debate [35–37].

The very first step in the photocycle has been studied by a special laser method of femtosecond resolution [38]. The absorption changes at $\lambda = 600$ nm after laser excitation are sketched in Fig. 9. The curve is composed of three different changes:

1) the photon-excited part of the ground state $BR_{568}$ does not absorb: it remains so until at the end of the cycle it is recovered. In the fs, ps range the negative absorption change from this source is constant.
2) Data show that BR is transformed into an other state called J which appears with a rate of 500 fs and has an absorption maximum at $\lambda = 600$ nm. Since the absorption of $BR_{568}$ is lower at $\lambda = 600$ nm than the absorption of the new form, a positive absorption change appears with 500 fs lifetime. This is the only light-driven step in the BR photocycle because the next form $K_{590}$ appears by thermal agitation.

As it is shown in Fig. 9 the absorption of the J form decays with a lifetime of 5 ps and the absorption of the K form increases with the same rate. Because the absorption of $K_{590}$ is smaller at $\lambda = 600$ nm than the absorption of $J_{600}$, a small decay with 5 ps appears. The $K_{590}$ form has a lifetime of $\simeq 1$ μs which appears constant at this scale.

In Fig. 10 our measurements show the lifetime of the $K_{590}$ state. A positive absorption change is shown: it rises within ~ 20 ns, and decays with $\tau_K = 1.2$ μs. It is known from the previous data with high time resolution that $K_{590}$ appears within 5 ps in reality. The slow rise therefore has its origin in the width of the exciting laser pulse (~20 ns) which determines the rise of the absorption signal. The response of the electric recording system also adds to the rise time in general. It is, however, negligible in our measurement (~ 3 ns).

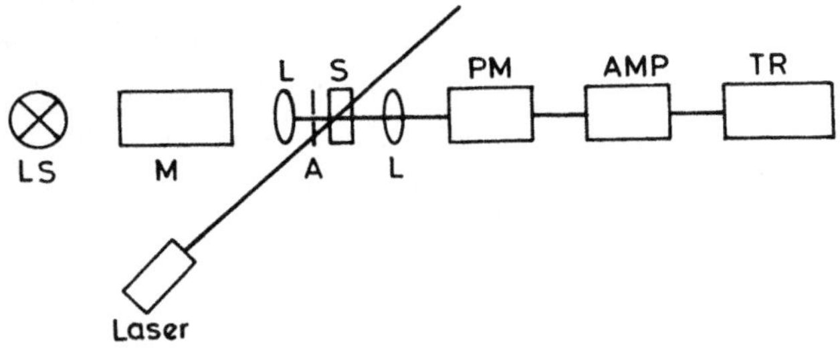


Fig. 8 – Apparatus to measure the light absorption changes after laser excitation (LS - light source, M-monochromator, L-lenses, A-aperture, S-sample, PM-photomultiplier, AMP-amplifier, TR-transient recorder).

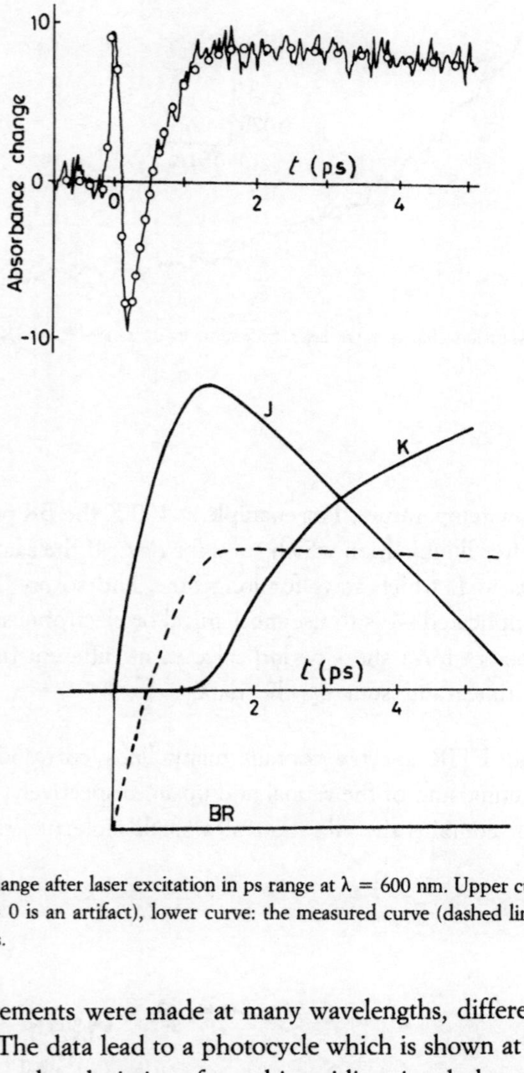


Fig. 9 – Light absorption change after laser excitation in ps range at $\lambda = 600$ nm. Upper curve: measured (from Ref. 38, the positive peak at $t = 0$ is an artifact), lower curve: the measured curve (dashed line) is the superposition of three different components.

Similar measurements were made at many wavelengths, different temperatures and different pH values. The data lead to a photocycle which is shown at Fig. 11. It should be emphasized once more that deviations from this unidirectional photocycle were suggested by many authors (like branchings at different intermediates, equilibration between them, *etc.*) but it is thought that a well defined photocycle called *restricted photocycle* is advantageous to correlate the data of the four cycles.

### *3.5. The retinal and opsin cycles*

In general the collection of RR and FTIR spectra takes much longer time than the lifetime of the different intermediates. The easy solution of the problem is to freeze the

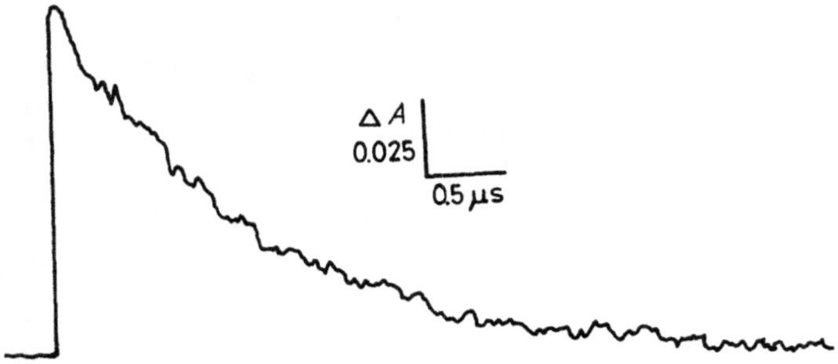


Fig. 10 – Light absorption change after laser excitation in μs range. λ = 632 nm, room temperature.

intermediates at low temperature. For example at 100 K the BR photocycle proceeds only until the K state after illumination which stays for days. If the sample is warmed to 150 K the K state decays to L which stays for long time, and so on. The other possibility — which is more complicated — is to use mechanical or electronic stroboscopic systems, *i.e.* to measure the spectra for a short period selected at different times after excitation but perform it many times and sum up the data.

The RR and FTIR spectra contain many lines corresponding to the different vibrations of the retinal and of the retinal and opsin, respectively. The analysis of RR data is easier because it contains the vibration of a small molecule, the retinal only. Even in

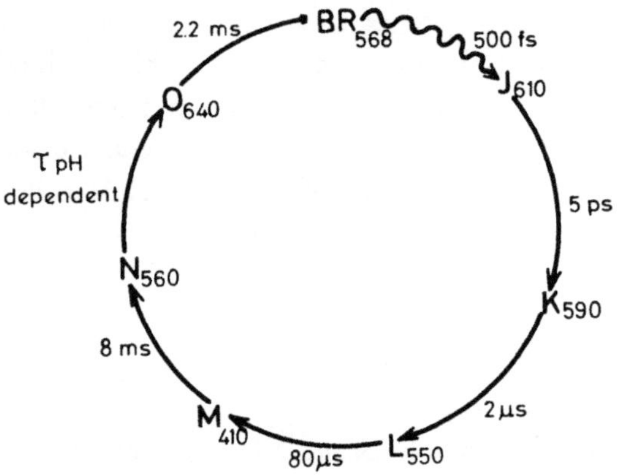


Fig. 11 – Photocycle of BR. Lifetimes at room temperature.

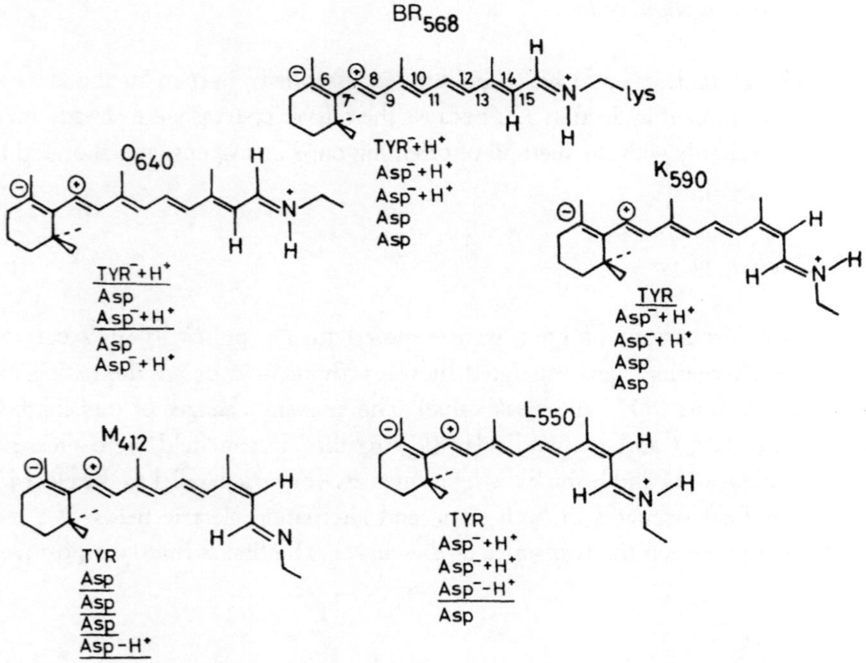


Fig. 12 – The retinal and opsin cycles of BR. The retinal cycle is determined by RR spectroscopy, changes in protonation of Tyr and Asp residues were detected by FTIR spectroscopy. The changed residues during the given transition are noted by underlining.

this case the isotopic substitution method should have been used: exchange of specific $^1H$ to $^2H$, $^{12}C$ to $^{13}C$ atoms helped to identify the vibrations and assign changes in the configuration of the retinal. The main findings are [25]:

— the K-state already has a 13-cis configuration. The retinal, however, is somewhat distorted which relaxes during the K→ L transition,

— the SCHIFF-base is deprotonated at the L→M transition, and gets reprotonated during the M→N transition,

— the N intermediate probably, the O surely, is all-trans.

The changes are shown in Fig. 12. The J state is 13-cis very probably but this intermediate has not yet been studied in detail.

The opsin cycle may be studied by FTIR and U.V. spectroscopy. The main results are: that tyrosine and aspartic acids deprotonate and reprotonate at different steps of the photocycle. Also these data are included in Fig. 12.

### 3.6. *The charge motion cycle*

The charge motion cycle has been studied essentially by two methods as it has already been mentioned in Section 3.3. Because the bilayer systems were already reviewed [39] here we deal only with the method of orienting pm-s in suspension elaborated in our laboratory [34, 40].

### 3.7. *Orientation of pm-s*

Aqueous suspensions of pm-s were exposed to *d.c.* pulses (0-60 V/cm) of 3 s duration with alternating signs separated by 6 s without field or an alternating electric field of 6 kHz of 0 to 350 V/cm (peak value). The intensity changes of transmitted light polarized parallel ($I_{II}$) and perpendicular ($I_{\perp}$) to the electric field were measured at $\lambda = 575$ nm. Figure 13a and b show the light intensity changes parallel to the electric field ($I_{II}$) at various field strengths in both static and alternating electric fields. We see two clear differences between the two series of the curves. The first is that $I_{II}$ is positive for a

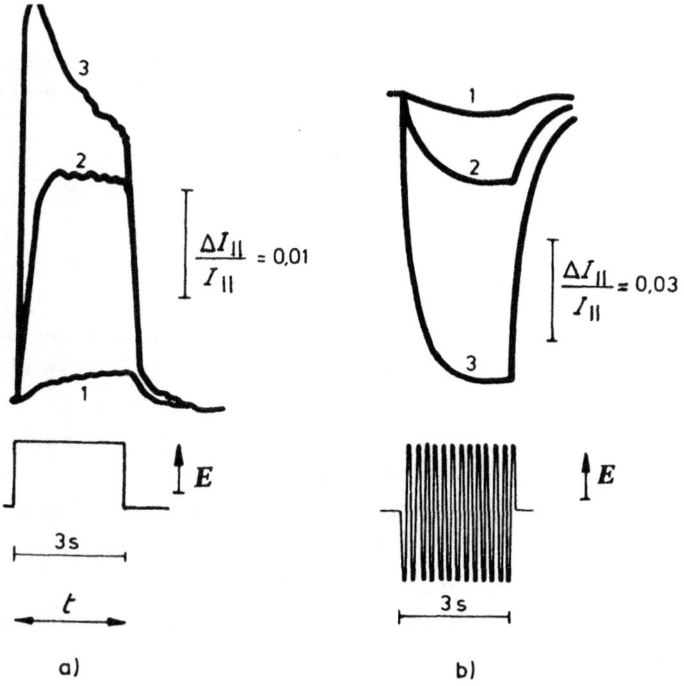


Fig. 13 – Light intensity changes (a) in a static electric field, $E_{1,2,3}$= 1.7, 5.2, 15.8 V/cm, and (b) in alternating electric field, $E_{1,2,3}$ = 31, 77, 165 V/cm (peak values). The light is polarized parallel with the direction of the electric field.

*d.c.* and negative for an *a.c.* field. This is due to the different spatial arrangement of the membrane fragments in the static and alternating electric fields. The second difference is in the shape of the curves at higher field strengths. The decay of the curves at *d.c.* fields is caused by an inevitable clustering of the membranes which was observed by microscope.

Light intensity changes are related to the reduced absorption change [41]:

$$\frac{\Delta A}{A} = - \frac{1}{A} \lg \left(1 + \frac{\Delta I}{I}\right) \tag{1}$$

where $\Delta A$ is the corresponding absorbance change, and $A$ is the absorbance in the absence of electric field. The rotational symmetry around the direction of the electric field predicts that

$$\Delta A_{II} = - 2\, \Delta A_{\perp} \tag{2}$$

The reduced dichroism at any field can be written as

$$\begin{aligned} \frac{\Delta A_{\perp}}{A} &= \left(\frac{3}{2} \sin^2 \theta - 1\right) \phi\,(\beta, \gamma) \\ \frac{\Delta A_{II}}{A} &= (3 \cos^2 \theta - 1)\, \phi\,(\beta, \gamma) \end{aligned} \tag{3}$$

The value of reduced dichroism depends upon the angle between the membrane normal and the transition moment of the chromophore ($\theta$) and upon the function, $\phi\,(\beta, \gamma)$. $\beta = \mu E/kT$; $\mu$, permanent dipole moment; $E$, electric field strength; $k$, BOLTZMANN constant; $T$, temperature in degrees Kelvin; $\gamma = \alpha E^2/2kT$; and $\alpha$, polarizability of the particles.

The general expression of the orientation function $\phi\,(\beta, \gamma)$, which denotes the degree of orientation for disk shaped particles having permanent and induced dipole moments perpendicular to each other, has been calculated by SHAH [42]. Using this equation we computed and plotted some curves of $\phi\,(E)$ at characteristic values of $\beta^2/\gamma$ (Fig. 14). SHAH suggested a simplified expression if $\beta^2/\gamma << 1$, then

$$\theta\,(\beta, \gamma) = \frac{1}{15} \beta^2 - \frac{2}{15} \gamma - \frac{\beta^4}{90} - \frac{2}{45} \gamma^2 + \frac{\beta^4 \gamma}{30} \ldots \tag{4}$$

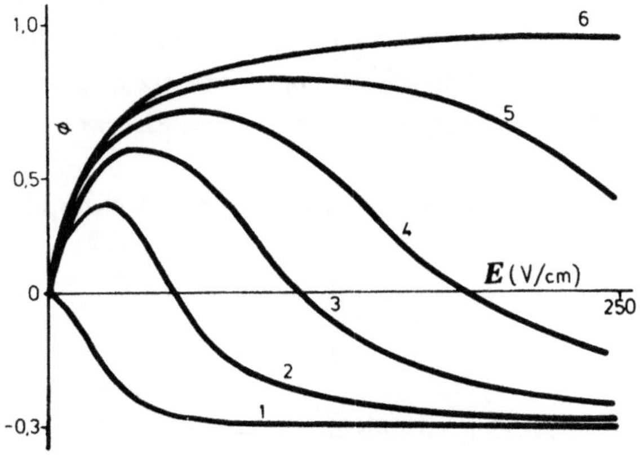


Fig. 14 – Calculated curves of $\phi(E)$ at different values of $\beta^2/\gamma$. The curves 1, 2, 3, 4, 5, 6 denote $\beta^2/\gamma$ = 0, 15, 30, 50, 90 and ∞ resp.

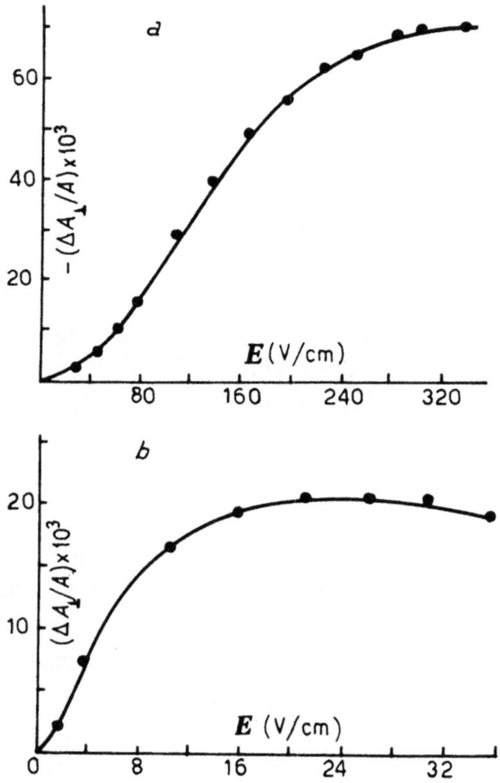


Fig. 15 – The reduced dichroism as a function of field strength for (a) alternating electric field and (b) static field.

For high electric field

$$\begin{aligned} &\phi(\beta, 0)_s \rightarrow \quad 1 \\ &\phi(0, \gamma)_s \rightarrow -1/2 \\ &\phi(\beta, \gamma)_s \rightarrow -1/2 \end{aligned} \tag{5}$$

To obtain values of $\mu$ and $\alpha$ the following procedures were used. Values of $\Delta A/A$ were plotted as a function of field strength. (Representative curves are displayed in Fig. 15*a* and *b*). Because only the induced moment participates in the orientation of the fragments in an *a.c.* field whose period is much greater than the relaxation time of the fragments, $\beta$ can be considered zero. Evaluation of $\gamma$ is straightforward: $\theta$ is obtained from the saturation value of the reduced dichroism, where $\phi(0, \gamma)_s = -1/2$. Using this $\theta$ value, we determine $\phi(0, \gamma)$ for any initial value of the $\Delta A/A$ data. Then $\gamma$ is calculated with the help of equation (4). When a *d.c.* field is used, both the permanent and induced dipole moments influence the orientation. The saturation at low fields is only apparent (Fig. 14). Therefore, $\phi(\beta, \gamma)$ and from it $\beta$ were calculated using equations (3) and (4) with $\theta$ and $\lambda$ determined from *a.c.* measurement. The permanent dipole moment was found to be ~ 70 Debye per one BR molecule in the pm; the polarizibility of pm-s is $\alpha = 3\times10^{-28}$ F $m^2$.

The most important finding of the experiments was that pm-s can be oriented practically to saturation by *d.c.* electric field of 15-20 V/cm. The orientation relaxes within 300-500 ms after the field is switched off. This way a sufficiently long period exists to perform measurements on oriented pm-s without external field. The orientation will stay for days if the sample is frozen to 100 K [43].

The oriented pm-s can be immobilized by methods worked out in our Institute. The pm suspension containing acrylamide and polymerizing agent is added during orientation by electric field [44]. Polymerization occurs within minutes. It is very easy to vary the bathing solution in the gel samples in which the BR molecules preserve their activity for years. This way the photocycle may be studied easily in very different conditions.

The pm-s move also electrophoretically in electric field providing the possibility to deposit them oriented on a transparent electrode [45]. After drying another electrode is evaporated on the other surface. This type of dried oriented sample works for many years.

### 3.8. *Protein electric response signals*

The samples containing oriented pm-s and consequently oriented BR molecules have enough material to excite ~ $10^{14}$ molecules by a typical laser flash. This quantity provides possibility to measure the light absorption changes and measure the current caused by the charges to be transported on the same sample. Because of the large number of moving charges a single flash is enough to obtain electric signals in the broad time range determined by the photocycle with an apparatus shown in Fig. 16.

A comparison of the electric and light absorption signals is presented in Fig. 17. We call this type of electric signals *protein electric response signals* (PERS). The first inspection of the lifetimes shows that the components of the PERS are associated with the transitions of the photocycle established by the three cycles previously described. Detailed investigations: temperature dependence (Fig. 18) and pH dependence (Fig. 19) demonstrated the correlation [46, 47]. The latter correspondence was confirmed and extended by MÜLLER *et al.* [48].

The origin of PERS is the displacement current caused by charge motion inside the protein. The protons emitted move randomly in a field-free solution and do not induce current in the external circuit. A short calculation will provide an understanding of the origin of the PERS. In the following we apply the plausible assumption that a single proton moves inside the BR considered as a dielectric. No real change would be needed if other charge(s) or dipole(s) were to move because the charge $Q$ can be specified later.

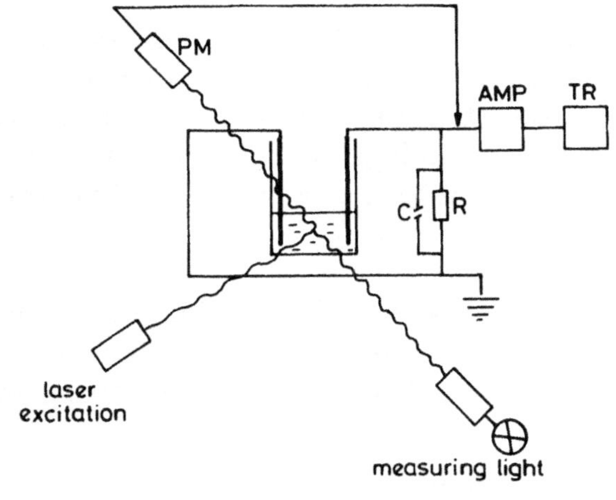


Fig. 16 – Scheme of the apparatus to measure the protein electric response signal and light absorption changes on the same sample. PM - photomultiplier, AMP-amplifier, TR-transient recorder.

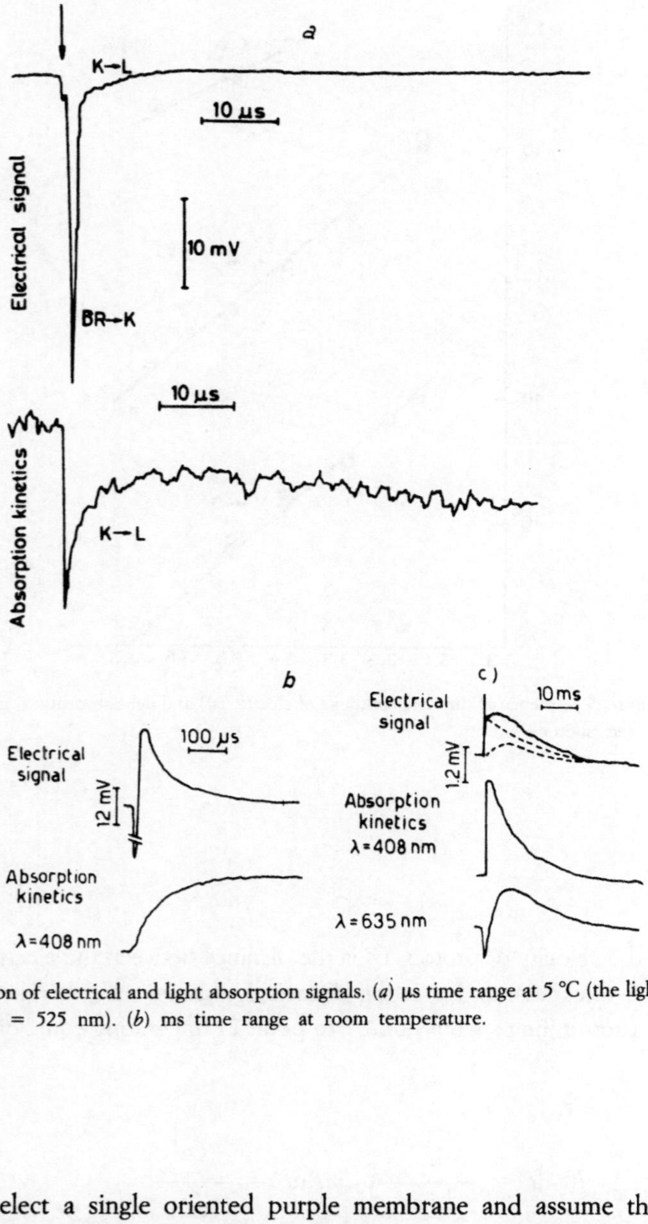


Fig. 17 – Comparison of electrical and light absorption signals. (*a*) µs time range at 5 °C (the light absorption signal was measured at λ = 525 nm). (*b*) ms time range at room temperature.

Let us select a single oriented purple membrane and assume that large planar electrodes are in contact with it (Fig. 20). A current *I* is induced in the external circuit if a charge moves between the electrodes:

$$I\,(t) = \frac{Qv(t)}{\varepsilon D'} \tag{6}$$

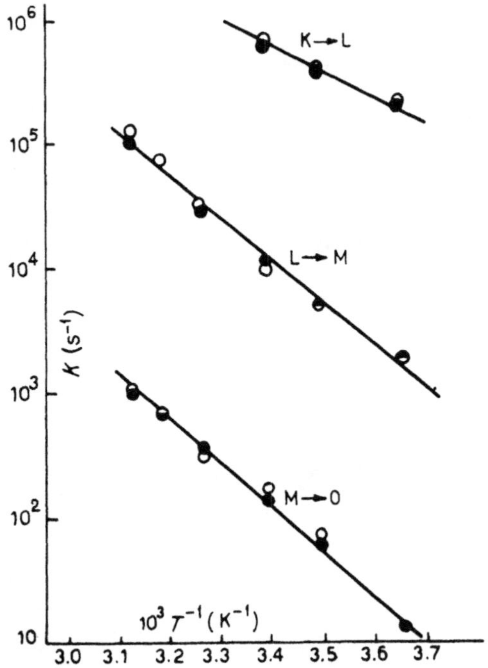


Fig. 18 – Temperature dependence of the time constants of electric (○) and light absorption signals (●) for different transitions during the photocycle.

where $v(t)$ is the velocity of proton, $D'$ is the distance between the electrodes, in this case the membrane thickness, and $\varepsilon$ is the dielectric constant. We assume that $v(t)$ is very large, *i.e.* the proton jumps from point 1 to point 2. Integrating equation (6) with respect to time:

$$Q_{ind} = \int_0^\infty I(t)\mathrm{d}t = \frac{Q}{\varepsilon D'}\int_0^\infty v(t)\mathrm{d}t = \frac{Qd}{\varepsilon D'} \qquad (7)$$

The proton jump induced a charge ($Q_{ind}$) proportional to the displacement $d$. $Q_{ind}$ charges the capacitance $C$ of the electrodes and that of the external circuit. The capacitance discharges through a resistance $R$. The potential difference for one charge displacement is

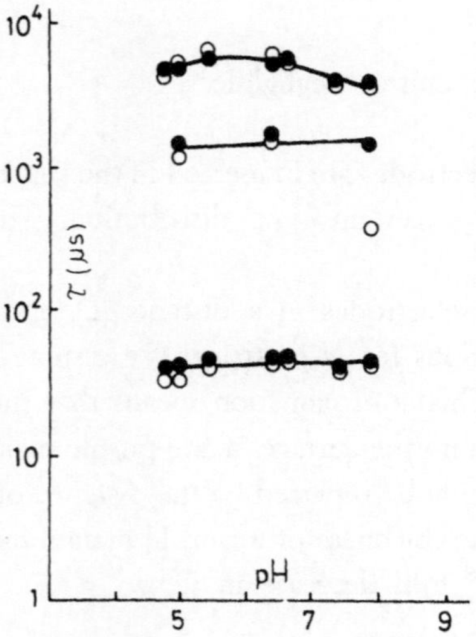


Fig. 19 – pH dependence of the time constants of electric (○) and light absorption signals (●) for different transitions during the photocycle (lower curve L→M, middle O→BR, upper M→O transition).

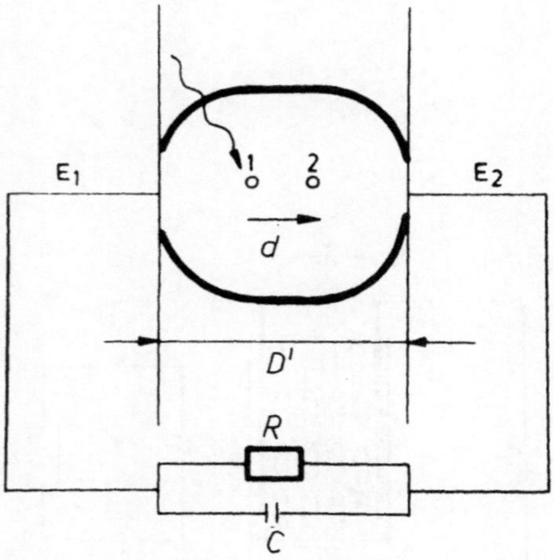


Fig. 20 – Elementary act to produce the displacement current. BR embedded in the membrane; charge moves from point 1 to 2; $E_1$ and $E_2$ hypothetical electrodes.

$$U_1\ (t) = \frac{Q_{ind}}{C} \exp\left(-\frac{t}{RC}\right) \tag{8}$$

if the inductance of the circuit is negligible.

In the real case the electrodes are immersed in the suspension in a distance $D$ and $N$ charges are moving with a certain time distribution. The calculation involves two processes:

(*I*) to determine $U_1(t)$ for electrodes in a distance $D$, and

(*II*) to sum the $U_1(t)$ functions for $N$ particles. We assume that the conductivity of the suspension is very small. This approximation means that the hypothetical electrodes in Fig. 20 are now removed from the surface of the purple membrane to a distance $D$ (Fig. 21). In this case $D'$ can simply be replaced by the distance of the immersed electrodes $D$ in equation (6). Because the thickness of a purple membrane $D' \sim 5$ nm and $D \sim 1$ cm, this means an approx. $10^6$ fold decrease in $Q_{ind}$.

Charge is induced in the external circuit only when a transition from state 1 to state 2 occurs. We assume a simple exponential decay of the $N$ states excited at $t=0$ then the number of states decaying in unit time is:

$$\varrho(t) = kN \exp(-kt) \tag{9}$$

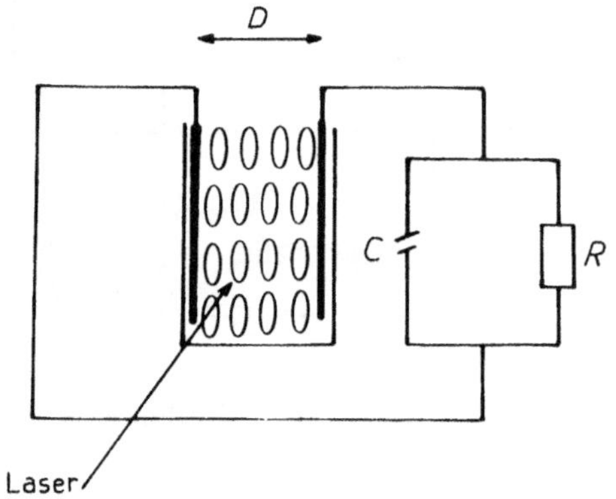


Fig. 21 – The displacement currents caused by elementary steps in BR-s in pm (the ellipses in the figure) are summed by the real electrodes $E_1$ and $E_2$ in a distance $D$ on the $RC$ circuit.

where $k$ is the rate constant. Every induced charge displacement produces a potential difference as given by equation (8). To obtain $U_N(t)$ we have to sum the $N$ uncorrelated $U_1(t)$ functions for all times $t' < t$ (Fig. 22). In calculation this means the folding of equations (8) and (9):

$$U_N(t) = \frac{NQdk}{\varepsilon DC} \int_0^t \exp(-kt') \exp\left(-\frac{t-t'}{RC}\right) dt'$$

$$= \frac{NQdk}{\varepsilon D} \frac{R}{1-kRC} \left[\exp(-kt) - \exp\left(-\frac{t}{RC}\right)\right) \tag{10}$$

If $k << 1/RC$, then:

$$U_N(t) = \frac{NQd}{\varepsilon D} Rk \exp(-kt) \tag{11}$$

*If* $k >> 1/RC$, then:

$$U_N(t) = \frac{NQd}{\varepsilon DC} \exp\left(-\frac{t}{RC}\right) \tag{12}$$

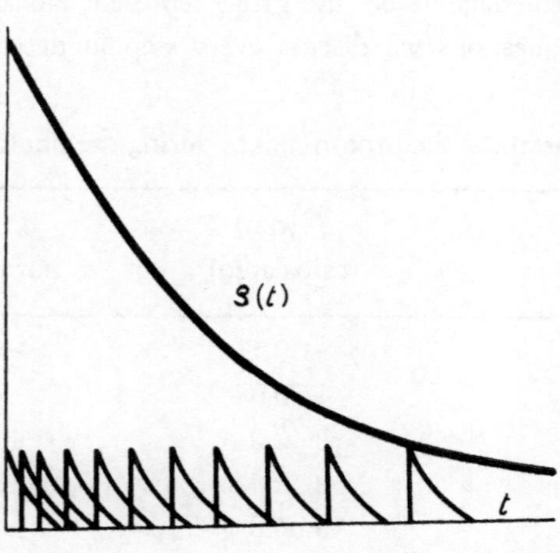


Fig. 22 – The $U_1(t)$ values (Equation 8) from the elementary acts are distributed in time with exponential density ($\varrho(t)$). $U_N(t)$ at a given value is obtained by summing all the $U_1(t)$ values from decays at times $t' < t$.

If there are many components of the charge movement in series, then we have to take into account that the expression for the number of decaying states [equation (9)] has several exponential components with different preexponential factors. These functions were first calculated by BATEMAN for the radioactive decay series and were recently given, specialized for the case of the BR photocycle [49]. Denoting this function by $f(k_1, \ldots, k_i, t)$ we obtain a simple expression for the $i$-th transitions (case $k << 1/RC$):

$$U_N^i\ (t) = \frac{NQR}{\varepsilon_i D} d_i\, k_i\, f\ (k_1, \ldots k_i,\ t) \tag{13}$$

where $U_N^i\ (t)$ is the measured component of the PERS, $k_i$, $D$, $R$ are measured too. The number of photocycles $N$ is determined from the light absorption signal, our assumption of one moving proton gives the value of $Q = 1.6 \times 10^{-19}\ C$. We introduce a new assumption that the dielectrics of the protein is homogeneous (it is probably wrong, but adequate data do not exist) and take $\varepsilon = 2$. The only unknown in equation (12) now is $d_i$. Calculations gave values for the different charge displacements as in Table 1. Naturally $\Sigma d_i$ should be equal to 5 nm, the thickness of the membrane. It was rather surprising that the above simplified treatment provided $\Sigma d_i = 20$ nm. *i.e.* the right order of magnitude. {A more elaborated theoretical approach resulted in expression similar to equation (12) [50]}. After normalization we obtain $d_i$ values corresponding to each step of the photocycle (Table 1, and Fig. 23).

Do the $d_i$ values have any meaning, *i.e.* do they correctly express that protons are translocated through the proteins in five different steps distributed in time according to the lifetimes of the intermediate? Does the PERS represent a charge motion cycle of the BR? To answer these questions we discuss every step in detail.

Table 1. – Distances the proton makes during the photocycle of BR

| Transition | $d_i$ (nm) (calculated) | $d_i$ (nm) normalized |
|---|---|---|
| BR - K | − 0.52 | −0.13 |
| K - L | − 0.08 | −0.02 |
| L - M | + 2.0 | +0.5 |
| M - O | +12.4 | +3.1 |
| O - BR | + 6.0 | +1.5 |
| $\Sigma d_i$ | 19.8 | 4.95 |

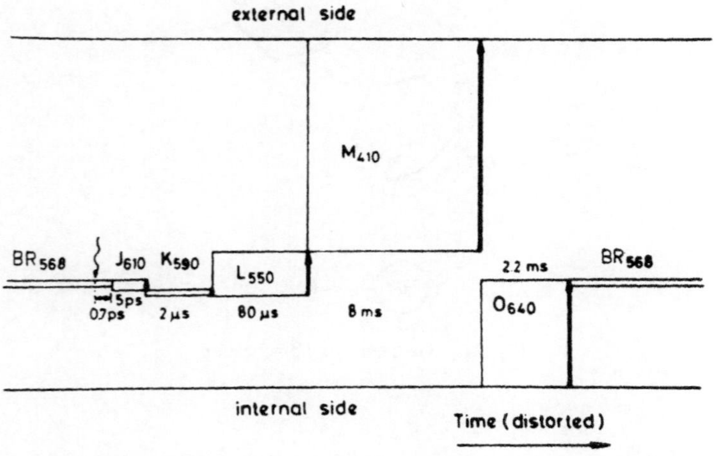


Fig. 23 – Charge motion cycle of BR. The length of the arrows is proportional to the calculated displacement of a single proton.

First of all it has to be emphasized that $d_r$-s determined form PERS represent displacements in the transmembrane direction. The lateral components do not induce current in the external circuit (laterally introduced electrodes would not pick up current either because of the cylindrical symmetry of the oriented pm-s).

$d_1$ is considered to be a negative signal because it is caused by displacement current in a direction opposite to proton translocation. Separate experiment with picosecond laser excitation and fast current recording (14 GHz sampling oscilloscope) demonstrated that this negative signal appears faster than 20 ps [51]. The signal is assigned to the first two transitions BR → J and/or to J → K because the time resolution of the recording is not sufficient to discriminate. It is known from the retinal cycle that the SCHIFF-base moves because of the all-trans → 13-cis isomerization. The proton stays on it, therefore a charge displacement is expected. The simple geometry of the isomerization suggests a motion of 0.16 nm (Fig. 24).

$d_2$ = – 0.02 nm may be a consequence of the relaxation of the 13-cis retinal as observed by resonance Raman spectroscopy. $d_1 + d_2 = -0.15$ nm is in good agreement with the distance calculated from the geometry of isomerization.

The third displacement $d_3$ = +0.5 nm occurs during the L → M transition. This first movement in the pumping direction is correlated with the deprotonation of the SCHIFF-base, known from the retinal cycle. The protonation of an aspartate is observed in

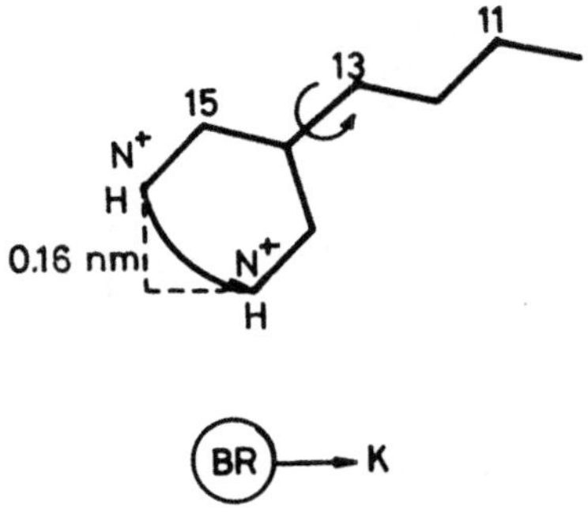


Fig. 24 – Geometry of isomerization.

the opsin cycle. Looking at the amino acid sequence (Fig. 3) Asp 212 is expected to serve as the acceptor of the SCHIFF-base proton.

Displacements $d_1$, $d_2$ and $d_3$ are deep inside the protein in the near neighbourhood of the SCHIFF-base. The two last components $d_4 = +3.1$ nm and $d_5 = +1.5$ nm connect the interior of BR with the exterior. $d_4$ is assigned to a long path for the proton to get to the external surface of BR and released, while $d_5$ corresponds to the uptake of proton from the internal side. The distances measured locate the SCHIFF-base 1/3 from the internal side of the membrane which coincides with the known geometry.

The distances $d_3$ and $d_4$ were confirmed by separate experiments. The electric potential difference affects the kinetics of a transition [52]:

$$\tau\ (U) = A\ \exp\ \{ -\ [(\Delta H + \frac{2\mathsf{F}dU}{D})/k'T\} \tag{14}$$

where $U$ is the potential difference across the membrane of thickness $D$, $A$ is the preexponential factor of the ARRHENIUS relation, $\Delta H$ is the activation enthalpy of the transition, F is Faraday's costant, $k'$ is the BOLTZMANN constant referred to 1 molecule and $d$ the thickness of the barrier in this case the distance the proton has to overcome.

External potential difference $U$ was applied to dried oriented samples and $\tau(U)$ of the L$\rightarrow$M transition was measured. The data in Fig. 25 show the expected exponential behaviour of $\tau$ *versus* $U$ and $d = 0.48 \pm 0.2$ nm is calculated from the slope in agreement

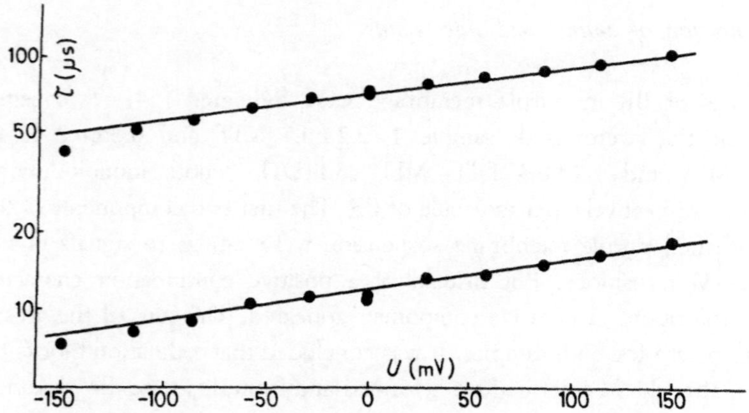


Fig. 25 – Dependence of the time constant of the electric signals (L-M transition, two components) from the externally applied potential difference, across the membrane.

with the PERS data (Fig. 23). A different approach provided data on the rate limiting step in creating membrane potential difference. Braun *et al.* [53] used the bilayer system to produce equilibrium membrane potential difference during continuous illumination in externally applied potential difference. It is evident that the lifetime of the rate limiting step which is regulated according to equation (13) determines the equilibrium potential difference. The authors calculate 60 % of the thickness of the membrane as the distance a proton makes during the rate limiting step which is evidently the M→O transition (Fig. 11). This value coincides with and conforms the value of $d_4$ = 3.1 nm $\simeq$ 0.6 × 5 nm.

## 4. Application

The study of PERS on differently modified BR exposed to different environmental effects provide valuable data on the mechanism of proton translocation. Since the sum of the area of the different components is equal to the total electric charge $\Sigma A_i = \Sigma \int_0^\infty U_i(t)dt$, $\Sigma A_i > 0$ means a real charge translocation, $\Sigma A_i = 0$ would mean a charge motion in one direction compensated by another motion in the opposite direction. $\Sigma A_i$ characterises the proton translocation. Therefore in the following the sum of the area of the electric signals will be investigated for BR with some of its amino acids modified, and also for BR in different environments. $\Sigma A_i$ is always related to the first negative signal ($A_1$) which is considered to be proportional to the number of excited cycles.

*4.1. Modification of amino acid side chains*

Tyrosines of BR in purple membrane were iodinated [54]. Two samples with different iodination were used: sample 1 : 2.2±0.5 MIT and 0.6±0.2 DIT; sample 2 : 3.1±0.5 MIT and 1.2±0.3 DIT. MIT and DIT denote monoiodotyrosine and diiodotyrosine, respectively, per molecule of BR. The first two components of the electric signals in oriented purple membrane suspension were similar to signals of normal BR (BR→K→L→M transitions) but instead of a positive continuation characteristic for M→O→BR transitions, a negative component appeared. The sum of the areas $\Sigma A_i$ was nearly equal to zero for both samples. It was concluded that iodination blocks the proton translocation after the M-intermediate is formed and from this state the proton returns to its original position [55].

There are 19 carboxyl groups in BR, one at the N-terminal, four in the C-terminal tail, seven in the helix–to–helix link regions protruding into the aqueous medium and the remaining seven are in the interior of the molecule. EEDQ (N-(ethoxycarboxyl)-2-ethoxy-1, 2-dihydroxyquinoline) is a hydrophobic, highly specific carboxyl-activating reagent for carboxyl groups in the interior of the protein. The modification by EEDQ means crosslinking of carboxyl groups inside the protein; they are suspected to play a role in proton translocation. Crosslinking between separate proteins is negligible [56].

The modified pm-s were oriented and immobilized in gel and their electric response signal and absorption kinetics were studied. The results have shown that the crosslinking of internal carboxyl group does not change the photocycle and the proton pumping activity of BR significantly [57]. Amidation of all the accessible ε-amino groups of lysine residues was performed by dimethyl-3,3'-dithiobispropionimidate [58]. Gel electrophoresis data have shown that the BR-s were partially crosslinked: the monomer content was 30-40 % which could be increased to 60-90 % by mercaptoethanol treatment.

The light absorption signals had two components of M-rise ($\tau_1$ = 20 μs, $\tau_2$ = 80 μs with equal amplitudes) and three components of M-decay ($\tau_3$ = 10 ms, $\tau_4$ = 50 ms, $\tau_5$ = 200 ms, relative amplitudes 0.25–0.5–0.25 respectively, the decay times were measured at ~22 °C). The time constants of the measured electric signals were in accord with those of absorption changes in a broad pH range (5-9) [58]. $\Sigma A_i$ was positive starting from a low value at pH 3 and becoming essentially constant between pH 5 and 9. It was interesting to observe in a limited pH range a small negative component after M-formation with a lifetime not found in absorption data. It is interpreted as a branching after the L-intermediate to an intermediate from which the protons fall back to their original position. The estimated loss is ~15 % at the maximum. The increase in the monomer

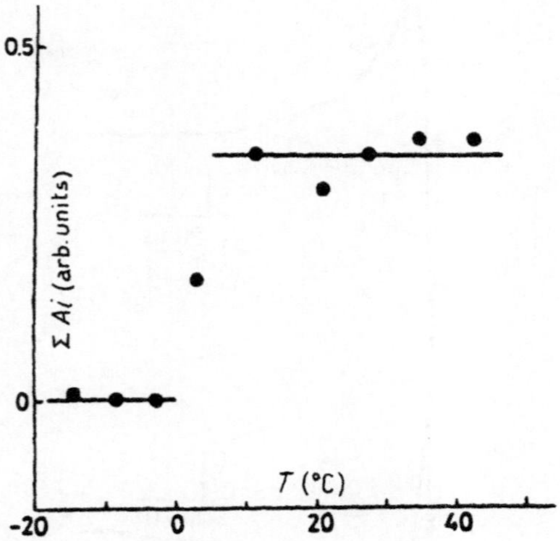


Fig. 26 – Temperature dependence of the sum of the area of the electric signal. Dried oriented sample, water content 0.12 g $H_2O$/g BR.

content does not influence the absorption and electric signals significantly [58]. It may be concluded that the lysine residues do not take part in proton translocation because $\Sigma A_i$ remained practically unchanged. This finding is in accord with previous data [59, 60] where the effect of succinylation and methylation of lysines was studied.

It has been observed in dried, oriented pm-samples that proton translocation does not occur if the water content of the sample is less than 0.06 g $H_2O$/g BR [61]; also the Arrhenius parameters and infrared spectra change at this water content [61-63]. From these data it may be concluded that water molecules are necessary for proton translocation. The temperature dependence of $\Sigma A_i$ was also measured in the presence of 0.12 g $H_2O$/g BR which is well above the limiting value. As Fig. 26 shows the area is zero below O °C demonstrating that the excess water should be in fluid state.

Recently an unexpected phenomenon was found in our laboratory: diamines, when added to pm-suspension may reverse the direction of the proton pump [64]. Electric signals measured on pm-s oriented in suspension with added N, N, N', N',-tetramethylethylenediamine (TEMED) are shown in Fig. 27. The main characteristics are the following:

*a*) the reversal depends on the concentration fo TEMED;

*b*) at ~ 1 TEMED molecule per BR the negative area equals the positive one without TEMED;

*c*) at higher TEMED concentrations the original pumping direction is reestablished;

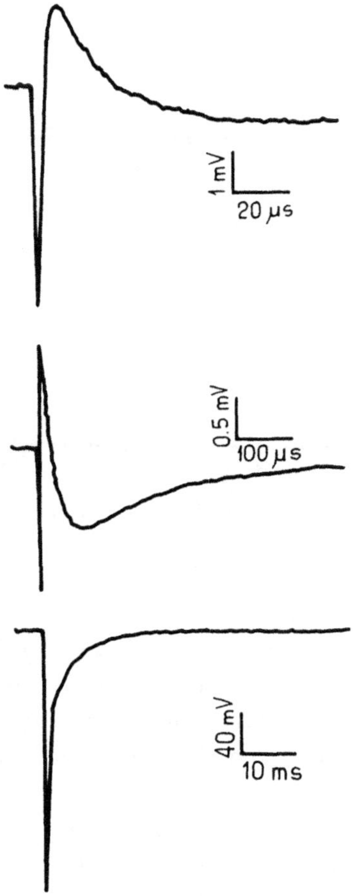


Fig. 27 – Electric signals from gel samples in presence of TEMED. Concentration ~ TEMED molecule /BR.

*d*) the effect is reversible: after washing, the unperturbed photocycle appears;
*e*) the TEMED is gradually inactivated by salt;
*f*) some other diamines cause similar variations, but monoamines do not.

The influence of diamines is interpreted as a charge effect: their presence alters the charge distribution within the protein (or the surface charges) which is needed to determine the direction of proton flow.

The normal and reversed proton flow can be sustained in gel samples at least for ~30 s by constant illumination [65]. In this case the light driven proton pump of BR behaves as a battery where the direction of the potential difference may be changed by adding TEMED in the proper concentration.

### 4.2. *Mechanism of proton conduction*

It was interesting to observe in all applications that first components of PERS were always present. Modifications and environmental effects influenced the two components representing contacts with the exterior of the protein. These two components are due to the transport of proton for long pathways (3.1 and 1.5 nm). The pathway may be constructed from proton–donating and –accepting side chains of amino acids through which the protons are conducted as in ice [66–68]. This theory of proton conduction adequately forecasts the speed of proton transfer [66]. The experiments, however, are not in favour of this mechanism:

(*a*) Cross-linking of possible proton donors and acceptors (lysines and carboxyls) varies the life times of the intermediates but does not decrease the number of protons per photocycle.

(*b*) Proton translocation ceases below a water content of 0.06 g $H_2O$/g BR; data show that even for higher water content, protons are pumped only above 0 °C, *i.e.*, in the presence of fluid water.

Based on the above data and on the general properties of proteins another simple model was proposed [46]. It is well known from $^1H$–$^2H$ exchange experiments that different parts of the protein are accessible to water because of the fluctuation in protein structure. This process is rather fast (in the millisecond range). We assume that the proton at the carboxyl group in the M-intermediate is approached by water molecules through temporarily opened channels, their dipole moment lowers the barrier for proton release, the resultant $H_3O^+$ is driven out to the external side and one $H_3O^+$ is taken up from the internal side, by the existing internal electric field due to the primary charge separation and surface charges.

Arguments in support of the model are points *a* and *b* above and also:

(*c*) Tryptophan fluorescence data obtained by PLOTKIN and SHERMAN [69] demonstrate the existence of intraprotein aqueous regions;

(*d*) Diamines reverse the direction of the proton flow. The direction of continuous currents generated by continuous illumination of BR oriented and immobilized in gels is reversed by the presence of TEMED.

(*e*) The model is general and is also applicable for other ions (for example, halorhodopsin, which pumps $Cl^-$ after light absorption [70]).

Data on the de- and reprotonation of aspartic acids do not contradict the model. It is highly possible that the conduction of the proton in the water-filled channel may be accelerated by these groups but their elimination, however, is not catastrophic and merely slows down the process. Recently, genetically modified BR-s were prepared and tested for proton translocation [7, 71]. Tyrosine, tryptophan, proline, serine and aspartic acid residues were changed by site-specific mutagenesis to other residues. The mutant apoproteins were regenerated by retinals and reconstitued into vesicles. The mutants pumped protons. The experiments point to a general pumping mechanism which is not influenced by many different changes in the protein.

## 5. Conclusions

It may be seen from the experimental data presented above that the understanding of the BR proton pump is still in the *hypothesis* state. We believe, nevertheless, that many features of this pump may have a *model value* in understanding other pumping mechanism:

(1) the primary energy uptake is of an electrostatic nature: a small motion picks up comparatively large amount of energy;
(2) significant changes occur in the charge states of the side chains of some amino acids;
(3) water is a decisive factor;
(4) surface charges are involved in pumping;
(5) many chemical and genetic changes do not abolish pumping, showing its dependence upon general features of the protein structure.

## References

[1] P. Boyer, B. Chance, L. Ernster, P. Mitchell, E. Racker and E. Slater, *Annu. Rev. Biochem.*, **46**, 966 (1977)

[2] W. Stoeckenius, R.H. Lozier and R.A. Bogomolni, *Biochim. Biophys. Acta*, **505**, 225 (1979).

[3] D. Oesterhelt and W. Stoeckenius, *Proc. Natl. Acad. Sci. USA*, **70**, 2853 (1973).

[4] Yu. A. Ovchinnikov, N.G. Abdulaev, M.Y. Feigina, A.V. Kiselev and N.A. Lobanov, *FEBS Lett.*, **100**, 219 (1979).

[5] H.G. Khorana, G.E. Gerber, W.C. Herlihy, C.P. Gray, R.J. Anderegg, K. Nihei and K. Biemann, *Proc. Natl. Acad. Sci. USA,* **76**, 5046 (1979).

[6] S. Das Sarma, K.L. Raj Bhandary and H.G. Khorana, *Proc. Natl. Acad. Sci. USA,* **81**, 125 (1984).

[7] T. Mogi, L.J. Stern, T. Marti, B.H. Chao and H.G. Khorana, *Proc. Natl. Acad. Sci. USA,* **85**, 4148 (1988).

[8] R. Henderson and P.N.T. Unwin, *Nature London*, **257**, 28 (1975).

[9] R. Henderson, J.M. Baldwin, K. Downing, J. Lepault and F. Zemlin, *Eur. J. Cell,* **44**, 23 (1987).

[10] F. Seiff, J. Wallant, P. Ermann and M.P. Heyn, *Proc. Natl. Acad. Sci. USA,* **82**, 3227 (1985).

[11] F. Seiff., J. Westerhausen, J. Wallat and M.P. Heyn, *Proc. Natl. Acad. Sci. USA,* **83**, 7746 (1986).

[12] D. Oesterhelt and W. Stoeckenius, *Meth. Enzymol.,* **31**, 667 (1974).

[13] J.K. Lányi and V. Vodyanoy, *Biochemistry,* **25**, 1465 (1986).

[14] J.L. Spudich and R.A. Bogomolni, in *Biophysical Studies of Retinal Proteins*, T.G. Ebrey, H. Frauenfelder, B. Honig and K. Nakanishi (Editors) University of Illinois Press, (1987), p. 24.

[15] E. Racker and W. Stoeckenius, *J. Biol. Chem.,* **249**, 1955 (1974).

[16] N.A. Dencher and M.P. Heyn, *FEBS Lett.*, **108**, 307 (1979).

[17] R. Henderson, J.M. Baldwin, T.A. Aske, R.M. Glaser and I.N. Tsygannik, in *Retinal Proteins*, Yu. A. Ovchinnikov (Editor), VNU Science Press, Utrecht, (1987), p. 153.

[18] P. Scherrer, W. Stoeckenius, M.K. Mathew and W. Sperling in *Biophysical Studies of Retinal Proteins*, T.G. Ebrey, H. Frauenfelder, B. Honig and K. Nakanishi (Editors), University of Illinois Press, (1987), p. 206.

[19] S.O. Smith, A.B. Myers, J.A. Pardoen, C. Winkler, P.P.J. Mulder, J. Lugtenburg and R. Mathies, *Proc. Natl. Acad. Sci. USA,* **81**, 2055 (1984).

[20] S.O. Smith, M.S. Braiman, A.B. Myers, J.A. Pardoen, J.M.L. Courtin, C. Winkel, J. Lugtenburg and R.A. Mathies, *J. Am. Chem. Soc.,* **109**, 3108 (1987).

[21] G.S. Harbison, S.O. Smith, J.A. Pardoen, J.M.L. Courtin, J. Lugtenburg, J. Hertzfeld, R.A. Mathies and R.G. Griffin, *Biochemistry,* **24**, 695 (1985).

[22] J.L. Spudich, D.A. McCain, K. Nakanishi, M. Okabe, N. Shimizu, H. Rodman, B. Honig and R.A. Bogomolni, *Biophys. J.,* **49**, 479 (1986).

[23] P. Hildebrandt and M. Stockburger, *Biochemistry,* **23**, 5539 (1984).

[24] R.H. Lozier, R.A. Bogomolni and W. Stoeckenius, *Biophys. J.,* **15**, 955 (1975).

[25] S.O. Smith, J. Lugtenburg and R.A. Mathies, *J. Membr. biol.,* **85**, 95 (1985).

[26] K. Bagley, G. Dollinger, L. Eisenstein, A.K. Singh and L. Zimányi, *Proc. Natl. Acad. Sci. USA,* **79**, 4972 (1982).

[27] B. Hess and D. Kuschmitz, *FEBS Lett.,* **100**, 334 (1979).

[28] G. Dollinger, L. Eisenstein, S.L. Lin, K. Nakanishi and J. Termini, *Biochemistry,* **25**, 6524 (1986).

[29] M. Engelhard and B. Hess, in *Retinal Proteins*, Yu. A. Ovchinnikov, (Editor), VNU Science Press, Utrecht, (1987), p. 259.

[30] H.W. Trissl and M. Montal, *Nature (London),* **266**, 655 (1977).

[31] L.A. Drachev, A.D. Kaulen and V.P. Skulachev, *FEBS Lett.,* **87**, 161 (1978).

[32] A. Fahr, P. Läuger and E. Bamberg, *J. Membr. Biol.,* **60**, 51 (1981).

[33] G.W. Rayfield, *Biophys. J.,* **48**, 111 (1985).

[34] L. Keszthelyi and P. Ormos, *FEBS Lett.,* **109**, 189 (1980).

[35] Zs. Dancsházy, R. Govindjee and T.G. Ebrey, *Proc. Natl. Acad. Sci. USA,* **85**, 6358 (1988).

[36] L.A. Drachev, A.D. Kaulen, V.P. Skulachev and V.V. Zorina, *FEBS Lett.,* **226**, 139 (1987).

[37] T. Kouyama, A. Nasuda-Kouyama, A. Ikegemi, M.K. Mathew and W. Stoeckenius, *Biochemistry,* **27**, 5855 (1988).

[38] H.J. Polland, M.A. Franz, W. Zinth, W. Kaiser, E. Kölling and D. Oesterhelt, *Biophys. J.,* **49**, 651 (1986).

[27] B. Hess and D. Kuschmitz, *FEBS Lett.,* **100**, 334 (1979).

[28] G. Dollinger, L. Eisenstein, S.L. Lin, K. Nakanishi and J. Termini, *Biochemistry,* **25**, 6524 (1986).

[29] M. Engelhard and B. Hess, in *Retinal Proteins,* Yu. A. Ovchinnikov, (Editor), VNU Science Press, Utrecht, (1987), p. 259.

[30] H.W. Trissl and M. Montal, *Nature (London),* **266**, 655 (1977).

[31] L.A. Drachev, A.D. Kaulen and V.P. Skulachev, *FEBS Lett.,* **87**, 161 (1978).

[32] A. Fahr, P. Läuger and E. Bamberg, *J. Membr. Biol.,* **60**, 51 (1981).

[33] G.W. Rayfield, *Biophys. J.,* **48**, 111 (1985).

[34] L. Keszthelyi and P. Ormos, *FEBS Lett.,* **109**, 189 (1980).

[35] Zs. Dancsházy, R. Govindjee and T.G. Ebrey, *Proc. Natl. Acad. Sci. USA,* **85**, 6358 (1988).

[36] L.A. Drachev, A.D. Kaulen, V.P. Skulachev and V.V. Zorina, *FEBS Lett.,* **226**, 139 (1987).

[37] T. Kouyama, A. Nasuda-Kouyama, A. Ikegemi, M.K. Mathew and W. Stoeckenius, *Biochemistry,* **27**, 5855 (1988).

[38] H.J. Polland, M.A. Franz, W. Zinth, W. Kaiser, E. Kölling and D. Oesterhelt, *Biophys. J.,* **49**, 651 (1986).

[39] E. Bamberg and K. Fendler, in this volume, Chapter 4.

[40] L. Keszthelyi and P. Ormos, *Biophys. Chem.,* **18**, 397 (1983).

[41] E. Frederiq and C. Houssier: *Electric Dichroism and Electric Birefringence.* Clarendon Press, Oxford, (1973), p. 81.

[42] M.I. Shah, *J. Phys. Chem.* **67**, 2215 (1963).

[43] P. Ormos, L. Reinisch and L. Keszthelyi, *Biochim. Biophys. Acta,* **722**, 471 (1983).

[44] A. Dér, P. Hargittai and J. Simon, *J. Biochem. Biophys. Meth.,* **10**, 295 (1985).

[45] G. Váró, *Acta Biol. Acad. Sci. Hung.,* **32** 301 (1982).

[46] L. Keszthelyi, in *Information and Energy Transduction in Biological Membranes,* L. Bolis. E.J. Helmreich and H. Passow, (Editors), Alan R. Liss, Inc. New York, (1984), p. 51.

[47] P. Ormos, S.G. Hristova and L. Keszthelyi, *Biochim. Biophys. Acta,* **809**, 181 (1985).

[48] K.H. Müller, H.J. Butt, M. Engelhard, B. Hess and E. Bamberg, in *Molecular*

*Physiology of Retinal Proteins*, T. HARA, (Editor), Yamada Sci. Found., (1988), p. 341.

[49] P. LÄUGER, R. BENZ, G. STARK, E. BAMBERG, P.C. JORDAN, A. FAHR and V. BROCK, *Q. Rev. Biophys.*, **14**, 513 (1981).

[50] L. KESZTHELYI and P. ORMOS, *Biophys. Chem.*, **18**, 397 (1983).

[51] G. GROMA, F. RÁKSI, G. SZABÓ and G. Váró, *Biophys. J.*, **54**, 77 (1988).

[52] H.D. ZWOLINSKY, H. EYRING and C.E. REECE, *J. Phys. Chem.*, **53**, 1426 (1949).

[53] D. BRAUN, N.A. DENCHER, A. FAHR, M. LINDAU and M.P. HEYN, *Biophys. J.*, **53**, 617 (1988).

[54] P. SCHERRER, L. PACKER and S. SELTZER, *Arch. Biochem. Biophys.*, **212**, 589 (1981).

[55] L. PACKER, P. SCHERRER, K.T. YUE, GY. VÁRÓ, P. ORMOS, K. BARABÁS, A. DÉR and L. KESZTHELYI, *Biochem. Int.*, **5**, 437 (1982).

[56] S. WU-CHOU, A.E. ROBINSON, E. HRABETA and L. PACKER, *Biochem. Biophys. Res. Comm.*, **124**, 556 (1984).

[57] L. PACKER, E. HRABETA-ROBINSON, S.G. TANEVA, R. TÓTH-BOCONÁDI and L. KESZTHELYI, *Biochem. Int.*, **14**, 977 (1987).

[58] R. TÓTH-BOCONÁDI, S.G. TANEVA, A.V. KISELEV, N.G. ABDULAEV and L. KESZTHELYI, *Arch. Biochem. Biophys.*, **260**, 726 (1988).

[59] A.V. KISELEV, N.G. ABDULAEV, R.G. VASILOV, I.R. NABIEV and YU.A. OVCHINNIKOV, in *Short Reports of Third European Bioenergetics Conference*, Congress Edition Hannover, **3A**, 27 (1984).

[60] N.G. ABDULAEV, N.A. DENCHER, A.E. DERGACHEV, A. FAHR and A.V. KISELEV, *Biophys. Struct Mech,* **10**, 211 (1984).

[61] G. VÁRÓ and L. KESZTHELYI, *Biophys. J.*, **43**, 47 (1983).

[62] G. VÁRÓ and L. KESZTHELYI, *Biophys. J.*, **47**, 243 (1985).

[63] G. VÁRÓ and L. EISENSTEIN, *Eur. Biophys J.*, **14**, 163 (1987).

[64] R. TÓTH-BOCONÁDI, S.G. HRISTOVA and L. KESZTHELYI, *FEBS Lett.*, **19**, 164 (1986).

[65] A. DÉR, R. TÓTH-BOCONÁDI and L. KESZTHELYI, *FEBS Lett.*, **228**, 313 (1988).

[66] J.F. NAGLE and S. TRISTRAM-NAGLE, *J. Membr. Biol.*, **74**, 1 (1983).

[67] A. BRUNGER, Z. SCHULTEN and K. SCHULTEN, *Z. Phys. Chem.*, **135**, 1 (1983).

[68] H. MERZ and G. ZUNDEL, *Biochem. Biophys. Res. Comm.*, **101**, 540 (1981).

[69] B.J. PLOTKIN and W.V. SHERMAN, *Biochemistry,* **23**, 5353 (1984).

[70] A. DÉR, K. FENDLER, S. SZÁRAZ, L. KESZTHELYI, D. OESTERHELT and E. BAMBERG, *Biochim. Biophys. Acta,* in the press (1988).

[71] P.L. AHL, L.J. STERN, N.R. HACKETT, K.J. ROTSCHILD and H.G. KHORANA, *Biophys. J.*, **51** 416a (1987).

# THE METALS OF CYTOCHROME *C* OXIDASE AND THEIR ROLE IN THE KINETICS OF ELECTRON TRANSFER AND PROTON PUMPING

FRANCESCO MALATESTA (•), GIOVANNI ANTONINI (•), PAOLO SARTI (#), MAURIZIO BRUNORI (+)

*(•) Department of Experimental Medicine and Biochemical Sciences, University of Rome "Tor Vergata", Via O. Raimondo 1, 00173 Rome, Italy*

*(#) Department of Biomedical Technologies and Biometry, University of L'Aquila, 67100 L'Aquila, Italy*

*(+) Department of Biochemical Sciences and CNR Center of Molecular Biology, University of Rome "La Sapienza", 00185 Rome, Italy*

Contents



*Bioelectrochemistry III*
Edited by G. Milazzo and M. Blank
Plenum Press, New York, 1990

## 1. INTRODUCTION

Cytochrome *c* oxidase (ferrocytochrome *c*:02 oxidoreductase, EC 1.9.3.1) is a complex oligomeric metalloprotein incorporated in the mitochondrial inner membrane of eukaryotic cells or in the plasma membrane of bacteria [1, 2, 3]. This enzyme catalyses the transfer of electrons between two substrates, namely ferrocytochrome *c* and molecular oxygen. Part of the redox energy involved in this reaction is used by the enzyme to transfer protons from the matrix aqueous space to the intermembrane space of mitochondria [4]. Thus, cytochrome *c* oxidase is a redox-linked proton pump, which couples redox energy to the endoergonic vectorial transport of protons in line with the chemiosmotic theory.

According to a classical point of view, electron transfer and energy transduction by cytochrome *c* oxidase is achieved by at least two different metals (copper and iron) organized in three distinct prosthetic groups. These may be classified [1] on the basis of the substrate with which they may interact directly:

a) an iron atom inserted in a heme *a* prosthetic group, bearing a formyl group and a long isoprenoid side chain, called heme *a*,
b) an e.p.r. detectable copper ion termed CuA or Cu *a*, since it is in relatively close proximity to the heme *a* site and parallels its kinetic behaviour. These two centers represent the sites of electron donation by cytochrome *c*.
c) The third site represents the prosthetic group which binds and reduces dioxygen to water; it is an heterobinuclear center made up of one iron and a copper ion.

The iron atom is inserted into an otherwise identical heme *a* prosthetic group but located in a protein environment which confers quite distinct and differing properties. This metal center is referred to as the heme $a_3$ site and is within 0.5 nm from a copper ion, called CuB or Cu $a_3$, which under ordinary conditions can not be detected by e.p.r.

It is the purpose of this chapter to review current concepts on the structure of metal centers in cytochrome *c* oxidase and the kinetic properties of the enzyme in order to critically discuss possible mechanisms of electron transfer and energy transduction. Open questions relative to any of the above topics will also be highlighted, since some generally accepted dogmas, such as metal composition, have been recently challenged.

## 2. The metals of cytochrome *c* oxidase

It is now largely accepted that cytochrome *c* oxidase from beef heart is composed of 13 different polypeptides in a 1:1 stoichiometry, with molecular weights ranging from *ca.* 57000 to *ca.* 5000 daltons. These polypeptides are assembled in the functional complex by non-covalent bonds. As expected for an intrinsic membrane protein, many of these polypeptides are quite hydrophobic allowing for proper interaction with the non-polar components of the phospholipid bilayer. The metals of cytochrome *c* oxidase must be accomodated and ligated by residues in one or more of these subunits. There is still considerable debate on this subject since it is quite difficult to handle this protein and there are no measurements to locate unequivocally the metals in oligomeric proteins. A clue to the understanding of this aspect may come from the comparative aspects of sequence information (which is rapidly growing, see Chapter by R. Bisson) and from the study of simpler oxidases such as those obtained from prokaryotic organisms. The oxidase extracted from *Paracoccus denitrificans*, (see for example Refs. 5 and 6), is made up of only 2 polypeptides (homologous to subunits I and II of the eukaryotic enzyme [7]); yet most, if not all, of its spectroscopic properties are very similar to those of the eukaryotic enzyme. Mitochondrial respiratory complexes, and in particular cytochrome *c* oxidase, have a biosynthetic peculiarity: the genes coding for the higher molecular weight subunits, *i.e.* subunits I, II and III, reside on the mitochondrial DNA and translation occurs on mitochondrial ribosomes (which resemble the bacterial ones). The remaining 10 subunits are coded for by the nuclear DNA. These observations indicate that all metals are likely to be bound to subunits I and II also in the mitochondrial cytochrome oxidase. Although this result dramatically simplifies the overall picture, and opens the question on the role of cytoplasmically-synthesized polypeptides, locating the metal ligands still remains a difficult task of major importance. As discussed by Steffens *et al.* [8], the metal composition of cytochrome *c* oxidase, has been quite controversial since the discovery of this protein by Warburg [9]. If general consensus is found for the number of heme irons (two: cytochrome *a* and cytochrome $a_3$ per mole of functional unit (which may be defined as that containing 13 polypeptides with an overall molecular mass of 200 000 Da), the same, however, is not true for the copper content. This is shown in Table 1, which reports the stoichiometry of the metals found in beef heart and *Paracoccus* cytochrome *c* oxidase by different groups. Although it is generally agreed that the Cu/Fe ratio is close to 1 and thus at least 2 copper ions are bound to the protein, several reports indicate that this ratio is sensibly higher than unity. In 1984 and 1985 Einarsdottir and Caughey [10, 11] found a slight excess of copper over the expected 1:1 Cu/Fe ratio using inductively coupled plasma-atomic emission spectroscopy (i.c.p.-a.e.s. a novel technique which allows simultaneous determination of the metals of interest). In these papers stoichiometric amounts of zinc and magnesium were also found. In 1987 Steffens *et al* [8], using the same technique, extended previous determinations also to the enzyme from *Paracoccus denitrificans*.

TABLE 1: METAL CONTENT OF CYTOCHROME *c* OXIDASE (*)

| Source | Cu | Fe | Zn | Mg | Cu/Fe | Ref |
|---|---|---|---|---|---|---|
| Beef | 2.50 | 2.00 | 1.00 | 1.00 | 1.25 | 10,11 |
| Beef | 2.83 | 1.94 | 1.01 | 0.98 | 1.46 | 8 |
| Beef (^) | 2.80 | 1.90 | — | — | 1.47 | 12 |
| Beef (^) | 2.90 | 2.70 | — | — | 1.07 | 13 |
| Beef | — | — | — | — | 1.70 | 14 |
| Beef | 2.70 | — | — | — | — | 15 |
| Pd (#) | 2.97 | 2.09 | 0.33 | 0.70 | 1.42 | 8 |

(*) Numbers represent stoichiometries, *i.e.* gr-atoms of metal bound per mole of functional unit.
(^) Calculated assuming a molecular weight of 200000 Da.
(#) *Paracoccus denitrificans* two-subunit enzyme.

Moreover, since the sulphur content of cytochrome oxidase is known from sequence analysis, its determination was included as an internal standard. Their result (Table 1) indicated a Cu/Fe ratio of almost 1.5 which would suggest the presence of three copper ions per mole of functional unit in both eukaryotic and prokaryotic cytochrome oxidase. Up to date spectroscopy has not provided a solution to this problem.

Optical spectroscopy of cytochrome *c* oxidase is dominated by heme A transitions [1, 3]. The visible spectrum of cytochrome oxidase in the oxidized and reduced state is shown in Fig. 1. The spectrum of the oxidized protein as prepared is characterized by broad

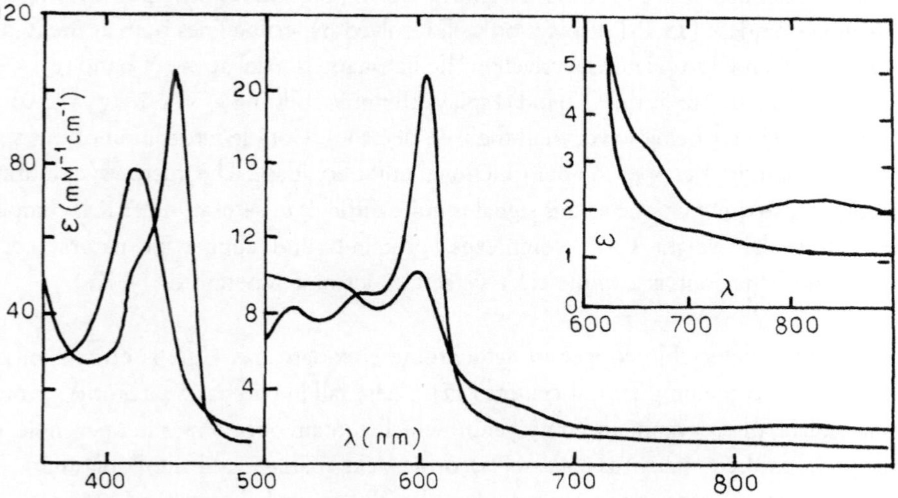


Fig. 1. – Absorption spectra of cytochrome *c* oxidase in the oxidized and reduced states.

absorption bands with peaks centered at around 418-420 nm (the SORET or γ band) and 598-600 nm (α band). Upon reduction of the enzyme (typically with sodium dithionite) the SORET band shifts to the red by as much as 28 nm increasing the molar absorbivity by about 37 %; the α band shifts to 605 nm almost doubling its extinction coefficient. The position of the SORET peak in the oxidized enzyme is particularly sensitive to the previous history of the protein and to the ligation state of cytochrome $a_3$ and in any case always shifts ipsochromically. It is now established that both cytochrome *a* and $a_3$ contribute to the optical transitions discussed above, although with different amplitudes. On the basis of the differential reactivity of cytochromes *a* and $a_3$ with ligands and the possibility of preparing various half-reduced derivatives of cytochrome *c* oxidase, it has been possible to deconvolute the optical properties of cytochromes *a* and $a_3$ [16-19], with the generally verified assumption that the redox state of either cytochrome does not perturb the spectroscopic properties of the other (but see Ref. 20). Other distinctive features of the oxidized enzyme are a broad low extinction near-infrared band centered at 830 nm and a shoulder at 655 nm. The latter is thought to arise from the interaction of CuB and cytochrome $a_3$ [21]. Previous studies on the 830 nm chromophore have indicated that it is due (largely if not exclusively) to copper [22-24]. It is now believed that the e.p.r. detectable CuA is, by and large, responsible for this transition.

A great deal of information on the metals of cytochrome *c* oxidase has been obtained by e.p.r. studies. The e.p.r. signals observed in this enzyme are due either to divalent copper or trivalent iron. Cupric copper has a $3d^9$ electronic configuration and one unpaired electron with spin 1/2. It is usually found in proteins in tetragonal or distorted tetragonal ligand fields, which yields *g*-values in the range 2.0-2.4*. Since the nuclear spin of copper is 3/2 one would expect four hyperfine lines in the e.p.r. spectrum. The copper e.p.r. signal of cytochrome *c* oxidase [13,15] displays no well resolved hyperfine lines both at the X-band and Q-band (9 and 35 GHz, respectively). The lineshape is axial at the X-band ($g_z = 2.18$ and $g_x = g_y = 2.00$) but at the Q-band displays rhombic splitting $g_z = 2.18$, $g_y = 2.03$ and $g_x = 1.99$; the latter being lower than the free electron). Double integration of this signal accounts for at most 1 copper ion per functional unit of oxidase. The microwave saturation behaviour is also quite unique as this signal is quite difficult to saturate at 93 K if compared to low molecular weight $Cu^{2+}$ complexes, protein-bound copper or natural copper proteins; the behaviour is completely reversed at lower temperatures (4 K).

The e.p.r.-detectable copper of cytochrome *c* oxidase, *i.e.* CuA is quite similar to copper proteins containing Type 1 centers [25]. These fall in two basic categories: proteins (such as azurin, plastocyanin and umecyanin) which contain one copper ion per mole, and multicopper oxidases (such as laccase, ascorbic acid oxidase and ceruloplasmin). The similarity in the spectroscopic properties may be also extended to the primary sequence level [26]. As a matter of fact sections of the primary structure of these proteins are homologous to the sequence of subunit II of the beef heart enzyme, which is rather conserved through evolution. This information strongly suggests that the binding site for CuA is on this polypeptide. The peculiar spectroscopic properties of CuA have been extensively investigated by CHAN and Colleagues [27-29]. The currently accepted model of this site involves coordination of the copper ion to at least one cysteine and histidine residues (and possibily an additional cysteine residue) on subunit II. This is based on spectroscopic observations on oxidase from yeast grown with either histidine or cysteine containing $^{15}N$-histidine or deuterium or $^{13}C$ in the beta methylene of cysteine. Both isotopically substituted derivatives yielded changes in CuA e.n.d.o.r. spectrum indicanting that these amino acids are in the coordination sphere of CuA. X-ray edge absorption studies of CuA indicated that the coordination charge at the copper ion does not change appreciably upon reduction [30] and established sulphur ligation to copper [31]; thus CuA in the oxidized state is a very covalent $Cu^{2+}$ and is formally equivalent to a $Cu^{1+}$ ion with the unpaired electron to a large extent residing on the cysteine sulphur, in agreement with the proposal of PEISACH and BLUMBERG [32]. This description of the CuA site, however, is not unanimoulsy accepted: GREENWOOD *et al.* [33] conclude that the optical properties of CuA point against an assignment of a $Cu^{1+}$ coupled to a thiyl radical as do HOFFMAN *et al.* [34] on the basis of e.n.d.o.r. experiments.

It is important to stress at this point that among the 2 or 3 copper ions per mole of functional unit only less than 1 is detected by e.p.r. techniques. If a second copper ion (CuB) can be taken into account on the basis of its magnetic interaction with cytochrome $a_3$ (see below), the *newly* discovered copper (if confirmed) still remains a mystery to be explained, unless it were constitutively cuprous.

$Fe^{3+}$ with a $3d^5$ configuration can have spin 5/2 or 1/2, which are called repectively high- and low- spin. Low-spin ferric heme has one umpaired electron. The ligand field symmetry is typically rhombic and the e.p.r. spectrum displays three $g$-values ranging from 0.8 to 3.5. Cytochrome *a* falls in this category [12, 13, 15]. It is a bis-imidazolate low-spin center in both oxidation states [35] (II and III). The shape of the e.p.r. signal and the e.p.r. parameters ($g_z$ = 3.03, $g_y$ = 2.24 and $g_x$ = 1.45) are typical for a magnetically isolated low-spin heme. The signal intensity closely matches the protein concentration. The nature of the proximal and distal ligands has been elucidated on the basis of comparison with the Raman [35, 36] and m.c.d. [37, 38] spectra of model compounds. The heme of cytochrome *a*, although easily accessible for electron transfer, is likely buried within an hydrophobic pocket due to its long isoprenoid side chain. Rapid kinetic experiments indicate that electrons from ferrocytochrome *c* enter the oxidase via cytochrome *a*, and are then passed rapidly to CuA (see below).

* The parameter $g$ is the spectroscopic splitting factor having a value function of the electron environment (close to 2 for a free electron). The symbols $g_x$, $g_y$ and $g_z$ represent the components of $g$ in the $x$, $y$ and $z$ directions.

The location of this prosthetic group in cytochrome *c* oxidase is not known. Available evidence (see chapeter by R. Bisson) indicates that in subunit II a total of three histidine and two cysteine residues (with one peculiar exception [39]) are conserved through evolution. Since spectroscopy has demonstrated that:

(*i*) cytochrome *a* is a low-spin bis-imidazolate complex and

(*ii*) CuA is ligated by two cysteine residues and by one or two histidine residues (see above), a model assuming subunit II to contain both the above-mentioned prosthetic groups may still be viable [1, 40, 41], although it is very unlikely.

The iron atom of cytochrome $a_3$ and its associated CuB form a magnetically coupled heterobinuclear center. The result of this interaction is the lack of e.p.r. signals from the individual metals even when in the oxidized state. It has been shown by magnetic susceptibility experiments [42, 43] and m.c.d. spectroscopy [44] that cytochrome $c_3$ ($Fe^{3+}$, spin 5/2) and CuB ($Cu^{2+}$, spin 1/2) are strongly antiferromagnetically coupled (as in type III copper centers [25]) to form a spin = 2 system which, using ordinary e.p.r. instrumentation, can not be detected. It has been proposed [45] that the oxidized resting cytochrome $a_3$-CuB center contains a ferryl ion coupled to a cuprous copper site. This interesting proposal would explain why this site is e.p.r. undetectable under ordinary conditions and would give a structural basis for the CO driven reduction of this site which yields the mixed-valence species [46, 47] (a derivative of cytochrome *c* oxidase in which cytochrome *a* and CuA are oxidized and the cytochrome $a_3$-CuB site is reduced and in combination with carbon monoxide). Fe (IV) may, indeed, be generated during one-turnover experiments [48] and is readily suitable for reaction with carbon monoxide being a two-electron acceptor. This proposal should be contrasted, however, with recent Mossbauer experiments [49] on beef heart and *Thermus thermophilus* cytchrome *c* oxidase, which suggest a high-spin ferric cytochrome $a_3$ in the resting, oxidized state.

The oxidized beef heart enzyme always displays a low intensity high-spin e.p.r. signal [13]. This signal may be due to partial autoreduction (possibly of CuB?) of the protein since it is abolished by addition of ferricyanide or is increased following the addition of cytochrome *c* and ascorbate (under these conditions the signal accounts at most for 23 % of the enzyme concentration) [15].

The spectroscopic properties of cytochrome $a_3$ and CuB have been studied [1-3] by selective reduction experiments. Since the lack of e.p.r. signals is due to the coupling of the two metal spins, by preferential reduction of one or the other metal centers, the interaction is disrupted and its partner metal may be induced to exhibit e.p.r. signals. In this perspective, a classical experiment is that carried out by Stevens *et al.* [50]. In this investigation nitric oxide (which bears one unpaired electron) was added to a strictly anaerobic solution of oxidized beef heart cytochrome *c* oxidase. No changes in the e.p.r.

signals from either cytochrome *a* or CuA were observed upon addition of NO; however, a rhombic high-spin e.p.r. signal was generated by this treatment ($g_z = 6.16$, $g_y = 5.82$, $g_x$ was obscured by the CuA signal) accounting for approximately 60 % of one heme (the amount being preparation-dependent). This experiment demonstrated that the high-spin signal arises from cytochrome $a_3$, since the binding of nitric oxide to the oxidized protein breaks the antiferromagnetic coupling between the iron and the copper, by reducing the latter metal exclusively.

The presence in oxidized resting oxidase of a ligand bridging cytochrome $a_3$ and CuB, and thereby mediating magnetic exchange interaction between the two metals, has been amply discussed; e.x.a.f.s. studies have proposed a sulphur atom as a bridging ligand [51]. Since the binuclear center is almost certainly located in subunit I and no cysteines are available in this subunit (by metionines are), the assignment of the bridging ligand has been actively debated. Recently [52], Fe-e.x.a.f.s. performed on a chloride-depleted cytochrome *c* oxidase preparation shows that the intensity of the peak attributed to the bridging sulphur (which e.x.a.f.s. can not distinguish from chlorine) becomes quite attenuated. This result would seem to indicate that the bridging ligand is a chloride ion, whose function, if any, remains to be established.

Reoxidation with dioxygen in the presence of cytochrome *c* led to the discovery [53] of a new type II [25] copper signal, detected as a transient. This signal is clearly rhombic ($g_z = 2.278$, $g_y = 2.052$) and was assigned either to an oxidized CuB ion with cytochrome $a_3$ in the reduced state (no $g = 6$ signal was present in some preparations) or coupled to a Fe (IV)-oxo complex originating after the cleavage of the O-O bond of dioxygen [54]. Further experiments on this intermediate [48,55] demonstrated that the rhombic CuB signal arises from an intermediate in which ferrous (low-spin) cytochrome $a_3$ is liganded either to carbon monoxide or molecular oxygen.

Additional e.p.r. signals from cytochrome $a_3$ have been obtained using the completely reduced protein in the presence of nitric oxide [56, 57]. These studies have suggested the presence of one endogenous axial nitrogen ligand on cytochrome $a_3$. Following incorporation of $^{15}$N-histidine in yeast cytochrome *c* oxidase, a histidine residue could be unambiguously identified as the fifth endogenous ligand to cytochrome $a_3$ [58]. Finally, Cu e.n.d.o.r. spectra of cytochrome oxidase [59] indicate the presence of three distinct nitrogenous ligands; of these at least one is a histidine. These results established the similarity between CuB and the Type III Cu site found in laccase [25, 53, 54]. Fig. 2 summarizes the present knowledge on the ligands to the metal centers in cytochrome oxidase.

A very unusual transient e.p.r. signal is observed when reduced cytochrome *c* oxidase is rapidly mixed with molecular oxygen ($g = 5$, 1.78, 1.69) [60]. As discovered by

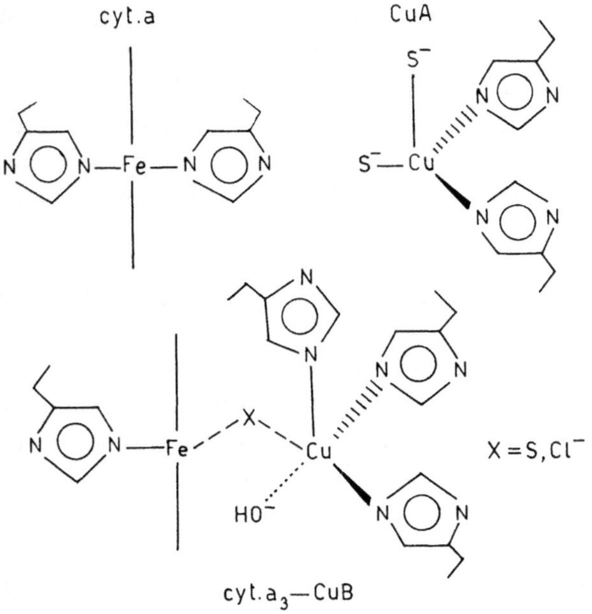


Fig. 2. – Schematic representation of the redox-active metal centers in cytochrome *c* oxidase.

Antonini *et al.* [61], pulsed oxidase (as it is called, see below) is a conformational state of the enzyme generated within milliseconds after mixing the fully reduced enzyme with dioxygen; in the fully oxidized state, pulsed oxidase decays back to resting with halftimes of many seconds to minutes. Expriments with $^{17}O_2$ [62] provided no evidence of oxygen being involved in the structure of this intermediate, in complete agreement with kinetic data showing that the pulsed state can be populated by oxidation with oxidants other than oxygen (see Ref. 1 for discussion and references).

In addition to the *canonical* redox-active metals, it appears that the functional unit of cytochrome *c* oxidase also contains one atom of zinc [8, 10, 11] as shown in Table 1. This metal has been recently studied by e.x.a.f.s. [63]. The conclusions are that zinc is in a distorted tetrahedral environment and is coordinated to three sulphurs and one nitrogen; this would seem to place this metal, with hitherto unknown function, in one of the cytoplasmically synthesized subunits.

## 3. Metals in Action: Kinetics of electron transfer and proton pumping

The reaction catalysed by cytochrome *c* oxidase is the following:

$$4 \text{ cyt } c^{2+} + 4 \text{ H}^+ + \text{O}_2 \rightarrow 4 \text{ cyt } c^{3+} + 2 \text{ H}_2\text{O} \quad (1)$$

which indicates that four electrons provided by cytochrome *c* are delivered to molecular oxygen with the uptake of four protons to produce two molecules of water. One of the most important and intriguing aspects of this reaction is the mechanism of dioxygen reduction. Considering that cytochrome *c* oxidase is responsible for more than 90 % of the oxygen consumption in respiring cells, this reaction must be carried out without the release in the bulk of oxygen radicals, such as the superoxide anion and $^{\bullet}$OH, or oxidants such as hydrogen peroxide.

The reaction of cytochrome *c* with the oxidase has been extensively studied both in the steady state and with rapid kinetic techniques [1-3]. The main features of transient and steady state kinetics will be briefly summarized. On mixing ferrocytochrome *c* with oxidized cytochrome oxidase, both at micromolar concentrations, spectral changes in the millisecond time range are observed. The reaction can be followed at wavelengths representing either cytochrome *c* (*i.e.* the α-band at 550 nm) or cytochrome *a* (605 nm) and the observed changes indicate a rapid electron transfer from reduced cytochrome *c* to oxidized cytochrome *a* [64-67]. The reaction is clearly concentration-dependent with second-order rate constants as high as $10^8$ $M^{-1}$ $s^{-1}$ [68], which decreases to *ca.* $10^6$ $M^{-1}$ $s^{-1}$ as the ionic strength of the medium increases. The electron transfer within the collisional complex must be very rapid since, with the stopped-flow apparatus (with upper limits of approximately 300 $s^{-1}$) no evidence for saturation behaviour at high protein concentration has been reported [65, 69]. This limit is increased to *ca.* 2000 $s^{-1}$ in temperature-jump experiments [69], showing bimolecular kinetics. This reaction is, by and large, insensitive to the redox and ligation state of cytochrome $a_3$, *i.e.*, the rate of reduction of cytochrome *a* in the completely oxidized protein is unaffected when starting with the cyanide-inhibited oxidase (in which the cytochrome $a_3$-CuB site is stabilized in the oxidized state) or with the mixed-valence enzyme in combination with carbon monoxide (in which the binuclear site is in the reduced state) [1]. When the reaction between ferrocytochrome *c* and oxidase is followed at a wavelength representing CuA (830 nm) [65] the time course in either identical or slower, indicating that reduction of CuA may lag behind the reduction of cytochorme *a*. Rapid mixing experiments have also been carried out starting from ferricytochrome *c* and reduced oxidase, *i.e.* the reverse electron transfer reaction [70]. The results indicate that the oxidation of cytochrome *a* is simultaneous with the reduction of cytochrome *c*, and no evidence for reduction of cytochrome *c* preceding oxidation of cytochrome *a* was reported. Absorbance recovery measurements [65] demonstrate that during the burst phase more than one electron enters the complex, *i.e.* both cytochrome *a* and CuA have been partially reduced (the amplitude of the burst phase may, however, depend on the presence of dioxygen [66]). This observation opens the question on which of these two sites is the primary electron acceptor from cytochrome *c*, a fundamental problem from the mechanistic point of view.

Temperature-jump experiments [69] carried out on an anaerobic equilibrium mixture of the mixed-valence CO-complex of cytochrome *c* oxidase and cytochrome *c* indicated that two processes were present. The faster relaxation process obeyed bimolecular kinetics and the slower process had a half-time 10 to 100 times longer. The data were interpreted with the following scheme:

$$a^{3+} - Cu_A^{2+} + c^{2+} \rightleftarrows a^{2+} - Cu_A^{2+} + c^{3+} \quad \text{(a)}$$

$$a^{2+} - Cu_A^{2+} \rightleftarrows a^{3+} - Cu_A^{+} \quad \text{(b)}$$

Accordingly, the two relaxations reflect the intermolecular electron transfer reaction between cytochrome *a* and cytochrome *c* [step (a) in the above scheme] and an intramolecular electron transfer between cytochrome *a* and CuA [step (b)], coupled to step (a).

These considerations may indicate that the dissociation of ferricytochrome *c* from the complex is not a rate-limiting step, since, if this were the case, diffusion and binding of a second molecule of reduced cytochrome *c* would be hindered by bound cytochrome *c* (see Ref. 1 for a complete discussion). For this reason it is quite important to design experiments which can exclude (or confirm) direct binding of cytochrome *c* to the CuA site. A first step in this direction comes from the preparation of a derivative of beef heart cytochrome *c* oxidase in which the CuA site has been selectively modified by treatment with mercurials [71]. As judged from optical and e.p.r. spectroscopy the treatment changes the CuA site to a Type II copper. The redox potential of this modified copper appears to be significantly lowered as to be non-reducible by cytochrome *c*. Preliminary kinetic experiments with this derivative indicate that the electron transfer reaction between cytochrome *a* and cytochrome *c* is unaffected. This result further supports the above scheme according to which electrons from cytochrome *c* are not donated to CuA, but rather to cytochrome *a*.

Once electrons have entered the oxidase, these are transferred intramolecularly to the binuclear center, where the oxygen reduction process can take place. Since CuB is spectroscopically non-detectable, only the reduction of cytochrome $a_3$ can be appropriately followed as a function of time. Experiments dealing with the dynamics of the electron transfer from the cytochrome *a*-CuA couple to the cytochrome $a_3$-CuB unit are quite complex. As for most hemeproteins, a characteristic property of ferrous cytochrome $a_3$ in combination with carbon monoxide is that the iron-CO bond is photosensitive, which permits the rapid photolytic removal of the ligand and determination of kinetic parameters for re-binding. If cytochrome $a_3$ is initially oxidized, as in resting state of cytochrome *c* oxidase (*i.e.* as purified), and is mixed with excess ferrocytochrome *c* in the absence of dioxygen, then the time course of appearance of photosensitivity will yield information on the rate of reduction of cytochrome $a_3$ [64]. Evidence that reduction occurs *via* a process internal to the oxidase is very compelling and is supported by the mechanism of reduction of cytochrome *c* oxidase by dithionite and other reductants studied by stopped-flow and flow flash techniques [72-75]. The results of these experiments indicate that the rate of reduction of cytochrome $a_3$ is so slow (0.5 $s^{-1}$) to appear incompatible with the turnover number of the enzyme. A possibile explanation for this apparent discrepancy has been offered by ANTONINI *et al.* [61] who showed that resting oxidase is converted into a more active, pulsed conformation by undergoing a complete reduction and reoxidation. The activation to the pulsed state may thus be achieved either starting from the fully reduced enzyme (as was done in the initial experiment [61, 67]) or during turnover, as determined from the shift of the steady state of cytochrome *a* during turnover, provided enough substrate is present. It has been demonstrated that pulsed oxidase is characterized by an internal electron transfer rate to the binuclear center which is considerably faster (2 to 10 fold) [76] than that of the resting state. Following exhaustion of the substrate the relaxation back to the resting state is so slow that in a continuous turnover situation the enzyme, although initially in the resting form, is always in the pulsed conformation.

Molecular oxygen reacts very rapidly with fully or partially reduced cytochrome *c* oxidase [77-83]. This reaction is usually studied by flow flash techniques by mixing the carbon monoxide derivative of the enzyme (either fully reduced or in the mixed valence state) with oxygen-containing buffer. Following the photolytic removal of bound CO, the subsequent reaction with dioxygen can be studied. GIBSON and GREENWOOD [78] proposed a sequential mechanism which involves oxygen binding ($k = 10^8\ M^{-1}\ s^{-1}$) synchronous to the oxidation of cytochrome $a_3$ and presumably CuB. The subsequent steps involve the oxidation of CuA (7000 $s^{-1}$) and finally, of cytochrome *a*. CHANCE and coworkers [84, 85] introduced the low-temperature triple trapping technique which opened the way to the discovery of a series of intermediates in the oxygen reduction reaction by cytochrome oxidase; these intermediates were characterized by optical and e.p.r. spectroscopy. These investigations together with further studies [48, 54, 55, 86, 87], established that there are independent electron transfer pathways within the enzyme. Thus electrons may be delivered, albeit at different rates, independently from cytochrome *a* and CuA to the cytochrome $a_3$-CuB site. Parallel electron transfer pathways may not just be fortuitous since *in vivo* these may be linked to specific functions such as proton pumping and control of energy transduction [88].

Many investigations have been carried out on the proton pumping function of cytochrome *c* oxidase (see Refs. 3 and 89 for reviews) since its discovery by WIKSTROM [4] and subsequent supporting experiments [90-93]. Cytochrome *a* oxidase contributes to the development and maintenance of a proton gradient in two ways:

1) by transfering electrons from ferrocytochrome *c* to molecular oxygen. These substrates react with the protein on different sides of the membrane and therefore provide a basis for the capacitive charging of the phospholipid membrane, since proton consumption for water production occurs from within the matrix space of mitochondria (or the lumen of artificial phospholipid vesicles). In other words, redox activity alone is capable of generating a proton electrochemical gradient which in the living cell, may sustain metabolic processes such as ATP synthesis and ion transport.
2) Cytochrome *c* oxidase provides an additional mechanism for transducing energy. This is in essence a vectorial process consisting in the redox-linked transport of protons (accepted stoichiometry 1 $H^+/e^-$) from the mitochondrial matrix or vesicle lumen to the external aqueous medium. Apparently this process severely depends on the presence of an efficient charge-compensating system (such as valinomycin and potassium) [90, 93, 94]. This simple experimental observation has deep consequences with respect to the mechanism of proton pumping. As a matter of fact in the absence of a charge-compensating system no proton ejection can be detected, indicating either that protons cannot exchange with the bulk because of the membrane potential build-up or that a slipping conformation of the enzyme becomes populated as a consequence of trans-membrane charge separation [88, 95]. There is no obligatory reason to believe that the coupling between the scalar and vectorial reaction in a proton pump such as cytochrome *c* oxidase need to be infinitely tight. Experimental evidence supporting the slip concept [96, 97] in cytochrome *c* oxidase is growing. BRUNORI *et al.* [95] showed that the kinetics of cytochrome *c* oxidation (and presumably proton pumping [93]) by cytochrome *c* oxidase reconstituted into phospholipid vesicles is controlled by the build-up of a membrane potential; a two-state model was proposed according to which regulation of catalytic activity and proton pumping is mediated by a membrane potential-triggered transition between a pumping and slipping conformation. The $H^+/e^-$ stoichiometry would be thus a variable, dependent on the membrane potential generated by electron flow. MURPHY and BRAND [98] indeed confirmed the above proposals by showing that the proton-electron stoichiometry of cytochrome *c* oxidase decreases at high (*ca.* 180 mV) steady state membrane potentials.

The mechanism of proton pumping is unknown, largely in view of the limited information concerning the three-dimensional structure of the enzyme. The functionally active redox centers (Fig. 2) must provide the free energy for the vectorial transport of protons against their electrochemical protential gradient, and one or more of these may represent the physical *trigger* in the process of energy transduction. Both cytochrome *a* [99,

100] and CuA [27] have been proposed as the redox metals involved in the coupling process, but evidence is as yet not conclusive. Other fundamental questions such as the topology of the proton circuits or how the chemical and vectorial researctions are coupled together, must await for futher research.

**References**

[1] M. Brunori, G. Antonini, F. Malatesta, P. Sarti and M.T. Wilson, *Adv. Inorg. Biochem.*, **7**, 93, (1987).
[2] See Volume **23** of *J. Inorg. Biochem.*, (1985).
[3] M.K.F. Wikstrom, K. Krab and M. Saraste, in *Cytochrome Oxidase. A Synthesis,* Academic Press, New York (1981).
[4] M.K.F. Wikstrom, *Nature (London),* **266,** 271 (1977).
[5] B. Ludwig and G. Schatz, *Proc. Natl. Acad. Sci. USA,* **77,** 196 (1980).
[6] B. Ludwig, *Biochim. Biophys. Acta,* **594,** 177 (1980).
[7] G.C.M. Steffens, G. Buse, W.A. Oppliger and B. Ludwig, *Biochem. Biophys. Res. Commun.,* **116,** 335 (1983).
[8] G.C.M. Steffens, R. Biewald and G. Buse, *Eur. J. Biochem.,* **164,** 295 (1987).
[9] O. Warburg, *Biochem. Z.,* **152,** 479 (1924).
[10] O. Einarsdottir and W.S. Caughey, *Biochem. Biophys. Res. Commun.,* **124,** 836 (1984).
[11] O. Einarsdottir and W.S. Caughey, *Biochem. Biophys. Res. Commun.,* **129,** 840 (1984).
[12] R.F. Van Gelder and H. Beinert, *Biochim. Biophys. Acta,* **189,** 1, (1969).
[13] C.R. Hartzell and H. Beinert, *Biochim. Biophys. Acta*, **368**, 318 (1974).
[14] T. Yonetani, *Biochem. Biophys. Res. Commun.,* **3,** 549 (1960).
[15] R. Aasa, S.P.J. Albracht, K.E. Falk, B. Lanne and T. Vanngard, *Biochim. Biophys. Acta,* **422,** 260 (1976).
[16] D. Keilin and E.F. Hartree, *Proc. R. Soc. London,* **Ser. B 127,** 167, (1939).
[17] T. Yonetani, *J. Biol. Chem.,* **235,** 845 (1960).
[18] S. Horie and M. Morrison, *J. Biol. Chem.,* **238,** 2859 (1963).
[19] W.A. Vanneste, *Biochemistry,* **5,** 838 (1966).
[20] D.F. Blair, D.F. Bocian, G.T. Babcock and S.I. Chan, *Biochemistry,* **21,** 6928 (1982).
[21] C.R. Hartzell, R.E. Hansen and H. Beinert, *Proc. Natl. Acad. Sci. USA,* **70,** 2477 (1973).
[22] D.E. Griffiths and D.C. Wharton, *J. Biol. Chem.,* **236,** 1857 (1961).
[23] D.C. Wharton and A. Tzagoloff, *J. Biol. Chem.,* **239,** 2036 (1964).
[24] H. Beinert, C.R. Hartzell and W.H. Orme-Johnson, in *Probes of Structure and Function of Macromolecules and Membranes,* B. Chance, T. Yonetani, A.S. Mildran (Editors), Academic Press, New York (1971), Vol. **2,** pp. 575-592.

[25] R.A. Malkin and B.G. Malmstrom, *Adv. Enzymol.,* **33,** 177 (1970).

[26] G.C.M. Steffens and G. Buse, *Hoppe Seyler's Z. Physiol. Chem.,* **360,** 613 (1979).

[27] S.I. Chan, D.F. Bocian, G.W. Brudwig, R.H. Morse and T.H. Stevens, in *Cytochrome Oxidase,* T.E. King, Y. Orii, B. Chance and K. Okunuki, (Editors), Elsevier North-Holland Biomedical Press, Amsterdam, (1979) pp. 177-188.

[28] T.H. Stevens, C.T. Martin, H. Wang, G.W. Brudvig, C.P. Scholes and S.I.

[29] C.T. Martin, C.P. Scholes and S.I. Chan, *J. Biol. Chem,* **263,** 8240 (1988).

[30] V.W. Hu, S.I. Chan and G.S. Brown, *FEBS Lett.,* **84.** 287 (1977).

[31] R.A. Scott, S.P. Cramer, R.W. Shaw, H. Beinert and H.B. Gray, *Proc. Natl. Acad. Sci. USA,* **78,** 664, (1981).

[32] J. Peisach and W.E. Blumberg *Arch. Biochem. Biophys.,* **165,** 691 (1974).

[33] C. Greenwood, B.C. Hill, D. Barber, D.G. Eglinton and A.J. Thomson, *Biochem. J.,* **215,** 303 (1983).

[34] B.M. Hoffman, J.E. Roberts, M. Swansson, S.H. Speck and E. Margoliash, *Proc. Natl. Acad. Sci. USA,* **77,** 1452 (1980).

[35] G.T. Babcock, P.M. Callahan, M.R. Ondrias and I. Salmeen, *Biochemistry,* **20** 959 (1981).

[36] G.T. Babcock, J. Van Steelan, G. Palmer, L.E. Vickery and I. Salmeen, in *Cytochrome Oxidase,* (T.E. King, Y. Orii, B. Chance, K. Okunuki (Editors), Elsevier North-Holland Biomedical Press, Amsterdam (1979), pp. 105-115.

[37] K. Carter and G. Palmer, *J. Biol. Chem.,* **257,** 13507 (1982).

[38] D.G. Eglinton, M.K. Johnson, A.J. Thomson, P.E. Gooding and C. Greenwood, *Biochem. J.,* **191,** 319 (1980).

[39] L. Bonen, P.H. Boer and M.W. Gray, *EMBO J.* **3,** 2531 (1984).

[40] M.K.F. Wikstrom, M. Saraste and I. Penttila, in *The Enzymes of Biological Membranes,* A.N. Martonosi (Editor), Plenum Publishing Corp., New York (1985), Vol. **4,** pp. 111-148.

[41] L. Holm, M. Saraste and M.K.F. Wikstrom, *EMBO J.,* **6,** 2819 (1987).

[42] K.E. Falk, T. Vanngard and J. Angstrom, *FEBS Lett.,* **75,** 23 (1977).

[43] M.F. Tweedle, L.J. Wilson, L. Garcia-Iniques, G.T. Babcock and G. Palmer, *J. Biol. Chem.,* **253,** 8065 (1978).

[44] A.J. Thomson, T. Brittain, C. Greenwood and J.P. Springall, *Biochem J.,* **165,** 327 (1977).

[45] C.H.A. Seiter and S.G. Angelos, *Proc. Natl. Acad. Sci. USA,* **77,** 1806 (1980).

[46] P. Nicholls and B.A. Chanady, *Biochim. Biophys. Acta,* **634,** 256 (1981).

[47] D. Bickar, C. Bonaventura and J. Bonaventura, *J. Biol. Chem.,* **259,** 10777 (1984).

[48] D.F. Blair, S.N. Witt and S.I. Chan, *J. Am. Chem. Soc.,* **107,** 7389 (1985).

[49] T.A. Kent, L.J. Young, G. Palmer, J.A. Fee and E. Munck, *J. Biol. Chem.,* **258,** 8543 (1983).

[50] T.H. STEVENS, G.W. BRUDVIG, D.F. BOCIAN and S.I. CHAN, *Proc. Natl. Acad. Sci. USA,* **76,** 3320 (1979).

[51] L. POWERS, B. CHANCE, Y. CHING and P. ANGIOLILLO, *Biophys. J.,* **34,** 465 (1981).

[52] R.A. SCOTT, P.M. LI and S.I. CHAN, *Ann. New York Acad. Sci.,* **550,** 53 (1988).

[53] B. REINHAMMAR, R. MALKIN, P. JENSEN, B. KARLSSON, L. ANDREASSON, R. AASA, T. VANNGARD and B.G. MALMSTROM, *J. Biol Chem.,* **255,** 5000 (1980).

[54] B. KARLSSON, R. AASA, T. VANNGARD and B.G. MALMSTROM, *FEBS Lett.,* **131,** 186, (1981).

[55] S.N. WITT, D.F. BLAIR and S.I. CHAN, *J. Biol. Chem.,* **261,** 8104 (1986).

[56] M.F.T. BLOKZIJL-HOMAN and B.F. VAN GELDER, *Biochim. Biophys. Acta,* **234,** 493 (1971).

[57] T.H. STEVENS, D.F. BOCIAN and S.I. CHAN, *FEBS Lett.,* **97,** 314 (1979).

[58] T.H. STEVENS and S.I. CHAN, *J. Biol. Chem.,* **256,** 1069 (1981).

[59] J. CLINE, B. REINHAMMAR, P. JENSEN, R. VENTERS and B.M. HOFFMAN, *J. Biol. Chem.,* **258,** 5124 (1983).

[60] B.V. SHAW, R.E. HANSEN and H. BEINERT, *J. Biol. Chem.,* **253,** 6637 (1978).

[61] E. ANTONINI, M. BRUNORI, A. COLOSIMO, C. GREENWOOD and M.T. WILSON, *Proc. Natl. Acad. Sci. USA,* **74,** 3128 (1977).

[62] B.V. SHAW, R.E. HANSEN and H. BEINERT, *Biochim. Biophys. Acta,* **548,** 386 (1979).

[63] A. NAQUI, L. POWERS, M. LUNDEEN and B. CHANCE, *Biophys. J.,* **51,** 312 (1987).

[64] Q.H. GIBSON, C. GREENWOOD, D.C. WHARTON and G. PALMER, *J. Biol. Chem.,* **240,** 888 (1965).

[65] M.T. WILSON, C. GREENWOOD, M. BRUNORI and E. ANTONINI, *Biochem. J.,* **147,** 145 (1975).

[66] L.E. ANDREASSON, B.G. MALMSTROM, C. STROMBERG and T. VANNGARD, *FEBS Lett.,* **28,** 297 (1972).

[67] M. BRUNORI, C. COLOSIMO, G. RAINONI, M.T. WILSON and E. ANTONINI, *J. Biol. Chem.,* **254,** 10769 (1979).

[68] K.J.H. VAN BUUREN, B.F. VAN GELDER, J. WILTING and R. BRAAMS, *Biochim. Biophys. Acta,* **333,** 421 (1974).

[69] C. GREENWOOD, T. BRITTAIN, M.T. WILSON and M. BRUNORI, *Biochem. J.,* **157,** 591 (1976).

[70] L.G. PETERSEN and L.E. ANDREASSON, *FEBS Lett.,* **66,** 52 (1976).

[71] J. GELLES and S.I. CHAN, *Biochemistry,* **24,** 3963 (1985).

[72] R.A. SCOTT and H.B. GRAY, *J. Am. Chem. Soc.,* **102,** 3219 (1980).

[73] G.D. JONES, M.G. JONES, M.T. WILSON, M. BRUNORI, A. COLOSIMO and P. SARTI, *Biochem. J.,* **209,** 175 (1983).

[74] F.G. HALAKA, Z.K. BARNES, G.T. BABCOCK AND J.L. DYE, *Biochemistry,* **23,** 2005 (1984).

[75] M. Brunori, D. Bickar, J. Bonaventura and C. Bonaventura, *J. Biol. Chem.*, **260,** 7165 (1985).
[76] G. Antonini, M. Brunori, A. Colosimo, F. Malatesta and P. Sarti, *J. Inorg. Biochem.,* **23,** 289 (1985).
[77] Q.H. Gibson and C. Greenwood, *Biochem. J.,* **86,** 51 (1963).
[78] Q.H. Gibson and C. Greenwood, *J. Biol. chem.* **240**, 2694 (1965).
[79] C. Greenwood and Q.H. Gibson, *J. Biol. Chem.,* **242**, 1782 (1967).
[80] J.K.V. Reichardt and Q.H. Gibson, *J. Biol. Chem.,* **257,** 9268 (1982).
[81] M. Brunori and Q.H Gibson, *EMBO J.,* **2,** 2025 (1983).
[82] B.C. Hill and C. Greenwood, *Biochem. J.,* **218,** 913 (1984).
[83] Y. Orii, *J. Biol. Chem.,* **259,** 7187 (1984).
[84] B. Chance, C. Saronio and J.S. Leigh, *Proc. Natl. Acad. Sci. USA,* **72**, 1635 (1975).
[85] B. Chance, C. Saronio and J. S. Leigh, *J. Biol Chem.,* **250**, 9226 (1975).
[86] G.M. Clore, L.E. Andreasson, B. Karlsson, R. Aasa and B.G. Malmstrom, *Biochem. J.,* **185,** 139 (1980).
[87] G.M. Clore, L.E. Andreasson, B. Karlsson, R. Aasa and B.G. Malmstrom, *Biochem. J.,* **185,** 155 (1980).
[88] F. Malatesta, G. Antonini, P. Sarti, B. Vallone and M. Brunori, *Ann. New York Acad. Sci.,* **550,** 269 (1988).
[89] M. Brunori, G. Antonini, F. Malatesta, P. Sarti and M.T. Wilson, *Eur. J. Biochem.,* **169,** 1, (1987).
[90] K. Krab and M.K.F. Wikstrom, *Biochim. Biophys. Acta,* **504,** 200 (1978).
[91] R.P. Casy, J. B. Chappel and A. Azzi, *Biochem. J.,* **182**, 149 (1979).
[92] E. Sigel and E. Carafoli, *Eur. J. Biochem.,* **111,** 299 (1980).
[93] P. Sarti, M.G. Jones, A. Antonini, F. Malatesta, A. Colosimo, M.T. Wilson and M. Brunori, *Proc. Natl. Acad. Sci. USA,* **82,** 4876 (1985).
[94] R.P. Casey, P.S. O'Shea, J.B. Chappel and A. Azzi, *Biochim. Biophys. Acta,* **765,** 30 (1984).
[95] M. Brunori, P. Sarti, A. Colosimo, G. Antonini, F. Malatesta, M.G. Jones and M.T. Wilson, *EMBO J.,* **4,** 2365 (1985).
[96] D. Pietrobon, G.F. Azzone and D. Walz, *Eur. J. Biochem.,* **117,** 389 (1981).
[97] D. Walz, in *Biological Structure and Coupled Flows,* A. Optalaka and M. Balaban (Editors), Academic Press, New York (1983), pp. 45-60.
[98] M.P. Murphy and M.D. Brand, *Eur. J. Biochem.,* **173,** 637 (1988).
[99] M.K.F. Wikstrom and R.P. Casey, *J. Inorg. Biochem.,* **23,** 327 (1985).
[100] P.M. Callahan and G.T. Babcock, *Biochemistry,* **22,** 452 (1983).

# MECHANISM OF $Ca^{2+}$ TRANSLOCATION AS STUDIED BY THE USE OF DETERGENT–SOLUBILIZED AND MEMBRANE PREPARATIONS OF SARCOPLASMIC RETICULUM $Ca^{2+}$–ATPase

JESPER MØLLER

*Institute of Medical Biochemistry, University of Aarhus,*
*8000 Aarhus C, Denmark*

Contents



*Bioelectrochemistry III*
Edited by G. Milazzo and M. Blank
Plenum Press, New York, 1990

## 1. Introduction

The sarcoplasmic reticulum (SR) forms a closed, membraneous network, consisting of tubules and terminal cisternae that surround the myofibrils of skeletal muscle. In these membranes the dominant protein component is a $Ca^{2+}$-ATPase that is involved in the uptake of $Ca^{2+}$ released from the terminal cisterna following neuromuscular excitation. The uptake occurs against a large concentration gradient of $Ca^{2+}$, maintaining a low concentration of $Ca^{2+}$ in the cytoplasma in the resting state of the muscle [1-4]. The sarcoplasmic reticulum $Ca^{2+}$-ATPase serves as a prototype of other $Ca^{2+}$-transporting

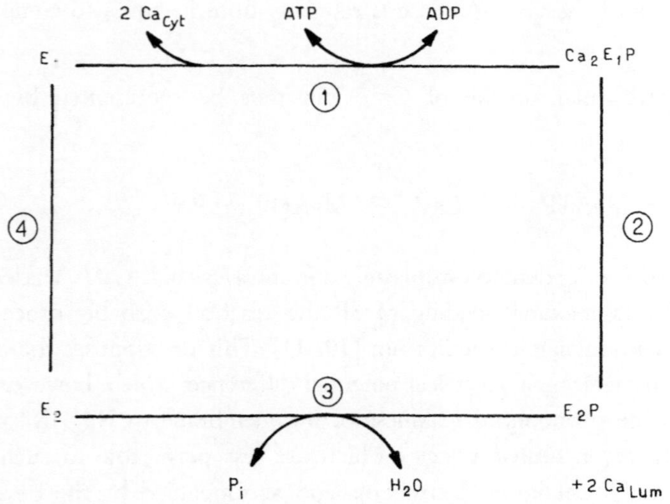


Fig. 1 - Minimal reaction scheme for sarcoplasmic reticulum $Ca^{2+}$-APTase. $Ca_{Cyt}$ ad $Ca_{Lum}$ denote $Ca^{2+}$ in the muscle citoplasma and lumen of SR, respectively.

ATPase's and is related to the family of $E_1P$-$E_2P$ aspartylphosphorylated transport ATPase's, which in addition to $Ca^{2+}$-ATPase comprises $Na^+$, $K^+$-ATPase, gastric and plasma membrane $H^+$-ATPase, and bacterial $K^+$-ATPases [5]. A minimal reaction scheme for sarcoplasmic reticulum $Ca^{2+}$-ATPase is shown in Fig. 1. In this scheme $E_1$

conformations denote enzyme with $Ca^{2+}$ translocating sites oriented outwards, while $E_2$ conformations denote conformations with $Ca^{2+}$ oriented inwards towards the intravesicular space. In the first step enzyme reacts with 2 $Ca^{2+}$ at the translocation sites on the cytoplasmic aspect of the membrane and then with MgATP to form phosphorylated enzyme, covalently modified at aspartate residue No. 351. The phosphorylation reaction is reversible, *i.e.* $E_1P$ can be dephosphorylated by addition of ADP ($Ca_2E_1P$ or $E_1P$ in short hand notation). Physiologically, MgATP is the substrate, but other divalent cations such as $Ca^{2+}$ [6] and $Mn^{2+}$ [7] can substitute, and many other energy-rich compounds than ATP can be used in place of ATP, *e.g.* guanosinetriphosphate and acetylphosphate [4]. Reaction 2 involves a conformational transition during which the enzyme is converted from the ADP reactive form into an ADP unreactive form, $E_2P$. During transition from $E_1P$ to $E_2P$ translocation of $Ca^{2+}$ is considered to take place, resulting in the displacement of $Ca^{2+}$ from the outer to the inner aspect of the membrane. The binding constant of $Ca^{2+}$-ATPase at the inner aspect of the membrane is decreased by a factor of $10^3$-$10^4$ [8, 9] and thus $Ca^{2+}$ can be released to the inner fluid phase against a much higher concentration than in the surrounding medium. In reaction 3 the enzyme is dephosphorylated by hydrolytic cleavage after the discharge of $Ca^{2+}$, and in reaction 4 the conformation of the $Ca^{2+}$-ATPase is returned from $E_2$ to $E_1$ to complete the cycle.

The intravesicular uptake of $Ca^{2+}$ can thus be represented by the following equation:

$$2\ Ca^{2+}_{Cyt} + MgATP = 2\ Ca^{2+}_{Lum} + MgADP + P_i$$

where $Ca^{2+}_{Cyt}$ and $Ca^{2+}_{Lum}$ denote cytoplasmic and intravesicular $Ca^{2+}$. Under physiological conditions with submaximal loading of SR the reaction probably proceeds to a state approaching thermodynamic equilibrium [10, 11]. This presupposes the absence of any appreciable transmembranal electrical potential difference, which, however, is unlikely to exist owing to the pronounced leakiness of the membrane to $Na^+$, $K^+$, and inorganic anions [12]. In reconstituted vesicles which are less permeable to such ions there is evidence that $Ca^{2+}$ uptake is electrogenic and accompanied by the development of a transmembranal potential difference [13, 14]. This does not rule out the possibility of partial charge compensation by countertransport of *e.g.* $K^+$ or $H^+$ as has been suggested [15, 16]. However, the only firm evidence of countertransport is by $Ca^{2+}$ itself. This phenomenon is observed as $Ca^{2+}$-$Ca^{2+}$ exchange at high levels of intravesicular $Ca^{2+}$ [17, 19].

The mechanistic features of the translocation process by $Ca^{2+}$-ATPase or by other ion pumps are not unravelled at the present time. However, sufficient structural information on sarcoplasmic reticulum $Ca^{2+}$-ATPase is now available at least to warrant speculations on this subject. A basic question concerns the mechanism by which $Ca^{2+}$ moves across the membrane. Does this involve passage through a channel or does it involve movement of the whole $Ca^{2+}$ binding site? By the application of various techniques to monitor overall conformational changes, including tritium exchange (20], circular dichroism [21, 23], neutron and X-ray diffraction [24], I.R. spectroscopy [25], and proteolytic cleavage [26, 27], the impression is gained that only minor conformational changes accompany the transport process. This may suggest that transmembrane passage of $Ca^{2+}$ after binding to the enzyme actually occurs through a channel. Is this putative channel situated between the subunits of an oligomeric ATPase complex, forming a water-filled channel? Or does $Ca^{2+}$ translocation take place *via* a channel centrally located in an ATPase monomer? To answer these questions the organization of $Ca^{2+}$-ATPase in the membrane and the effect of protein-protein interactions on functional aspects of $Ca^{2+}$- ATPase need to be known. These issues are difficult to study in the native membrane. However, useful information on the effect of protein-protein interactions has been obtained after solubilization of the protein with detergent in various states (monomeric and oligomeric) which can be analyzed precisely by analytical ultracentrifugation [28, 29] and by h.p.l.c. [30].

## 2. Properties of detergent-solubilized $Ca^{2+}$-ATPase

### 2.1. *General considerations*

Integral membrane proteins are amenable to solubilization in a native-like state by treatment with a variety of non-ionic detergents, deoxycholate, and cholate. However, solubilization in a native-like state (as contrasted to the unfolding effect of *e.g.* SDS) does not necessarily imply preservation of functional properties. For sarcoplasmic reticulum $Ca^{2+}$-ATPase care should be taken in the choice of detergent and other environmental variables to avoid inactivation. In a systematic study non-ionic detergents with an intermediate hydrophobic chain length ($\approx C_{12}$) and having a polyoxyethylene glycol or carbohydrate polar group [octaethyleneglycol dodecylether ($C_{12}E_8$) or dodecylmaltoside] were found to be optimal for retention of $Ca^{2+}$-ATPase function and stabilization, while detergents with short chain lengths ($C_{8-10}$) or bulky head groups [*e.g.* Tween detergents, (acyl polyethyleneglycol sorbitan ester)] and many zwitterionic detergents (*e.g.* dodecyl-dimethyl-diamineoxide) led to rapid inactivation of delipidated enzyme [31]. Stability after detergent-solubilization is best accomplished in the $E_1$ state (presence of $Ca^{2+}$), while the enzyme is considerably less stable in the $E_2$ and $E_2P$ state, and also during enzymatic turnover [31]. A non-specific increase in stability is provided by addition of polyols like sucrose and glycerol [29]. Two other conditions are known to confer stabilization of $Ca^{2+}$-ATPase, *viz.* solubilization of the enzyme in an oligomeric state and retention of lipid. These features are the background for the original success in the use of Tween 80 to prepare $Ca^{2+}$-ATPase with long term stability [32]. The resulting gel chromatographically prepared $Ca^{2+}$-ATPase is oligomeric (trimeric/tetrameric) and is only partially delipidated [33, 34]. By contrast, most long chain detergents, including Tween 80, despite their more lipid-like structure, have a markedly inactivating effect if added to completely delipidated $Ca^{2+}$-ATPase [31]. After solubilization with $C_{12}E_8$ or dodecylmaltoside $Ca^{2+}$-ATPase activity can be maintained in the completely delipidated state, especially if a high $Ca^{2+}$ concentration is present [30, 35].

Non-ionic detergents of appropriate size such as $C_{12}E_8$, dodecylmaltoside, and Triton X-100 readily solubilize $Ca^{2+}$-ATPase in predominantly monomeric form, provided the detergent-to-protein ratio is sufficiently high, *e.g.* for $C_{12}E_8$ 10:1 (*w/w*). Deoxycholate also efficiently solubilizes $Ca^{2+}$-ATPase in monomeric form [36], while cholate predominantly solubilizes the protein in oligomeric form. With the advent of the h.p.l.c. (high performance liquid chromatography) technique it has been possible to demonstrate the existence of two different forms of oligomerization:

(*i*) irreversible aggregation of inactive $Ca^{2+}$-ATPase that is either present in the sarcoplasmic reticulum membrane, or is formed slowly after detergent solubilization, and

(*ii*) reversible oligomerization that represents active enzyme [30].

As expected the extent of reversible oligomerization depends on protein and detergent concentration (Fig. 2) and, also, to some extent, on the amount of lipid associated with $Ca^{2+}$-ATPase (△, in Fig. 2). Hydrophobic interactions are probably involved in these protein-protein interactions as evidenced by decreased binding of detergent by oligomeric $Ca^{2+}$-ATPase [30]. We found the association constant to be of the order $10^5$-$10^6$ $M^{-1}$, much lower than in a similar study by SILVA and VERJOVSKI-ALMEIDA [37].

### 2.2. *Functional properties of monomeric* $Ca^{2+}$-ATPase

Regardless of non-ionic detergent used, solubilized $Ca^{2+}$-ATPase initially exhibits $Ca^{2+}$-dependent ATP hydrolysis. This activity can be maintained for a considerable period of time if care is taken to stabilize the protein as described above. Expression of full activity requires retention of a lipid layer around the hydrophobic region of $Ca^{2+}$-ATPase, but in the case of $C_{12}E_8$ high activities are observed even after complete delipidation [38]. The depressing effect of $C_{12}E_8$ and especially of dodecylmaltoside on ATP hydrolysis rate may be attributable to interaction of detergent with $Ca^{2+}$-ATPase at

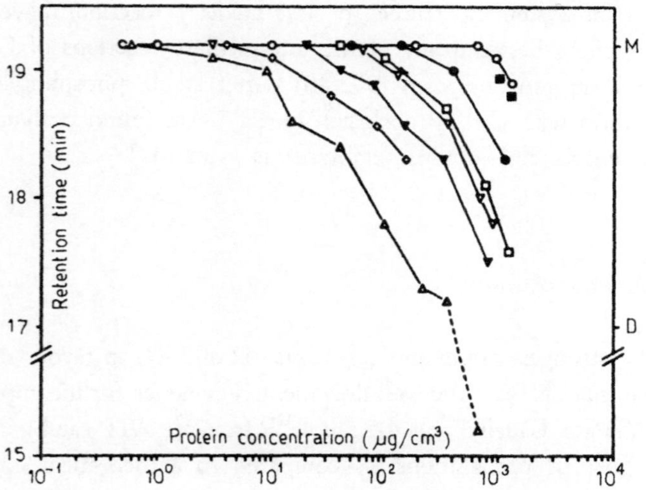


Fig. 2 - Reversible monomer-oligomer equilibria of active $Ca^{2+}$-ATPase as a function of protein concentration studied by large zone h.p.l.c. at the following concentrations of $C_{12}E_8$: ◊, 0.10 mg $C_{12}E_8$/cm³; ▼, 0.25 mg $C_{12}E_8$/cm³; ▽, 1 m*M* ADP and 0.25 mg $C_{12}E_8$/cm³; □, 0.50 mg $C_{12}E_8$/cm³; ●, 2.0 mg $C_{12}E_8$/cm³; ○, 10 mg $C_{12}E_8$/cm³; △, only partially delipidated protein (0.08 g phospholipid/g protein, 0.10 mg $C_{12}E_8$/cm³; ■, endogenous lipid present (0.5 g/g protein), 10 mg $C_{12}E_8$/cm³. In addition the eluant contained 20 m*M* Tes (pH 7.0), 0.1 *M*, NaCl, 10 m*M* $MgCl_2$, 1.5 m*M* $CaCl_2$, 1 m*M* EGTA and 5 m*M* DDT. M and D indicate elution positions of stable monomeric and dimeric $Ca^{2+}$-ATPase, respectively, as determined by analytical ultracentrifugation. Taken from Ref. 26.

a few critical sites, since a similar decrease in activity is observed when a much smaller amount of detergent is bound to protein after incorporation of detergent in the membrane at non-solubilizing concentrations [38].

The $Ca^{2+}$-dependence of ATP hydrolysis by detergent-solubilized $Ca^{2+}$-ATPase indicates that essential functional properties are maintained in the monomeric state, but does not exlude the possibility that detergent-solubilized $Ca^{2+}$-ATPase is deficient in $Ca^{2+}$ transport. This question is difficult to approach directly, owing to the non-compartmentalized nature of the system. However, a closer study has revealed that after detergent-solubilization the two forms of phosphorylated intermediate ($E_1P$ and $E_2P$), characteristic of membraneous $Ca^{2+}$-ATPase (Fig. 1), can still be distinguished on the basis of their reactivity with ADP. Furthermore, the two forms exhibit high and low affinity for $Ca^{2+}$, respectively, as found in leaky membranes [9, 26]. These features are exactly those expected if $Ca^{2+}$ translocation is present in the detergent-solubilized preparations. So far systematic studies of intermediary reactions have revealed only quantitative differences between any of the intermediary reactions of $Ca^{2+}$-ATPase in monomeric and membraneous form [39, 40]. After stable phosphorylation of $Ca^{2+}$-ATPase in $E_1P$ form with CrATP occlusion of $Ca^{2+}$ was found both in the detergent solubilized, monomeric and in the membraneous form [41].

## 2.3. *Half-of-the-sites reactivity*

One of the strongest arguments (*e.g.* Refs. 42 and 44) in favour of an oligomeric transport mechanism has been the well documented evidence for incomplete phosphorylation of $Ca^{2+}$-ATPase. Usually, only 4-5 nmol $^{32}P$ from $^{32}P$ ATP can be incorporated per mg protein in purified preparations, as compared to a theoretical upper value of 9 nmol/mg by phosphorylation of all polypeptide chains in a pure preparation of $Ca^{2+}$-ATPase (*m.w.* 110 000). This is the case also when phosphorylation is performed under conditions where dephosphorylation is the absolutely rate-limiting reaction. Thus, only about half of the $Ca^{2+}$-ATPase is available for phosporylation, suggesting a phenomenon of negative cooperativity between polypeptide chains in a functional dimer of $Ca^{2+}$-ATPase. Evidence for half-of-the-sites reactivity for probes covalently attached to specific sites has also been reported: fluorescein isothiocyanate which inhibits ATP binding [45] and dicyclohexylcarbodiimide [46] which inhibits $Ca^{2+}$ transport.

In terms of $Ca^{2+}$ transport half-of-the-sites reactivity was previously proposed to be based on an aggregate rearrangement mechanism [47] in which the $Ca^{2+}$ translocation sites do not move appreciably, but which during phosphorylation change their exposure from contact with the outer fluid phase to the inner fluid phase by a slight movement of the polypeptide chains with respect to each other (see Fig. 3). This theory is amenable to testing by the use of detergent-solubilized preparations. After solubilization of active $Ca^{2+}$-ATPase in monomeric form we found no increase in the phosphorylation level [40, 48] as it would have been expected if phosphorylation was under allosteric control in an oligomeric complex. Rather, the data indicate that low phosphorylation levels are the result of other factors such as the presence of inactive $Ca^{2+}$-ATPase, some contamination with other membrane proteins, and possibly also some systematic error in LOWRY protein determinations. As mentioned above inactive $Ca^{2+}$-ATPase is regularly found in the oligomeric fraction of detergent-solubilized $Ca^{2+}$-ATPase. Accordingly, a higher level of active sites (6-7 nmoles/mg detergent-solubilized $Ca^{2+}$-ATPase) has been reported after h.p.l.c. fractionation of monomeric and oligomeric $Ca^{2+}$-ATPase [49, 50]. Similarly, in a reinvestigation of the problem, fluorescein isothiocyanate was found to react with the active fraction of $Ca^{2+}$-ATPase alone and in stoichiometric amounts [51].

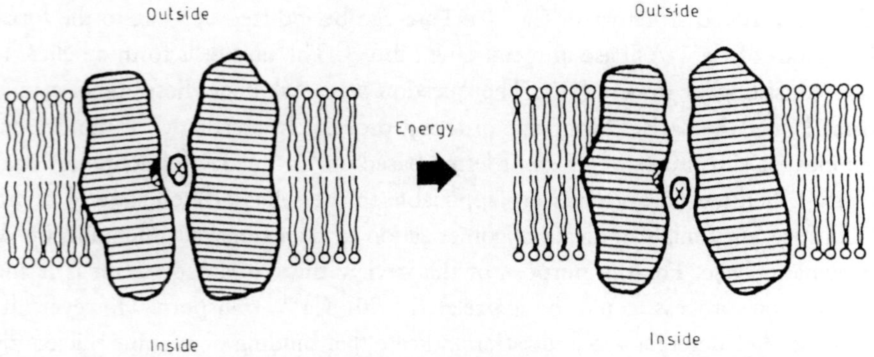


Fig. 3 - Schematic illustration of $Ca^{2+}$ translocation according to the aggregate rearrangement mechanism. Taken from DUTTON *et al.* [47].

## 2.4. *Kinetic heterogeneity*

Evidence of biphasic phosphorylation of membraneous $Ca^{2+}$-ATPase that disappears after solubilization of $Ca^{2+}$-ATPase with $C_{12}E_8$ was interpreted in terms of an oligomeric mechanism [52]. In a reinvestigation slow dephosphorylation of $Ca^{2+}$-ATPase was found to reflect reaction with CaATP instead of MgATP [39]. Phosphorylation of $C_{12}E_8$ solubilized $Ca^{2+}$-ATPase with CaATP also occurred after solubilization with $C_{12}E_8$, but to a lesser extent than for membraneous enzyme [39].

The stoichiometry between $Ca^{2+}$ binding at the translocation sites and phosphorylation is found to be 2:1, both for membraneous and detergent-solubilized $Ca^{2+}$-ATPase. At alkaline pH, HILL coefficients in the binding of $Ca^{2+}$ considerably larger than 2 have been reported, suggesting an effect of protein-protein interactions [53]. However, the same high cooperativity was observed after solubilization of $Ca^{2+}$-ATPase in monomeric form [54].

# 3. Studies on membraneous $Ca^{2+}$–ATPase

## 3.1. *Organization of $Ca^{2+}$-ATPase in the membrane*

The existence of $Ca^{2+}$-ATPase oligomers in the membrane has been inferred from ultrastructural data on the basis of a relatively low density of freeze-fracture particles and, to some extent, by the use of various spectroscopic probes to monitor rotational motion and peptide interactions (see previous reviews, Refs. 55 and 56). In the presence of vanadate an ordered structure of $Ca^{2+}$-ATPase can be induced, leading to the formation of 2D-crystals of $Ca^{2+}$-ATPase in membrane tubules. The unit cells form a helical repeat pattern, consisting of dimers [57]. The question arises whether these dimeric units are preformed or if they arise during the ordering process. Interestingly, in the presence of lanthanides (as substitutes of $Ca^{2+}$) ordering based on $Ca^{2+}$-ATPase monomers has been found [58]. If these observations are applicable to the selfassociated state of unordered $Ca^{2+}$-ATPase, they might indicate oligomerization to be a transient phenomenon during the enzymatic cycle. For the purpose of this review the relevant question is if such an oligomerization process could be correlated with $Ca^{2+}$ transport. However, h.p.l.c. esperiments after detergent-solubilization indicate that binding of adenine nucleotide and phosphorylation favours dissociation of oligomeric $Ca^{2+}$-ATPase [50]. Similarly, WATANABE and INESI [59] found that phosphorylation led to dissociation of $Ca^{2+}$-ATPase which had been poised between the detergent-solubilized and membraneous state by addition of an appropriate amount of $C_{12}E_8$.

Even if the detergent-solubilization studies emphasize the functional competence of monomeric $Ca^{2+}$-ATPase, the question remains if the putative channel for $Ca^{2+}$ transport in the membrane is located between individual polypeptide chains. In order to approach this question we have examined the properties of $Ca^{2+}$-ATPase, reconstituted by a cholate procedure [60], with an excess of lipid (weight ratio of lipid to protein 100:1), followed by fractionation on a sucrose gradient [61]. Stereological analysis indicates that $Ca^{2+}$-ATPase is evenly dispersed in the light fraction which has a high transport capacity and a high coupling ratio (1-1.5). Furthermore, the number of intramembraneous particles corresponds to the number of $Ca^{2+}$-ATPase polypeptide chains in this preparation. On this basis a $Ca^{2+}$-ATPase monomer would seem capable not only of hydrolyzing ATP, as indicated by the detergent-solubilization studies, but also to effect $Ca^{2+}$ transport in a membraneous environment.

It seems possible that, despite the high concentration of $Ca^{2+}$-ATPase in the sarcoplasmic reticulum membrane, an appreciable fraction exists in monomeric form. This conclusion can be derived from e.s.r. spectroscopy of spin-labeled $Ca^{2+}$-ATPase in which chemical polypeptide cross-limking is used to gauge the effect of oligomerization on rotational motion [62]. After rotary evaporation-shadowing FRANZINI-ARMSTRONG and FERGUSON [63] reported that the number of identifiable intramembraneous particles is much higher than previously observed by the standard freeze-fracture technique, using unilateral shadowing. In our study there was little difference between the size of the intramembraneous particles in the reconstituted and sarcoplasmic reticulum vesicles: both preparations were characterized by a preponderance of particle diameters around 6-7 nm [60].

### 3.2. *Structure of the $Ca^{2+}$-ATPase polypeptide*

At the end of the 1970s around 2/3 of the $Ca^{2+}$-ATPase polypeptide chain had been sequenced through the efforts of ALLEN and GREEN (see Ref. 64). By the use of recombinant DNA methodology the whole amino acid sequence of rabbit skeletal muscle $Ca^{2+}$-ATPase is now deduced [65, 66]. In conjunction with other advances on the structure of the $Ca^{2+}$-ATPase, *e.g.* obtained by image reconstruction from 2D-crystals [67, 68] and diffraction techniques [69], as well as the topological relationships between active sites, studied by spectroscopic techniques [70, 72], important clues concerning $Ca^{2+}$-ATPase structure and structure-function relationships can be obtained.

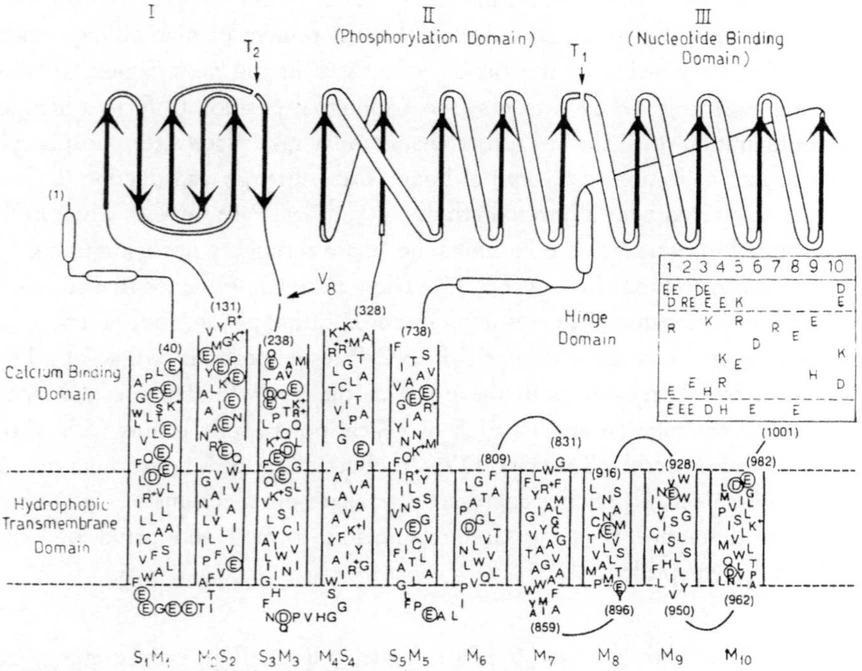


Fig. 4 - Model for sarcoplasmic reticulum $Ca^{2+}$-ATPase, proposed by BRANDI *et al.* [66]. The $Ca^{2+}$-ATPase polypeptide chain is seen to be composed of a $Ca^{2+}$ binding *neck region* ($S_{1-5}$) and a membrane traversing region ($M_{1-10}$) plus 3 cytoplasmic domains (I-III). The figure shows the location of the first two tryptic splits ($T_1$ and $T_2$) and of the major $Ca^{2+}$ -dependent V8 split [81].

Before the advent of these new data it was known from a large number of diverse studies that $Ca^{2+}$-ATPase insertion into the membrane is markedly asymmetric, a large fraction of the mass projecting from the outer aspect of the SR membrane (reviewed *e.g.* in Refs. 55 and 56). On the basis of the amino acid sequence predictions of secondary structure have been made [ 5, 65]. Hydropathy plots have suggested 10 probable stretches of transmembrane helices [65, 66], see Fig. 4. The major part of the extramembraneous portion of these is connected with 5 of the transmembrane stretches. MacLennan *et al.* [65] suggest that the extramembraneous portion forms three interacting domains (I, II and III in Fig. 4). Domains II and III contain the phosphorylation site and adenine binding site of ATP, respectively. The 3 domains are considered to be connected with the intramembraneous portion of $Ca^{2+}$-ATPase via a *neck-region*, consisting of 5 peptide chains that contain many glutamate residues and which are proposed to have a basically α-helical structure. The juxaposition of many carboxylate groups to the intramembraneous portion of the $Ca^{2+}$ATPase has not been found in other homologous ion transporting ATPases [5]. It is suggested that the neck region forms the locus for initial binding of transported $Ca^{2+}$. It should be noted that high-affinity binding sites are not preformed, but arise during the $E_2$ to $E_1$ transition (Fig. 1). Brandl *et. al* have considered that the structural basis for the inducement of high-affinity binding of $Ca^{2+}$ is rotation or twisting of the bundle of helices in the neck region. In this way carboxylate groups are envisaged to move to an internal position to form a high-affinity binding configuration for 2 $Ca^{2+}$. Further changes may take place after phosphorylation, giving rise to internalization (*occlusion*) of bound $Ca^{2+}$. Energy transduction ($E_1P$ to $E_2P$ transition) is associated with a lowered affinity of $Ca^{2+}$-ATPase for $Ca^{2+}$. Released $Ca^{2+}$ is then suggested to escape to the interior fluid phase through a narrow channel formed between the transmembrane helices [66, 73]. There is kinetic evidence that release of the 2 bound $Ca^{2+}$ occurs sequentially, resulting in translocation of $Ca^{2+}$ one at a time [74]. A model according to which transition of $E_1P$ to $E_2P$ results in both change of a localized binding region from exposure with the outer to the inner fluid phase and decreased binding of $Ca^{2+}$ by rotation and twisting of helical polypeptide chains with respect to each other was presented by Tanford [75]. In effect these ideas concerning internalization of $Ca^{2+}$ resemble the previous aggregate rearrangement mechanism (Fig. 3), but with the important difference that the $Ca^{2+}$ channel is formed and gated by only one $Ca^{2+}$- ATPase polypeptide.

In accordance with the $Ca^{2+}$-ATPase model (Fig. 4) fluorescence energy-transfer experiments indicate that the phosphorylation site is distant from both the $Ca^{2+}$ binding sites and the membrane [70-72]. Energy transduction is thus carried over a large distance that must involve conformational changes of the peptide chain and interactions between amino acid residues in the various domains. Time-resolved neutron diffraction experiments [24] and electron microscopic examination of vanadylated $E_2$ [76, 77] suggest that $Ca^{2+}$ -ATPase to some extent *sinks* into the membrane during $Ca^{2+}$ transport. Interestingly, amino acid residues corresponding to helices 4 and 5 in the neck region were found to be embedded into membrane lipid as indicated by an increased labeling with hydrophobic azido derivatives during $Ca^{2+}$ binding and phosphorylation [78]. Other evidence for the involvement of domain I in the translocation process is provided by disruption of $Ca^{2+}$ accumulation and decreased binding of $Ca^{2+}$ (79, 80] after tryptic cleavage at $T_2$ (see Fig. 4). Using staphylococcal V8 to cleave peptide bonds adjacent to acidic amino acid residues [81] we found that $Ca^{2+}$ binding induced accelerated cleavage of a particular peptide bond between amino acid residues 231 and 232, localized in the outer portion of the neck region (Fig 4). This is evidence of a $Ca^{2+}$ induced conformational change close to the putative $Ca^{2+}$ binding region. Binding of $Ca^{2+}$ persisted after breakage of this bond, stabilizing the enzyme against further degradation by V8. Concerning the possible role of domain I in the transduction process it is of interest that tryptic cleavage of $Ca^{2+}$-ATPase at $T_2$ under some conditions was found to increase $Ca^{2+}$ efflux of accumulated $Ca^{2+}$. Modification of -SH groups by $Ag^{+}$ was also found by GOULD *et al.* [82] to induce $Ca^{2+}$ release, while $La^{3+}$ causes retention of accumulated $Ca^{2+}$ [83]. These observations may signify the existence of a $Ca^{2+}$ channel in the intact $Ca^{2+}$-ATPase which can be opened or closed by certain experimental manipulations, also in the absence of energy transducing processes.

## Addition in proof

As noted above the suggested mechanism for $Ca^{2+}$ translocation is speculative. We have recently obtained evidence after proteolytic digestion of $Ca^{2+}$-ATPase with V8 and trypsin that the carboxyterminal end of $Ca^{2+}$-ATPase has a large potential for $Ca^{2+}$ binding (M. LE MAIRE, S. LUND, P. CHAMPEIL and J.V. MØLLER, manuscript in preparation). It is therefore a moot point if the $Ca^{2+}$ translocation sites are in fact located in the stalk region at the N-terminal end of $Ca^{2+}$-ATPase, or whether some other mechanisms for $Ca^{2+}$ translocation than described in the text should be sought, based on $Ca^{2+}$ translocation sites at the C-terminal end.

## References


[1] W. HASSELBACH and M. MAKINOSE, *Biochem. Z.,* **333,** 518 (1961).

[2] W. HASSELBACH, *Top. Curr. Chem,* **78,** 1 (1979).

[3] M.L. ENTMAN and W.B.V WINKLE, in *Sarcoplasmic Reticulum in Muscle Physiology,* M.L. ENTMAN and W.V. Winkle, (Editors), CRC Press, FLA, (1986), p. 1.

[4] L. DE MEIS, in *The Sarcoplasmic Reticulum, Transport and Energy Transduction,* J. WILEY & SONS, NY, (1981).

[5] N.M. GREEN, W.R. TAYLOR, C. BRANDL, B. KORCZAK and D.H. MACLENNAN, *Ciba Fdn. Symp.,* **122,** 93 (1986).

[6] S. YAMADA and N. IKEMOTO, *J. Biol. Chem.,* **260,** 3108 (1980).

[7] M. CHIESI and G. INESI, *Biochemistry,* **19,** 2912 (1980).

[8] N. IKEMOTO, *J. Biol. Chem.,* **251,** 7275 (1976).

[9] J.P. ANDERSEN, K. LASSEN and J.V. MØLLER, *J. Biol. Chem.,* **260,** 371 (1985).
[10] C. TANFORD, *Gen. J. Physiol,* **77,** 223 (1981).
[11] W. HASSELBACH and H. OETLIKER, *Annu. Rev. Physiol.,* **45,** 325 (1983).
[12] G. MEISSNER, *Mol. Cell Biochem.,* **55,** 65 (1983).
[13] P. ZIMNIAK and E. RACKER, *J. Biol. Chem.,* **253,** 4631 (1978).
[14] T. MORIMOTO and M. KASAI, *J. Biochem. (Tokyo),* **99,** 1071 (1986).
[15] M. YANAGUCHI and T. KANAZAWA, *J. Biol. Chem.,* **260,** 4896 (1985).
[16] D.H. HAYNES, *Arch. Biochem. Biophys.,* **215,** 444 (1982).
[17] Y. TAKAKUWA and T. KANAZAWA, *J. Biol. Chem.,* **257,** 10770 (1982).
[18] J.J. FEHER and F.N. BRIGGS, *Biochim. Biophys. Acta,* **727,** 389 (1983).
[19] U. GERDES and J.V. Møller, *Biochim. Biophys. Acta,* **734,** 191 (1983).
[20] G. FRANCOIS, *Biochim. Biophys. Acta,* **173,** 86 (1969).
[21] W.H.F.M. MOMMAERTS, *Proc. Natl. Acad. Sci. USA,* **58,** 2476 (1967).
[22] R. K. NAKAMOTO and G. INESI, *FEBS Lett.,* **194,** 258 (1986).
[23] P. CSERMELY, C. KATOPIS, B.A. WALLACE and A. MARTONOSI, *Biochem. J.,* **241,** 663 (1987).
[24] J.K. BLASIE, L.G. HERBETTE, D. PASCOLINI, V. SKITA, D.H. PIERCE and A. SCARPA, *Biophys. J.,* **48,** 9 (1985).
[25] I.L.R. ARRONDO, H.H. MANTSCH, N. MULLNER, S. PIKULA and A. MARTONOSI, *J. Biol. Chem.,* **262,** 9037 (1987).
[26] J.P. ANDERSEN, P.L. JØRGENSEN and J.V. MØLLER, *Proc. Natl. Acad. Sci. USA,* **82,** 4573 (1985).
[27] J.P. ANDERSEN, B. VILSEN, J.H. COLLINS and P.L. JØRGENSEN, *J. Membr. Biol.,* **93,** 85 (1986).
[28] C. TANFORD, Y. NOZAKI, J.A. REYNOLDS and S. MAKINO, *Biochemistry,* **13,** 2369 (1974).
[29] J.V. MØLLER, M. LE MAIRE and J.P. ANDERSEN in *Progress in Protein-Lipid Interactions,* A. WATTS, and J.J.H.H.M. DE PONT (Editors), Elsevier, Amsterdam, (1986), Vol. **2,** Chap., 5, p. 147.
[30] J.P. ANDERSEN, B. VILSEN, H. NIELSEN and J.V. MØLLER, *Biochemistry,* **25** , 6439 (1986).
[31] S. LUND, S. ORLOWSKI, B. DE FORESTA, P. CHAMPEIL, M.LE MAIRE and J.V. MØLLER (unpublished observations).
[32] J.V. MØLLER, M. LE MAIRE and J.P. ANDERSEN, *Meth. Enzymol.,* **157,** 261 (1988).
[33] M. LE MAIRE, J.V. MØLLER and C. TANFORD, *Biochemistry,* **15,** 2336 (1976).
[34] M. LE MAIRE, K.E. LIND, K.E. JØRGENSEN, H. RØIGAARD-PETERSEN and J.V. MØLLER, *J. Biol. Chem.,* **253,** 7051 (1978).
[35] L. DUX, S. PIKULA, N. MULLNER and A. MARTONOSI, *J. Biol. Chem.,* **262,** 6439 (1987).
[36] K.E. JØRGENSEN, K.E. LIND, H. RØIGAARD-PETERSEN and J.V. MØLLER, *Biochem. J.,* **169,** 489 (1978).

[37] J.L. Silva and S. Verjovski-Almeida, *J. Biol. Chem.,* **260,** 4764 (1985).

[38] B. de Foresta, M. le Maire, S. Orlowski, P. Champeil, S. Lund, J.V. Møller, F. Michelangeli and A.G. Lee, *Biochemistry* (in press).

[39] S. Lund and J.V. Møller, *J. Biol. Chem.,* **263,** 1654 (1988).

[40] B. Vilsen and J.P. Andersen, *Eur. J. Biochem.,* **170,** 421 (1987).

[41] B. Vilsen and J.P. Andersen, *Biochim. Biophys. Acta,* **855,** 429 (1986).

[42] Y. Dupont, Y. Chapron and R. Pougeois, *Biochem. Biophys. Res. Comm.,* **106,** 1272 (1982).

[43] J.P. Froehlich and E.W. Taylor, *J. Biol. Chem.,* **251,** 2307 (1976).

[44] T. Watanabe, D. Lewis, R. Nakamoto, M. Kurzmack, C. Fronticelli and G. Inesi, *Biochemistry,* **20,** 6617 (1981).

[45] U. Pick and S.J.D. Karlish, *Biochim. Biophys. Acta ,* **626,** 255 (1980).

[46] U. Pick and E. Racker, *Biochemistry,* **18,** 108 (1979).

[47] A. Dutton, E.D. Rees and S.J. Singer, *Proc. Natl. Acad. Sci. USA,* **73,** 1532 (1976).

[48] J.P. Andersen, J.V. Møller and P.L. Jørgensen, *J. Biol. Chem.,* **257,** 8300 (1982).

[49] H. Barrabin, H.M. Scofano and G. Inesi, *Biochemistry,* **23,** 1542 (1984).

[50] J.P. Andersen and B. Vilsen, *FEBS Lett.,* **189,** 13 (1985).

[51] C. Mitchinson, A.F. Wilderspin, B.J. Trinnaman and N.M. Green, *FEBS Lett.,* **146,** 87 (1982).

[52] N. Ikemoto and R.W. Nelson, *J. Biol. Chem.,* **259,** 11790 (1984).

[53] T.L. Hill and G. Inesi, *Proc. Nat. Acad. Sci. USA,* **79,** 3978 (1982).

[54] B. Vilsen and J.P. Andersen, *Biochim. Biophys. Acta,* **898,** 313 (1987).

[55] J.V. Møller, J.P. Andersen and M. le Maire, *Mol. Cell Biochem.,* **42,** 93 (1982).

[56] M. le Maire and J.V. Møller, in *Sarcoplasmic Reticulum in Muscle Physiology,* M.L. Entman and W.B.V. Winkle (Editors), CRC Press, FLA., (1986), Vol. **1,** Chap. 4, p. 101.

[57] K. Taylor, L. Dux and A. Martonosi, *J. Membr. Biol.,* **174,** 193 (1984).

[58] L. Dux, K.A. Taylor, H.P. Ting-Beall and A.N. Martonosi, *J. Biol. Chem.,* **260,** 11730 (1985).

[59] T. Watanabe and G. Inesi, *Biochemistry,* **21,** 3254 (1982).

[60] J.P. Andersen, E. Skriver, T.S. Mahrous and J.V. Møller, *Biochim. Biophys. Acta,* **685,** 355 (1983).

[61] M. le Maire, C. Heegaard, J.P. Andersen, T. Gulik-Krzwycki and J.V. Møller (unpublished observations).

[62] T.C. Squire, S.E. Hughes and D.D. Thomas, *J. Biol. Chem.,* **263,** 9162 (1988).

[63] C. Franzini-Armstrong and D.G. Ferguson, *Biophys. J.,* **48,** 607 (1985).

[64] G. Allen, B.J. Trinnaman and N.M. Green, *Biochem. J.,* **187,** 591 (1980).

[65] D.H. MacLennan, C.J. Brandl, B. Korczak and N.M. Green, *Nature (London),* **316,** 696 (1985).

[66] C.J. Brandl, N.M. Green, B. Korczak and D.H. MacLennan, *Cell,* **44,** 597 (1986).

[67] K.A. Taylor, L. Dux and A. Martonosi, *J. Mol. Biol.,* **187,** 417 (1986).

[68] L. Castellani, P.M.D. Hardwicke and P. Vibert, *J. Mol. Biol.,* **185,** 579 (1985).

[69] L. Herbette, in *Sarcoplasmic Reticulum in Muscle Physiology,* M.L. Entman and W.B.V. Winkle (Editors), CRC Revs., CRC Press, FLA., (1986), Vol. **1,** Chap. 5, p. 127.

[70] S. Highsmith and A.J. Murphy, *J. Biol. Chem.,* **259,** 14651 (1984).

[71] T.L. Scott, *J. Biol. Chem.,* **260,** 14421 (1985).

[72] C. Gutierrez-Merino, F. Munkonge, A.M. Mata, J.M. East, B.L. Levinson, R.M. Napier and A.G. Lee, *Biochim. Biophys. Acta,* **897,** 207 (1987).

[73] J.P. Froehlich and P.F. Heller, *Biochemistry,* **24,** 126 (1985).

[74] G. Inesi, *J. Biol. Chem.,* **262,** 16338 (1987).

[75] C. Tanford, in *Structure and Function of Sarcoplasmic Reticulum,* S. Fleischer and Y. Tonomura (Editors), Academic Press, NY, (1985), p. 259.

[76] C. Perracchia, L. Dux and A.N. Martonosi, *J. Mus. Res. Cell. Motil.,* **5,** 431 (1984).

[77] W. Hasselbach, P. Medda, A. Migala and B. Agostini, *Z. Naturforsch.,* **38C,** 1015 (1983).

[78] J.P. Andersen, B. Vilsen, J.H. Collins and P.L. Jørgensen, *J. Membr. Biol.,* **93,** 85 (1986).

[79] T.L. Scott and A.E. Shamoo, *J. Membr. Biol.,* **64,** 137 (1982).

[80] T.L. Scott and A.E. Shamoo, *Europ. J. Biochem.,* **143,** 427 (1984).

[81] M. le Maire, S. Lund, P. Champeil and J.V. Møller (unpublished observations).

[82] G.W. Gould, J. Colyer, J.M. East and A.G. Lee, *J. Biol. Chem.,* **262,** 7676 (1987).

[83] U. Gerdes, A.M. Nakhla and J.V. Møller, *Biochim. Biophys. Acta ,* **734,** 191 (1983).

# THE STRUCTURE OF THE ATP-SYNTHASE FROM CHLOROPLASTS

PETER GRÄBER, BETTINA BÖTTCHER and EGBERT J. BOEKEMA*

*Biologisches Institut der Universität Stuttgart*
*Pfaffenwaldring 57, 7000 Stuttgart 80, F.R.G.*

** Biochemisch Laboratorium, Rijksuniversiteit*
*Nijenborgh 16, 9747 AG Groningen, Nederlande*

Contents



*Bioelectrochemistry III*
Edited by G. Milazzo and M. Blank
Plenum Press, New York, 1990

## 1. Introduction

Biological membranes have different functions:

1. They separate cells into different compartments. This allows the occurrence of otherwise interfering chemical reactions.
2. They form selective barriers for molecules and ions.
3. They actively transport molecules and ions into the different compartements (*pumps*).

The latter process is carried out by different membrane–bound proteins. One important class of such transport–proteins are the 10 called *$F_0F_1$–ATPases ($H^+$–ATPase* or *ATP-synthase).* These enzymes catalyze proton–transport–coupled ATP–synthesis/ hydrolysis. ATP–hydrolysis leads to a transmembrane proton–transport and thereby to generation of a transmembrane pH–difference, ΔpH, and an electric potential difference, $\Delta\phi$. In the opposite direction the proton flux driven by ΔpH and $\Delta\phi$ provides the free energy for ATP–synthesis from ADP and phosphate. This implies that these enzymes must have a structure (and mechanism) which allows to tranform the free energy derived from the proton flux into the energy of a phosphate anhydrate bond. $F_0F_1$–ATPases occur in the membranes of bacteria, mitochondria and chloroplasts. The enzymes from the different sources are very similar; they consist of a membrane–integrated, hydrophobic part, $F_0$, which is supposed to act as a proton channel through the membrane and a hydrophilic part, $F_1$, which contains the nucleotide–binding sites, This device is able to use the free energy of a chemical reaction to transport protons across the membrane (*ion pump*).

Functional studies on the mechanism of the ATP–synthase from chloroplasts are summarized in chapter 12. The basis for the interpretation of the functional studies and for the development of a biochemical feasable mechanism, of course, requires the knowledge of the structure of the enzyme. Therefore, in this chapter the current structural knowledge about the ATP–synthase from chloroplasts is summarized. The data are based mainly on electron-microscopic and biochemical investigation.

## 2. Subunit composition of $F_0F_1$ ATP-synthases

The subunit composition of the $F_0F_1$–ATP–synthases from the different sources has been investigated by SDS–gel electrophoresis [1–10]. Although the function and the shape of the enzymes are similar, there are differences in subunit composition. The enzyme from *E. coli* has 8 different subunits, that from chloroplasts has 9, and that from mitochondria at least 13. In Table 1 the subunit compositions of these different enzymes are summarized, together with their stoichiometries. There is general agreement that there are three copies of α– and β–subunits; however, the stoichiometry of the other subunits is controversial. Biochemistry and genetics of $F_0F_1$–ATP–synthases are covered in several reviews (see *e.g.* Refs. 1-11).

Table 1. Subunit composition and stoichiometry of different $F_0F_1$–ATP–synthases, $EF_0F_1$ from *E. coli*, $CF_0F_1$ from chloroplasts and $MF_0F_1$ from mitochondria. The subunits above the dashed line belong to the $F_1$ part, those below to the $F_0$ part.

| Subunits (stoichiometry) | | | suggested function |
|---|---|---|---|
| $EF_0F_1$ | $CF_0F_1$ | $MF_0F_1$ | |
| — | — | $F_1$ | inhibition of enzyme |
| α (3) | α (3) | α (3) | regulatory nucleotide binding |
| β (3) | β (3) | β (3) | catalytic nucleotide binding |
| γ (1) | γ (1) | γ (1) | |
| δ (1) | δ (1) | OSCP (2) | binding of $F_1$ to $F_0$ |
| ε (1) | ε (1) | δ (1) | ? |
| — | — | ε (1) | ? |
| — | — | F6 | ? |
| a (1) | IV (1) | a (1) | proton conduction |
| b (2) | I, II (1,1) | b (2) | binding of $F_1$ to $F_0$ |
| c (9–12) | III (6–12) | c (6–10) | proton conduction |
| — | — | d (?) | ? |
| — | — | A6L (?) | ? |

Usually, the subunits are designated according to their decreasing molecular mass in the SDS–gels. However, recent knowledge of the amino acid sequences derived from the corresponding DNA sequences shows that similarities exist between differently named subunits [12]. Therefore, the subunits are arranged in Table 1 in such a way that each line shows the subunits with the highest similarity in amino acid sequence and which, consequently, should have the same function in the enzyme. The reason for the different subunit composition of the $F_0F_1$–ATP–synthases is not clarified yet. However, the enzymes developed latest in evolution have the largest number of subunits. Presumably, these more complex enzymes allow better regulation of the enzyme activity by different physiologically relevant substances.

## 3. Subunits of $CF_0F_1$

The subunit composition of the ATP–synthase from chloroplasts is shown in Fig. 1 as revealed by sodium dodecylsulfate polyacrylamide gel–electrophoresis (SDS–p.a.g.e.). Fig. 1A, lane 1 shows $CF_0F_1$, the five subunits of $CF_1$ are designated by Greek letters, the four subunits of the $CF_0$–part by Roman numbers. In this case, the enzyme was treated with the SDS buffer for 5 min at 95 °C. If the enzyme is incubated in SDS buffer for 5 min at room temperature, the pattern on the SDS–gel is different (lane 2): subunit III is not observed; however, an additional band at the mole mass of 100 kDa can be seen.

In order to investigate the 100 kDa–protein, $CF_0F_1$ was subjected to a preparative gel electrophoresis, and the different protein bands were cut from the gel and electroeluted. Lane 3 shows the SDS–gel of the isolated 100 kDa–protein. No other protein bands can be detected. If the 100 kDa–protein is incubated in SDS buffer at 95 °C for 5 min, the 100 kDa band disappears and the band of subunit III at 8 kDa is observed (lane 4). Since no other protein band can be seen on lanes 3 and 4, we can conclude that the 100 kDa complex contains only subunit III. From the observed molecular masses we can conclude that the 100 kDa complex contains at least 12 copies of subunit III [13].

Another unusual phenomenon was the faint band designated IV in the SDS gel (see lanes 1 and 2). In order to check whether this band is an impurity, we isolated the band by quantitative gel–electrophoresis. Lane 5 shows the isolated subunit IV. From this protein, the amino acid sequence of the first 16 amino acids from the N–terminus was determined. The sequence was compared with the amino acid sequence deduced from the chloroplast gene atpI [14]. It turned out that the isolated subunit is identical to the gene derived sequence; however, the first 18 amino acids are obviously removed proteolytically before the subunit IV is assembled into $CF_0F_1$ [13]. Comparison of the amino acid sequence of subunit IV with that of *E. coli* subunit a shows strong homologies (14, 15). This result was already indicated in Tab. 1.

Figure 1B shows SDS–gels of the isolated $CF_0$–part, $CF_1$–part, and, for a comparison, again $CF_0F_1$. $CF_1$ shows the well-known five subunits (lane 3). $CF_0$ (lane 1) shows four subunits. Subunit III is observed as its 100 kDa complex. Slight impurities by

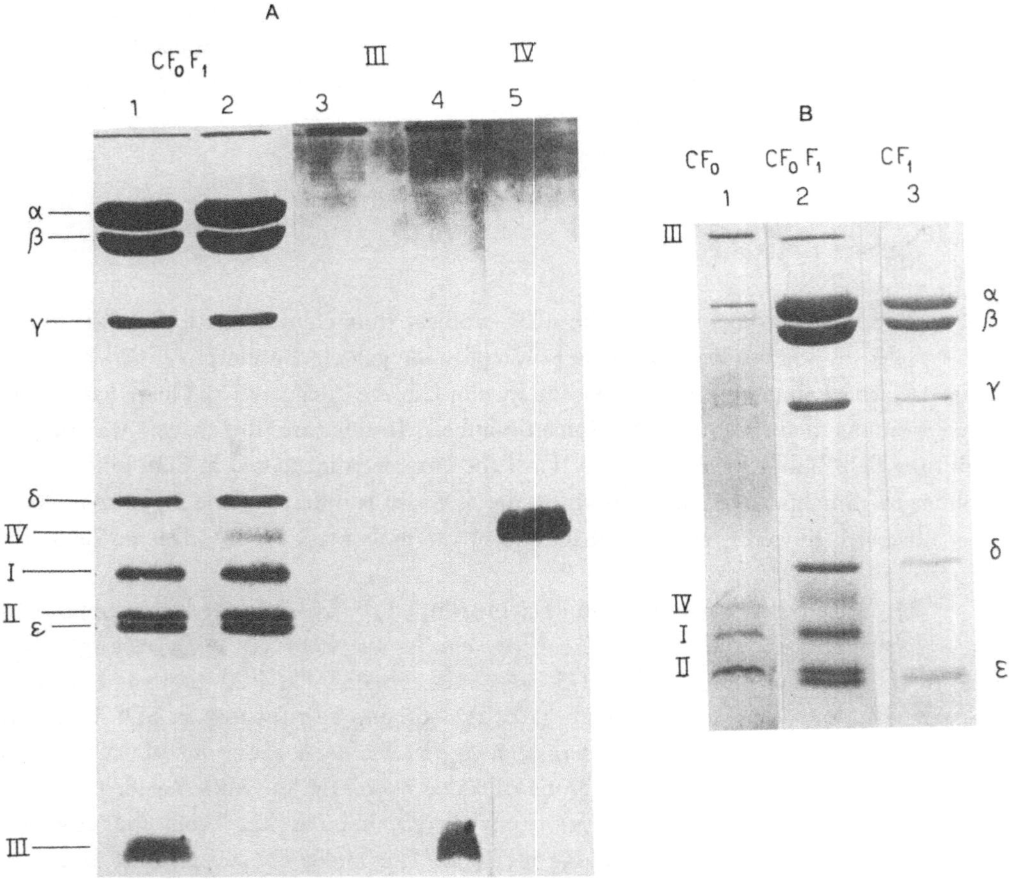


Fig. 1. – Subunit composition of $CF_0F_1$ as revealed by SDS–gel- electrophoresis and separation of different structures of the enzyme. A: Lane 1–$CF_0F_1$ (30 μg protein, enzyme treated for 5 min at 95 °C (in SDS–buffer); lane 2–$CF_0F_1$ (30 μg protein); lane 3–isolated subunit III–complex (8 μg; protein); lane 4–isolated subunit III–complex (8 μg protein) treated for 5 min at 95 °C in SDS–buffer; lane 5–isolated subunit IV (8 μg protein). Lanes 1–2 stained with Coomassie Blue, lanes 3–5 stained with silver. B: Lane 1–isolated $CF_0$ (20 μg protein); lane 2–isolated $CF_0F_1$ (30 μg protein); lane 3–isolated $CF_1$ (20 μg protein), stained with Coomassie Blue.

the subunits α, β, and γ can be seen (< 5 %). From the nine different subunits of $CF_0F_1$ the subunits α, β, ε, I, III and IV are encoded in the chloroplast DNA; the subunits γ, δ and II are encoded in the nuclear DNA. This is indicated in Table 2. In addition, the mole masses are given as calculated form the amino acid sequence and the mole masses found by SDS–gel–electrophoresis [16, 17].

Table 2. Molecular masses of the subunits from $CF_0F_1$ obtained from SDS–gel-electrophoresis and calculated from the DNA–derived amino acid sequence [14–17, 44, 45]

| subunit | α | β | γ | δ | ε | I | II | III | IV | $CF_0$ I II $III_{12}$ IV | $CF_1$ $\alpha_3\beta_3\gamma\delta\varepsilon$ | $CF_0F_1$ |
|---|---|---|---|---|---|---|---|---|---|---|---|---|
| mole mass (kDa) from sequence | 56.8 | 53.9 | — | — | 14.9 | 19.001 | 16.499 | 7.968 | 24.6 | | | |
| mole mass (kDa) from SDS-gel | 62 | 57 | 38 | 21 | 14 | 15.5 | 13.5 | 7.7 | 20 | 141.4 | 430 | 571.4 |
| encoded in chloroplast (c) nucleus (n) | c | c | n | n | c | c | n | c | c | | | } |

## 4. The shape of $CF_0F_1$

The shape of $CF_0F_1$–ATP–synthases has been investigated by electron microscopy of detergent–solubilized enzyme after negative staining with uranylacetate [18-21]. Figure 2 shows a gallery of electron micrographs of $CF_0F_1$. Single $CF_0F_1$ molecules and different aggregates (*strings*) can be seen. Additionally, a schematic interpretation of the observed structure is given [21]. The $F_0$–parts (dashed areas) are strongly hydrophobic and, therefore, have a tendency to aggregate. At the ends of a string the hydrophobic parts are surrounded by detergent as indicated schematically in Fig. 2. It is obvious that the $F_0$–parts are smaller than the $F_1$–parts. This influences the shape of the strings. During preparation of electron microscopy, the $F_1$–parts become attached to the carbon support film. Since they are so big their positions will mostly alternate along the strings, avoiding overlap or friction. In Fig. 2 one can see that in most of the strings adjacent $F_1$–parts are alternating and not neighbouring.

The contours of the $F_1$–parts attached to the strings can be seen clearly. In some cases the length of $F_1$ parallel to the string was longer than that vertical to the string. In other cases both dimensions were quite similar. In the average, it resulted 11 nm parallel to the string and 8.3 nm vertical to the string. The $F_1$-parts are connected to the strings by a stalk. Its length is 3.7 nm, its diameter 2.7 nm. If the $CF_0F_1$ is reconstituted into asolectin liposomes the length of the stalk is found to be 3.2 nm. The diameter of the

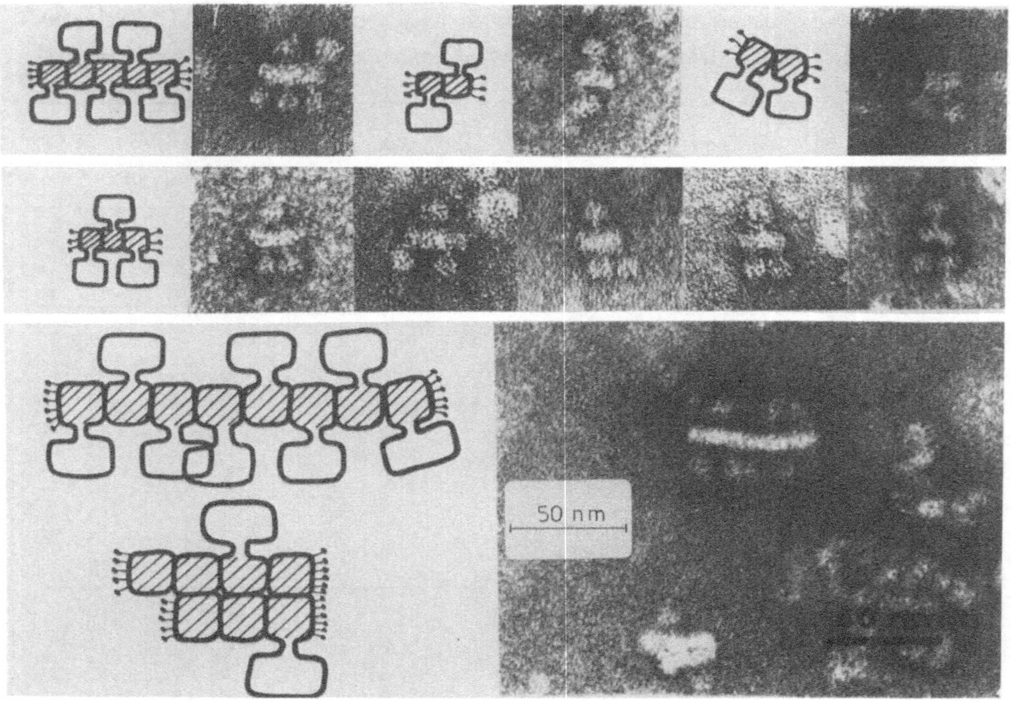


Fig. 2. – Electron micrographs of isolated $CF_0F_1$ showing different aggregates and schematic interpretation of this structure. Dashed areas represent the $F_0$–parts.

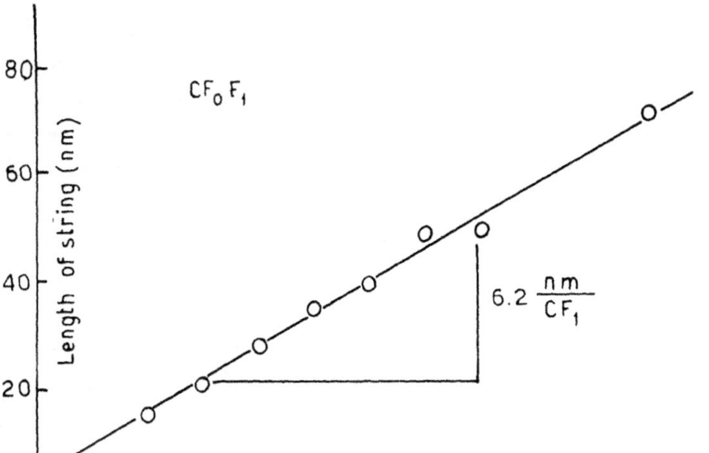


Fig. 3. – String length as a function of the number of attached $CF_1$–parts. The slope gives the string length per $CF_1$; this represents the maximum dimension of $CF_0$ in the direction of the string axis.

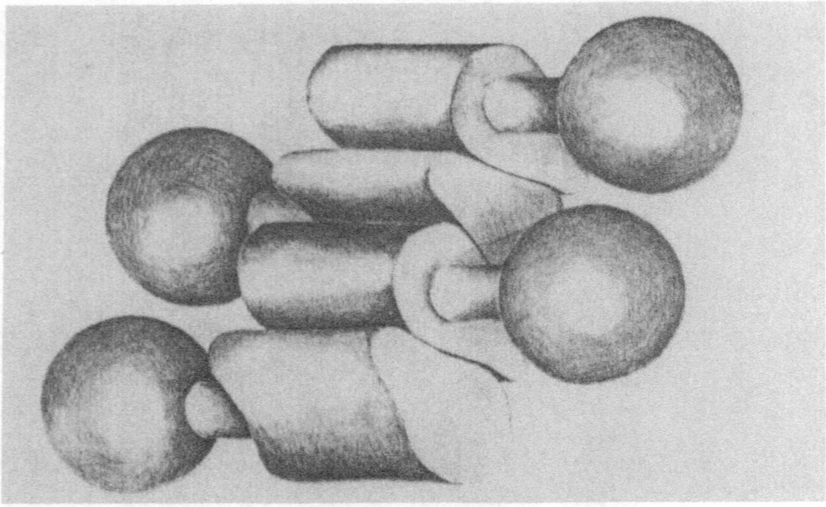

Fig. 4 – Three–dimensional view of a string with attached $F_1$–part on both sides. The $F_0$–part was drawn pear–shaped perpendicular to the string axis.

stalk could not be measured with accuracy in the reconstitued system. The thickness of the strings was 8.3 nm, and this obviously constitutes the transmembrane length of $CF_0$, *i.e.*, the $CF_0$–part can easily span a bilayer membrane which has a typical thickness of about 5 nm.

The dimensions of the $F_0$–part parallel to the strings cannot be measured directly from the micrographs as these parts stick together forming smaller or longer strings. Additionally, detergent is present at least at both ends of the strings. In such a case, the length of one $F_0$–part can be determined as follows: the length of well preserved strings was measured and plotted *versus* the number of attached $F_1$–parts. Figure 3 shows a plot of such data. The slope of this curve gives the increase in length of the string per $CF_1$. A length of 6.2 nm resulted and this represents the maximal dimension of the $F_0$–part along the string. If there are additional molecules (*e.g.* detergent) present in the string, the diameter of $F_0$ should be smaller. If the curves are extrapolated, a value of 3.6 nm results. This is exactly the length expected when the detergent octylglucoside (1.8 nm) is present at both ends of the string. It is, however, not clear from these results whether the $F_0$–part has a circular or an ellipsoidal shape. In the latter case, it must be assumed that along the string axis the smaller dimension is found and the larger one perpendicular to the string axis. This would result in a bigger area of contact between hydrophobic $F_0$–parts with a corresponding lowering of the free energy of the string–water system. Figure 4 shows an artist's view of such a string with attached $F_1$–parts, where the $F_0$–parts are shown in an ellipsoidal shape. The aggregation of $CF_0F_1$ into strings can be regarded as a one–dimensional crystallization. From these *one–dimensional* crystals the diameters of the $F_0$–parts were obtained.

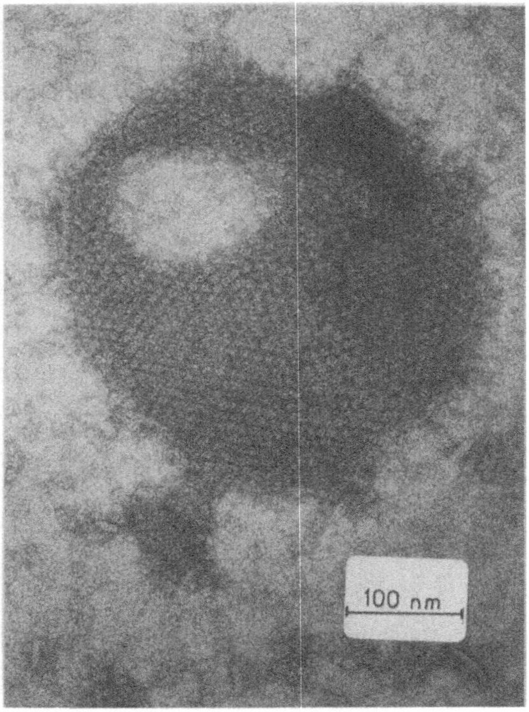


Fig. 5 – Electron micrograph of a two–dimensional ordered aggregation of isolated $CF_0F_1$ after negative staining (*two–dimensional crystal*).

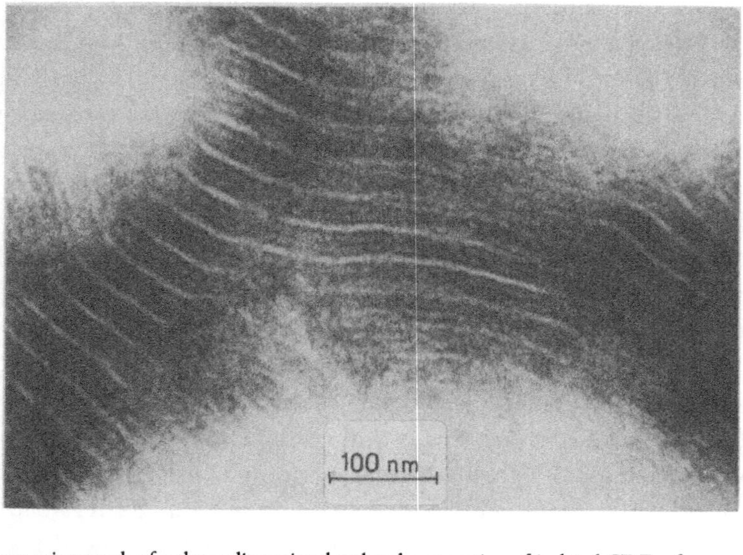


Fig. 6 – Electron micrograph of a three–dimensional ordered aggregation of isolated $CF_0F_1$ after negative staining.

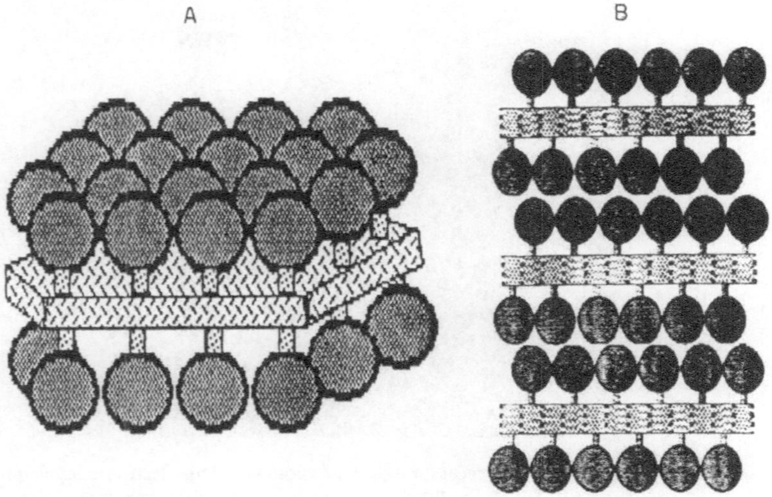


Fig. 7 – Schematic interpretation of the two–dimensional and three–dimensional aggregation of $CF_0F_1$. Definition of side view and top view.

Since from ordered structures more detailed information can be obtained, it has been investigated whether further aggregations are possible [22]. Figure 5 shows an electron micrograph of a two–dimensional ordered aggregation of $CF_0F_1$ molecules, *i.e.*, a *two–dimensional* crystal in a top view. However, this is not the last step: the two–dimensional crystalline sheets can also be aggregated to three–dimensional ordered structures. Figure 6 shows an electron micrograph of such a three–dimensional structure. In contrast to Fig. 5, in this case a side of this structure is shown. Figure 7 presents an interpretation of these structures. Figure 7A shows schematically a two–dimensional crystalline sheet with the $F_1$–parts on alternating positions on the sheet. This structure is seen in a top view in Fig. 5. Different sheets can be aggregated with the $F_1$–parts oriented head–on–head, as indicated in Fig. 7B. Such a structure is seen in a side view in Fig. 6.

All these electron microscopic investigations have been carried out with negatively stained samples. It has been argued that negative staining of the biological material might lead to artefacts. For example, the appearance of a stalk has been attributed to such staining artefacts. In order to investigate this possibility, cryoelectron microscopy has been applied [23]. In this case, the sample of biological material is frozen very rapidly so that the water does not crystallize in the ice structure, but the structure of the liquid water is frozen. This is called *frozen hydrated*. These samples are then investigated in the electron microscope at liquid nitrogen temperature. The sample is only about 30–400 μm thick so that the electron beam can traverse it. The contrast of the image is in this case merely a phase contrast of the electron waves which result from slight differences in the velocity of the electrons in water, protein and lipid. After taking the electron micrograph,

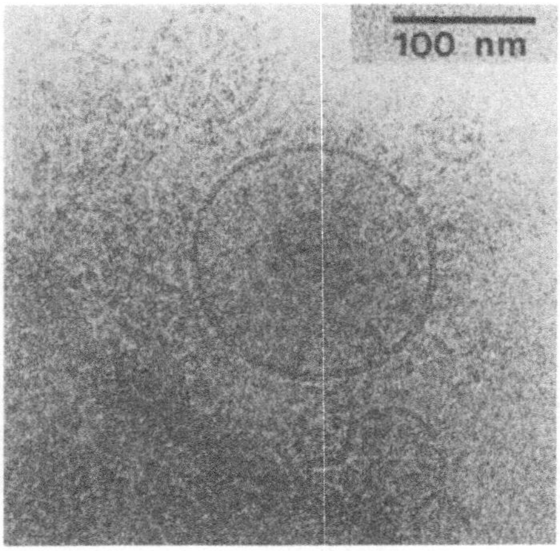


Fig. 8 – Electron micrograph of $CF_0F_1$ after reconstitution into liposomes. This electron micrograph was taken without any staining under frozen hydrated conditions (cryo electron microscopy). Details see text. Electron micrograph provided by Dr. U. LÜCKEN.

the sample can be thawed again and still show its functional activity. Also in such samples a stalk has been observed [23]. Also when $CF_0F_1$ is reconstituted into liposomes a stalk is observed in the frozen hidrated state.

Figure 8 shows an electron micrograph of $CF_0F_1$ reconstituted into liposomes in the frozen hydrated state. A stalk is clearly visible. Figure 9 shows a model of the shape of $F_0F_1$–ATP–synthases in a lipid bilayer. The different measured parameters are defined in the figure. In Table 3 the measured parameters for different $F_0F_1$–ATP–synthases are summarized. Structural investigation of bacterial $F_1$, isolated and reconstituted $F_0F_1$ by neutron small angle scattering, gave parameters which are in accordance with the electron microscopic data [24, 25].

Table 3. Comparison of the shapes of $F_0F_1$–ATP–synthases obtained by different authors. Parameters are defined in Fig. 9.

| length (nm) | $CF_0F_1$ [21] | $CF_0F_1$ [22] | $MF_0F_1$ [21] | $MF_0F_1$ [61] | $EF_0F_1$ [23] |
|---|---|---|---|---|---|
| a | 11.0 ± 1.1 | 9.2 ± 1.4 | 11.7 ± 0.9 | 11 ± 10 | 9.5 ± 0.5 |
| b | 8.3 ± 1.0 | 7.6 ± 1.4 | 9.0 ± 0.8 | 8.5 ± 1.0 | 9.0 ± 0.7 |
| c | 2.7 ± 0.6 | 2.8 ± 0.7 | 4.3 ± 1.0 | 3.0 ± 0.6 | 2.7 ± 0.3 |
| d | 3.7 ± 0.7 | 3.1 ± 0.6 | 4.3 ± 0.6 | 3.5 ± 0.5 | 4.0 |
| e | 8.3 ± 0.8 | 6.0 ± 0.7 | 8.9 ± 0.8 | 6.5 ± 0.8 | 5.8 ± 0.4 |
| f | 6.2 ± 0.4 | — | 6.4 ± 0.7 | — | — |

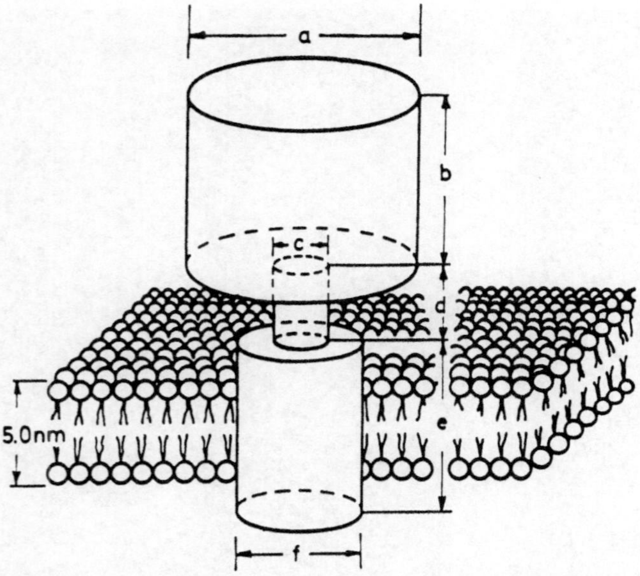


Fig. 9 – Schematic drawing of the shape of $F_0F_1$–ATP–synthases in a lipid bilayer. The parameters used in Table 3 are defined in this figure.

## 5. The structure of $CF_1$

The hydrophilic $F_1$–part can be separated from the membrane–integrated $F_0$–part by several treatments. Since no detergents are present in such $F_1$ preparations, the investigation of the $F_1$–parts is greatly facilitated. It has been even possible to obtain three–dimensional $F_1$–crystals which allow X–ray analysis [26, 27]. We have investigated the structure of isolated $CF_1$ by electron microscopy with negatively stained samples [28]. Figure 10 shows an electron micrograph of negatively stained $CF_1$. The $CF_1$ molecules show a clear subunit structure. Most of the molecules are oriented on the carbon support film in such a way that a hexagonal subunit structure can be seen. This view is called the *hexagonal* or *top* view (so also Refs. 29 and 30). Some examples of this orientation are surrounded by a hexagon. In some cases a different orientation is seen: the molecules show a small protrusion on one side. This view is called a *side* view, and some examples are surrounded by a square with a protrusion. The low signal–to–noise ratio does not allow interpretation of the finer details of the single molecules directly — or we have to select at random one of the molecules which seems to look *good*. This approach is, however, very arbitrary. Therefore, the signal–to–noise ratio has been enhanced by averaging the images of similarly oriented molecules. Figure 11 shows a scheme of this procedure. First, the electron micrograph (*e.g.* Fig. 10) is scanned with a microdensito-

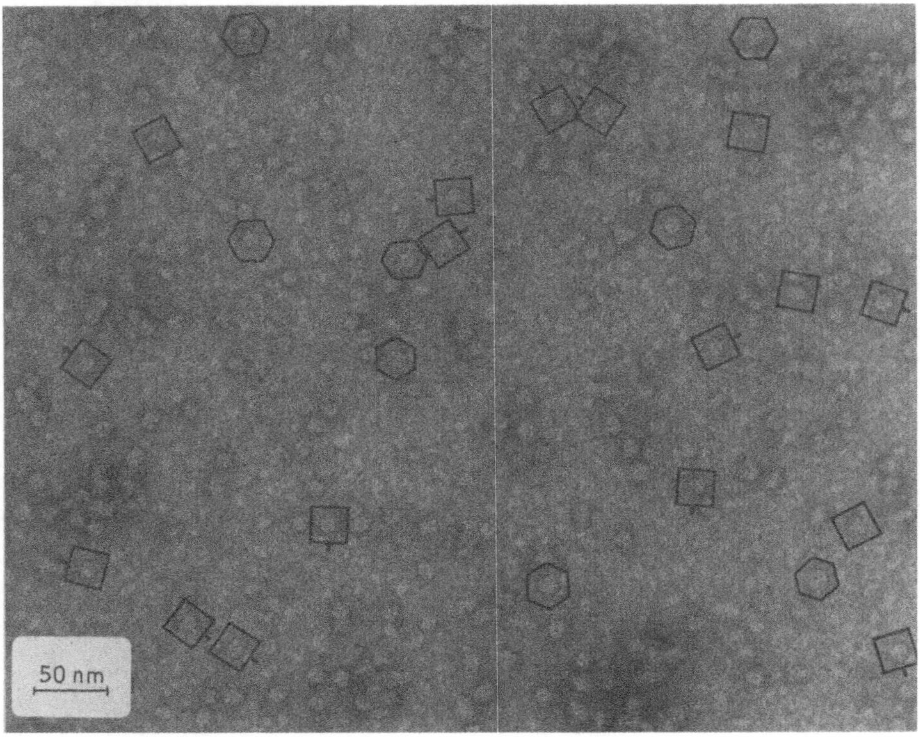


Fig. 10 – Electron micrograph of isolated $CF_1$. Typical top views (*hexagonal view*) are surrounded by hexagons, typical side views are surrounded by squares indicating also the stalk. The sample was negatively stained with 1 % uranyl acetate.

meter and the light/dark–pattern of the different molecules is measured. Each molecule is resolved in 64 x 64 elements (*pixel*). One pixel is a square with an area of 0.36 nm x 0.36 nm. These digitized images are stored in a computer and one of the images is chosen as a first–reference molecule. The variance of the light/dark pattern of each image is normalized. Then these images are oriented with regard to the reference molecule; *i.e.*, they are rotated over an angle and also shifted translationally until there is an optimum fit of the light/dark–pattern with respect to the reference. Finally, the oriented images are added to give a summation image which now has an increased signal–to–noise ratio. This summation image is then taken as a new reference molecule and the whole sequence is started again until there is no further change. This procedure is carried out with different reference molecules. It must be mentioned that the result of the classification depends on the appropriate choice of the reference.

Figure 12 shows the result of such a procedure. First, the digitized image of one $CF_1$ molecule is shown. Then 10 images are added, then 30 images and, lastly, 160 images

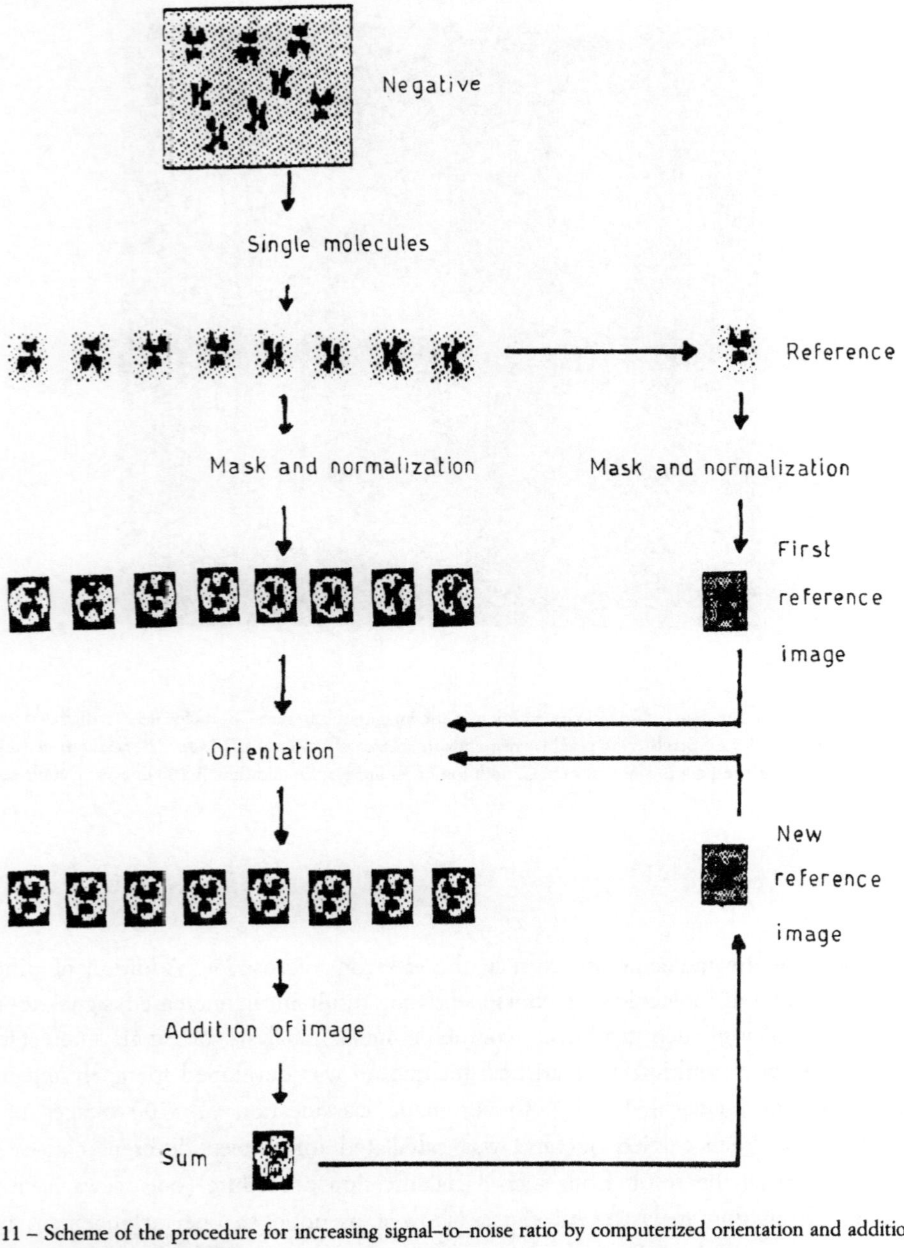


Fig. 11 – Scheme of the procedure for increasing signal–to–noise ratio by computerized orientation and addition of images from different molecules.

of different molecules are added. A significant increase of the signal–to–noise ratio is obtained which permits distinction of the different subunits in the molecule. However, not all molecules are oriented in the same way on the carbon film so that different

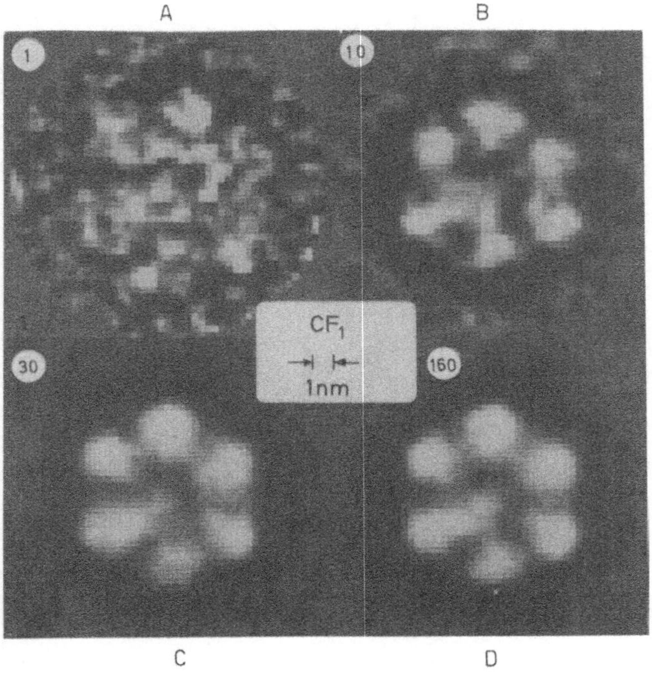


Fig. 12 – Increase of signal–to–noise ratio by addition of the images of different $CF_1$ molecules. A: digitized image of one $CF_1$ molecule 64 x 64 pixels. One pixel corresponds to an area of 0.36 nm x 0.36 nm. B: addition of 10 images after orientation with respect to the first one. C: addition of 30 images; D: addition of 160 images. Details see text.

projections of the molecule are seen in the electron microscope. Addition of different projections of the molecules, of course, did not result in an increased signal–to–noise ratio. For solving such problems computer image analysis of single molecules in combination with multivariate statistical techniques was developed for analyzing mixed populations of images [31, 33]. An automatic classification of 3300 images of $CF_1$ molecules has been carried out and was calculated for sixteen different classes [28]. Figure 13 shows the result from such a classification procedure (side views have been excluded from this analysis). All classes show a pseudohexagonal arrangement of six stain–excluding masses. These are the α– and β–subunits. With these data it is not possible to distinguish between α– and β–subunits, since they have practically the same molecular mass. In order to elucidate their arrangement $CF_1$ has been labeled with a monoclonal antibody against subunit α and the antibody–tagged $CF_1$ was investigated by electron microscopy [34, 36]. As a maximum, three antibodies are bound per $CF_1$ in

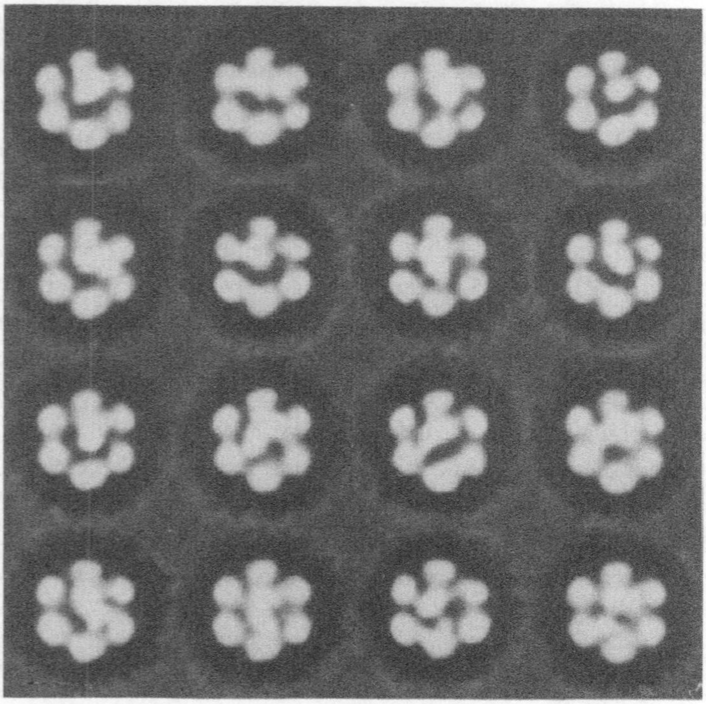

Fig. 13 – Molecular projections of $CF_1$ as determined by multi–reference alignments and classification of 3300 molecules into 16 classes. Number of molecules per class is about 150.

accordance with stoichiometry. The angle between the antibodies is 120 degrees; *i.e.* the α– and β–subunits are arranged in alternating positions.

The classes shown in Fig. 13 differ mainly in the configuration of the central mass which represents the smaller subunits γ, δ and ε. These differences are due to different stain accumulation in the central cavity. They probably reflect differences in handedness, *i.e.* projections of the molecule oriented *upside–up* and *upside–down* with regard to the supporting carbon film. Eventually, these different projections might reflect different conformational states of the enzyme; however, there is yet no convincing evidence for this hypothesis. In some of the classes in Fig. 13 the central mass is split into two parts. Such a splitting has never been observed in mitochondrial $F_1$ lacking the subunit OSCP [37]. Since OSCP is homologous to subunit δ from $CF_1$, one can tentatively assign the smaller mass to subunit δ and the larger one to subunits γ and ε (28). This hypothetic assignment is shown in the scheme of Fig. 14A. The same assignment has been found in recent work with $EF_1$ [36].

The three–dimensional arrangement of subunits in $F_1$ is not yet clarified. From X–ray analysis of $MF_1$ at 0.9 nm resolution it resulted that $F_1$ has a dimension of 12 x 11 x 8 nm and consists of six approximately spherical masses with the dimensions 6 x 5 x 4 nm [26, 27]. Electron microscopy of single crystals resulted in a pseudohexagonal structure of subunits with the dimensions 11.5 x 11.5 x 7 nm [38]. Electron microscopy of negatively stained single $CF_1$ molecules revealed six spherical masses arranged in a flattened trigonal antiprism and a central seventh mass. The dimensions were 11.5 x 11.5 x 6 nm [34]. Similar results were reported for $MF_1$. Also in this case a trigonal antiprism was observed with the dimensions of 10.5 x 10.5 x 6 nm [39]. Three–dimensional image reconstruction from negatively stained two–dimensional $EF_1$ crystals and investigation of frozen hydrated $EF_1$ by cryoelectron microscopy revealed also a hexagonal arrangement of the large subunits: however, the subunits are elongated rather than spherical with a diameter of 3.5 nm and a length of 9.0 nm [29, 40]. Depending on the shape of the α– and β–subunits the structural arrangements in $F_1$ is described either by an *apple* — or a *banana* — model. These results are summarized in Table 4. The top view of these models is similar (Fig. 14A); however, the side views are different (Fig. 14, B and C.).

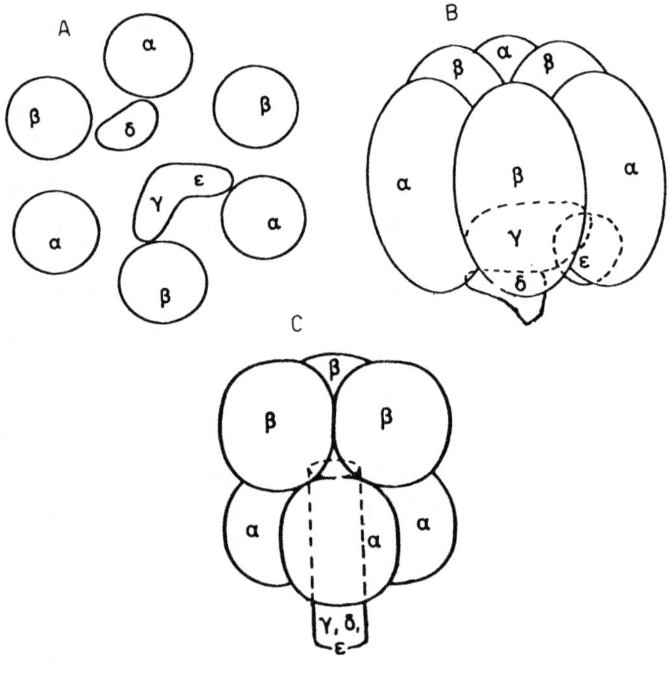


Fig. 14 – Scheme of subunit arrangement in $CF_1$ in the top view (A) and in a side view as *apple*–model (B) or *banana*–model (C).

Table 4. Comparison of the shape of the $F_1$–part and the α– and β–subunits obtained by different authors. Parameters are defined in Fig. 9. Length in nm.

| | $MF_1$ [27] | $CF_1$ [34] | $CF_1$ [38] | $MF_1$ [39] | $EF_1$ [29] |
|---|---|---|---|---|---|
| a x a | 12 x 11 | 11.5 x 11.5 | 11.5 x 11.5 | 10.5 x 10.5 | 9.5 x 9.5 |
| b | 8 | 6 | 7 | 6 | 9.0 |
| α, β | approx. spherical | spherical | spherical | spherical | ellipsoidal |
| | 6 x 5 x 4 | 6.5 | | 6.0 | 3.5 x 3.5 x 9.0 |

## 6. The structure of $CF_0$

The hydrophobic $CF_0$–part can be isolated from the reconstituted $CF_0F_1$ [41]. It contains the subunits I, II, III and IV [41, 42]. Electron microscopy of isolated $CF_0$ shows string–like structures. Presumably, these strings represent aggregates of $CF_0$ with their hydrophobic sides. Individual molecules cannot be distinguished; however, the thickness of the strings can be measured. It results 7.5 nm and this, presumably, represents the membrane spanning length of $CF_0$ [43].

As described in Section **3**, the subunits III form an oligomeric complex that can be isolated by SDS–gel–electrophoresis with subsequent electroelution of the gel. The subunit III–complex has an apparent molecular mass of 100 kDa. This implies that it is formed from 12 subunits III since one subunit III has a molecular mass of 8 kDa. Figure 15 shows the results of electron microscopic investigations of this complex [13]. The complex forms roll–like structures with up to 35 units of the 100 kDa complex arranged along the axis of one stack. The complexes are lying on their side. The diameter of the stack varies, indicating that one (Fig. 15, left, *monomers*), two (Fig. 15, center, *dimers*) or three (Fig. 15, right, *trimers*) complexes can form the repeating unit of a stack. At the bottom of Fig. 15 a schematic interpretation of the structure is depicted, and the different parameters are defined which can be measured: $a_1$, $a_2$ and $a_3$ are the diameters of the stack units, b is the repeating distance along the string, c is the diameter of one 100 kDa subunit III–complex, and d is the length of the detergent, SDS.

Figure 16 shows a plot of the diameter of the stack units *versus* the number of units. The number of units can be distinguished by a small space between two complexes (see arrows in Fig. 15). From the slope of the curve the diameter c of one complex can be obtained; c = 6.2 nm resulted. Extrapolation to zero gives 4.4 nm in agreement with the

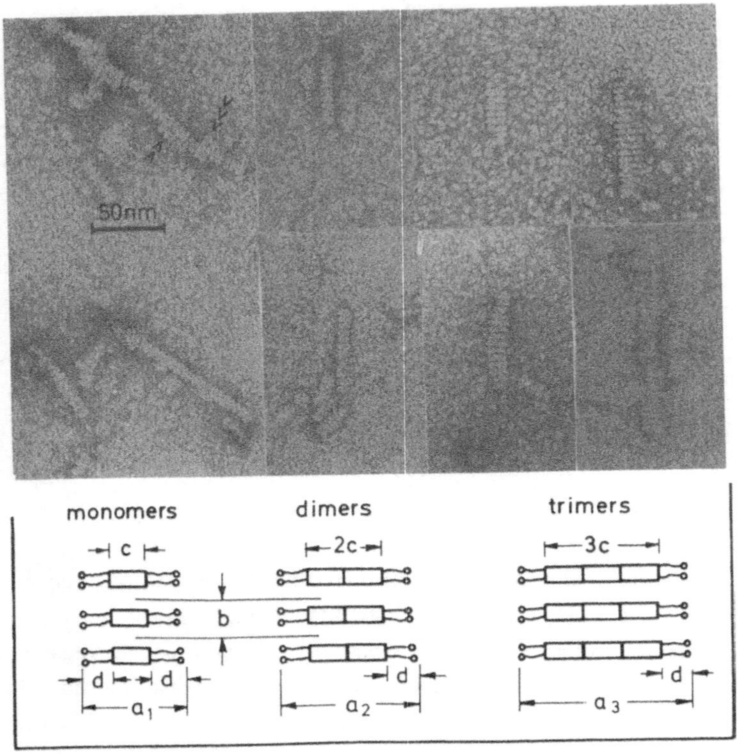


Fig. 15 – Electron micrograph of the subunit III–complex, negatively stained with uranyl acetate. Left: string composed of single complexes; center: string composed of two complexes; right: string composed of three complexes. At the bottom a schematic interpretation of the structure is shown.

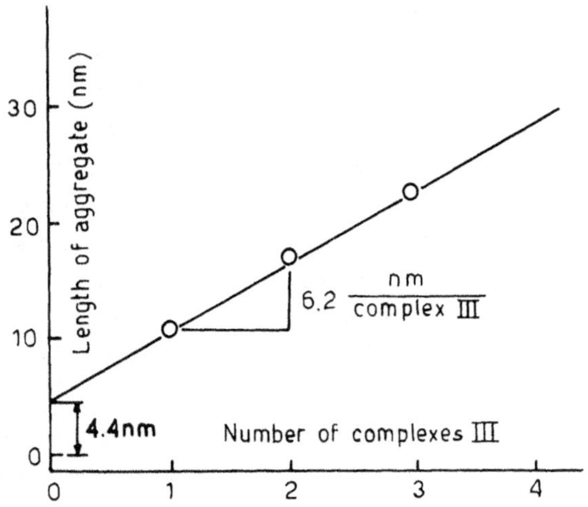


Fig. 16 – Diameter of subunit III–aggregates (strings) as a function of the number of complexes in the aggregate.

Figure 18 shows a hydropathy plot for the different proteins using the KITE–DOOLITTLE procedure. In this case each amino acid is given a number which reflects its hydrophilicity *e.g.* arginine –4.5 (strongly hydrophilic), isoleucine +5 (strongly hydrophobic). The basis for these estimated numbers is the free energy of transfer for the side presence of SDS (d = 2.2 nm) on both sides of the complex. Along the stack axis the hydrophilic parts of the complexes are aggregated. Therefore, the parameter c= 6.1 nm represents the membrane spanning length of the complex.

Figure 17 shows the subunit III–complex after increasing the signal–to–noise ratio by image processing of 96 images from digitized micrographs. The aggregation of three subunit III–complexes can be clearly seen; however, within one complex no subunit structure is detected. Presumably, this is due to SDS molecules which cover the hydrophobic parts of the protein thereby shielding the subunit structure. The data described here are presently the only structural information available about $CF_0$. However, biochemical data can be used for construction of a more refined model. The minimal stoichiometry for $CF_0$ is I II $III_{12}$ IV. The amino acid sequence of the four subunits is known [14, 15, 44-46]. From the primary structure predictions about the secondary structure can be made. Several algorithms have been developed for such structural prediction [47-50].

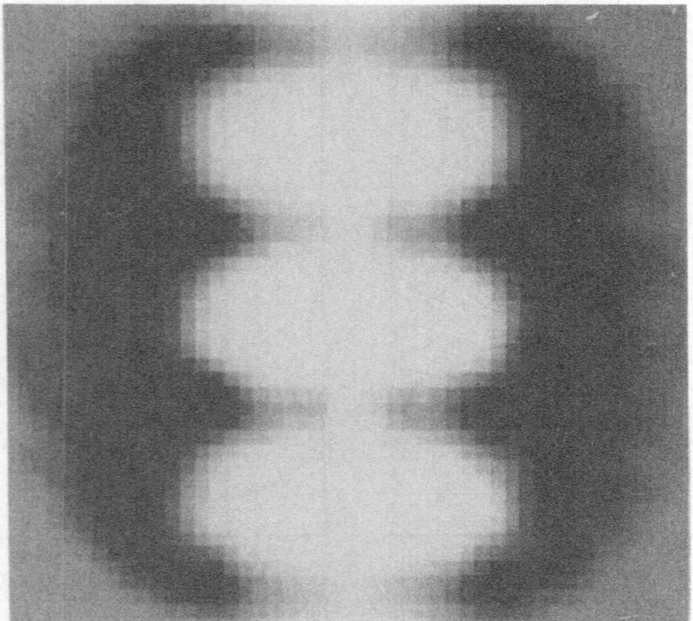

Fig. 17 – Average image of 96 oriented images of subunit III–aggregates with one complex in the repeating unit of the string.

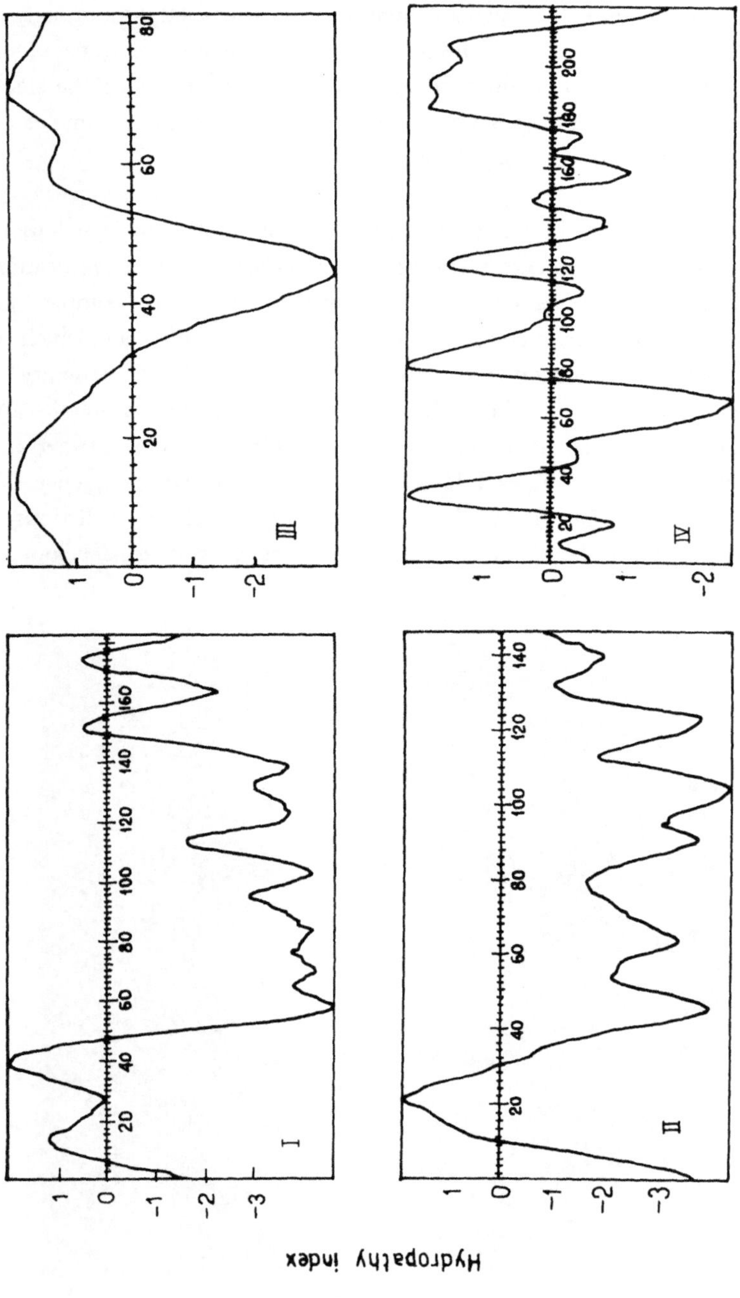


Fig. 18 – Hydropathy plots of the subunits from $CF_0$ using KYTE–DOOLITTLE procedure.

chains of the amino acids between water into condensed vapor [47]. Then an appropriate span of amino acids is chosen (usually between 7 and 11), the average hydropathy index of the span is calculated and the span is moved through the sequence. In Fig. 18 these average numbers are plotted as a function of the number of the amino acids in the corresponding sequence starting with N–terminal end (span was seven amino acids).

In subunit I a hydrophobic sequence of amino acids is seen between amino acids 22 and 48. If such a hydrophobic sequence is longer than approximately 23 amino acids, it could span a bilayer membrane in form of an α–helix. The α–helix is the most probable structure in a hydrophobic environment like a membrane since intramolecular hydrogen bonds strongly favor this secondary structure. On this basis structural predictions can be made from hydropathy patterns. (It should be noted that much more detailed information can be obtained from these structural predictions. For the following conclusions these simplified considerations are sufficient). It can be seen from the hydropathy plots in Fig. 18 that subunits I and II have one transmembrane helix each, subunit III has two and subunit IV has five helices. Since twelve subunits III are found, we have 24 helices from subunit III and at least seven helices from subunits I, II and IV. Therefore, within $CF_0$ — which has a diameter of 6.2 nm — a total of 31 helices has to be arranged. Each helix has a diameter between 0.7 and 0.9 nm. $CF_0$ has practically the same diameter as the isolated subunit III–complex (compare Figs. 3 and 16). Therefore, the following plausible hypothetical arrangements can be constructed which are shown in Fig. 19 in a top view: In model (a) the subunits III are arranged in identical positions in a ring–like structure, the seven helices from the other subunits are located in the center. In model (b) the subunits III are arranged in non–identical positions (dashed and dotted circles). In this case the helices from the other subunits are located mainly outside the subunit III complex. In model (c) it is assumed that $CF_0$ has not a circular shape in the top view. In this case the helices from the other subunits are arranged at one side of the subunit III complex. Model (c) has been proposed from the labeling pattern of the $F_0$ subunits from *E. coli* with 3–(trifluoromethyl)–3–(*m*($^{125}$J)–iodophenyldiazirine [51, 52) and is also preferred by other authors [11]. This model would also explain that the subunit III–complex can be very simply dissociated from the subunits I, II and IV; *i.e.*, that their interaction with subunit III is much weaker than the interaction between subunits III. All three models are consistent with the electron microscopic data. In case of model (c) the string–like structure of $CF_0$ seen in electron micrographs must be interpreted as shown in Fig. 4. The transmembrane arrangement of the different subunits is shown in a scheme in Fig. 20. The orientation of the N–terminal ends of the subunits is drawn according to the results obtained for the $F_0$–part from *E. coli* (for review see Refs. 11 and 53).

The subunit III (and the homologous subunits, c, from other $F_0$–parts) has a hair pin–like structure [54] and only a small part of the protein is hydrophilic and thus located

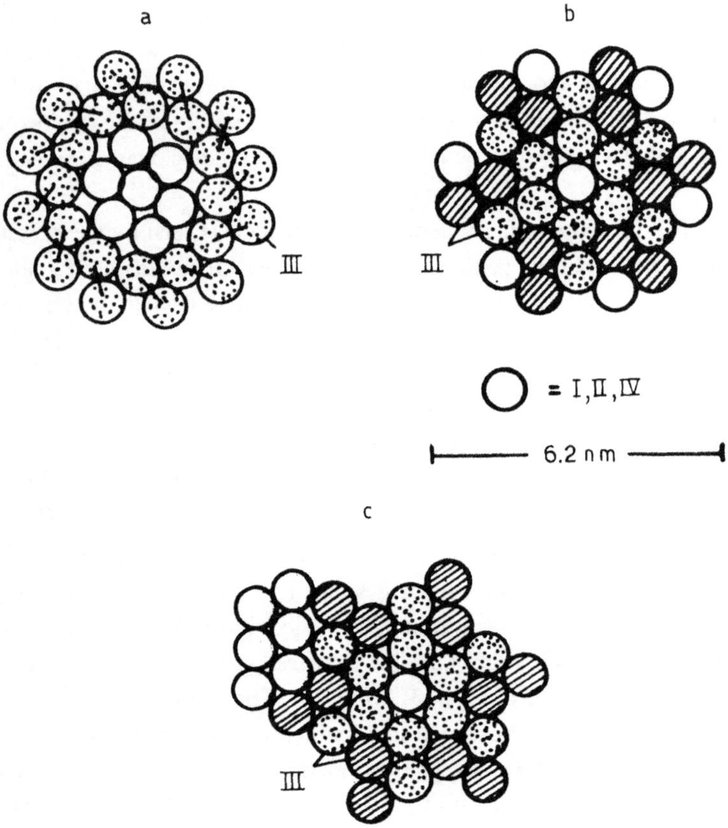


Fig. 19 – Top view of three different models for subunit arrangement in $CF_0$. The diameter of $CF_0$ is 6.2 nm, the diameter of one helix is 0.9 nm.

outside the membrane. This is in accordance with the electron microscopic results: the membrane-spanning length of the subunit III–complex (see Fig. 15) is 6.1 nm and the thickness of a bilayer membrane is about 5 nm. Also, the subunit IV (and the homologous subunits, a, from other $F_0$–parts) has only a small hydrophilic part, *i.e*, only a very small part of the protein is located outside the membrane. The situation is different for subunits I and II: they have only one transmembrane helix and a large hydrophilic part.

The volume of the hydrophilic part can be estimated as follows: subunit I has a molecular mass of 15.5 kDa, subunit II 13.5 kDa. The transmembrane span from the N–terminus to amino acid 30 has a molecular mass of about 3 kDa each. Therefore, the hydrophilic protein mass of subunit I and II together is about 23 kDa. From the molecular mass the molecular volume can be calculated, *i.e.*,

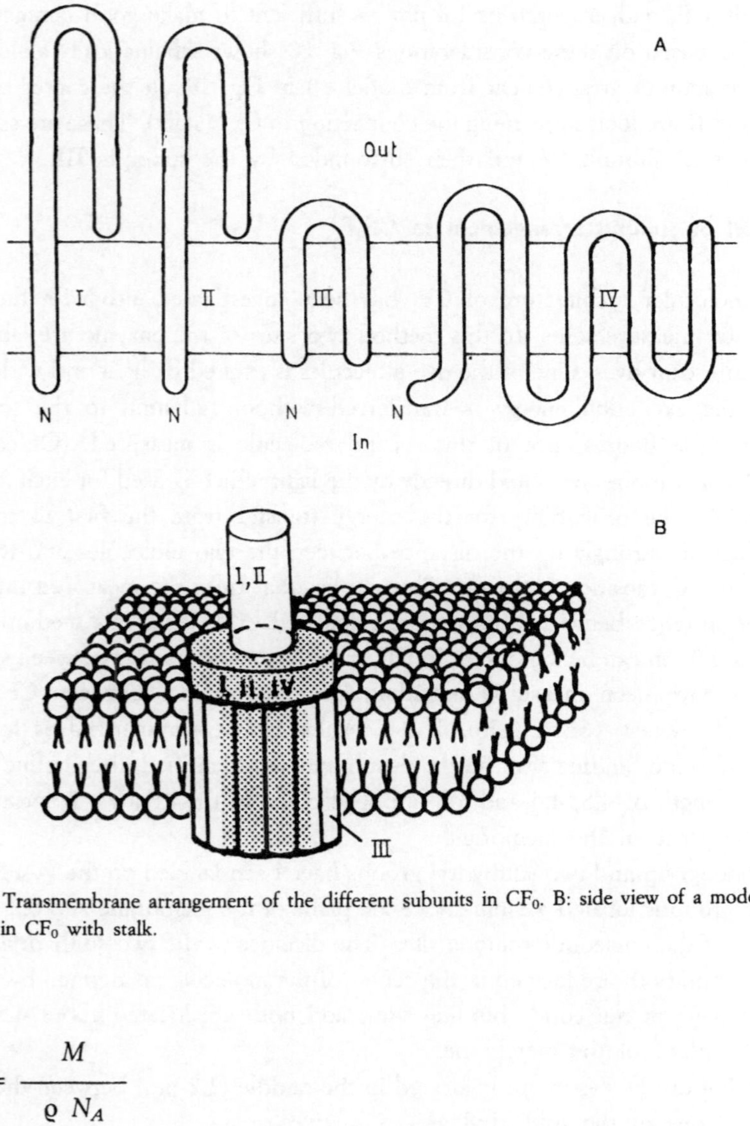


Fig. 20 – A: Transmembrane arrangement of the different subunits in $CF_0$. B: side view of a model for subunit arrangement in $CF_0$ with stalk.

$$\nu_M = \frac{M}{\varrho \, N_A}$$

where $\nu_M$ molecular volume; $M$ mole mass; $\varrho$ density of the protein and $N_A$ AVOGADRO's number. Assuming a protein density of 1.3 g $cm^{-3}$ a molecular volume of 30 $nm^3$ is calculated. The thickness of the stalk is 2.8 nm (see Tab. 3). If we assume that the hydrophilic parts of subunits I and II form a cylinder with a diameter of 2.8 nm, a length of 4.9 nm is calculated for this cylinder. The measured length of the stalk is 3.1 nm (see Tab. 3). Therefore, subunits I and II alone can form the stalk and, additionally, can be buried within $CF_1$ with a length of 1.8 nm — sufficient to make good contact with the $CF_1$ subunits. Based on these considerations Fig. 20 shows schematically a side view of $CF_0$ with the subunit arrangement from model (a) in Fig. 19. In the center of $CF_0$ the subunits I and II are located forming the connection to $CF_1$ (stalk). These are surrounded by the helices of subunit IV and then surrounded by the subunits III.

## 7. Model of subunit arrangement in $CF_0F_1$

The molecular architecture of $CF_1$ has been investigated also with fluorescence energy transfer measurements. In this method two sites of the enzyme are labeled with different fluorescent dyes. One of the dye molecules is excited by light and with a certain probability the excitation energy is transferred without radiation to the second dye molecule and the fluorescence of this second molecule is measured. (Of course, the second molecule cannot be excited directly by the light which is used for excitation of the first molecule). The probability for the energy transfer from the first to the second molecule depends strongly on the distance between the two molecules and the relative orientation of the transition moments. This implies that from the measured fluorescence intensity the distance between both can be calculated [55]. With this method distances between 1 and 9 nm can be measured. In $CF_1$ more than 30 distances between specifically labeled sites have been measured including also the distance between $CF_1$ and the membrane (for review see Ref. 56). These results can be summarized as follows:

1. Three nucleotide binding sites on the β–subunits are identified, they define a triangle with side length of 4.8, 4.4 and 3.6 nm. The plane of this triangle is located 10 nm above the plane of the membrane.
2. A disulfide group and two sulfhydryl groups have been labeled on the γ–subunit. The disulfide group is located 3.1 nm above the plane of the membrane and 6.2 nm below the plane of the nucleotide binding sites. The distance to the two sulfhydryl groups is about 4.5 nm, both are located in the center of the molecule (as defined by the center of the triangle of nucleotide binding sites) and both are located about 4.5 to 5 nm above the plane of the membrane.
3. A sulfhydryl on the ε–subunit is located in the middle (2.2 nm) between the disulfide group and one of the sulfhydryl groups on γ.
4. An amino group on one of the α–subunits has been labeled. This is located about 6.4 nm above the plane of the membrane and about 2.6 nm below the plane of the nucleotide binding sites.

From these measurements finally a mapping of the specific sites on the enzyme has been obtained [56]. Using the electron microscopic results and those from the fluorescence energy transfer measurements it can be tried to construct a model of subunit interaction. It is clear that one must take into account that different parameters are

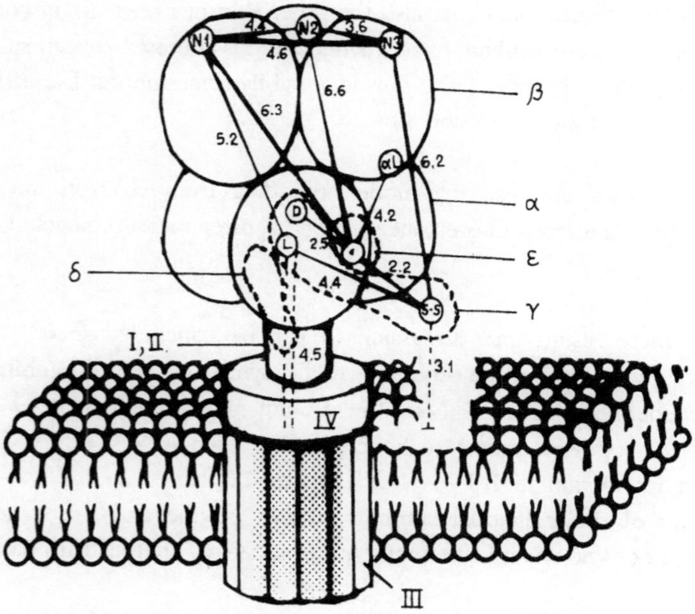


Fig. 21 – Model for subunit arrangement in $CF_0F_1$ based on electron microscopic investigations and fluorescence energy transfer measurements. The hypothetical location of the sites labeled according to fluorescence transfer measurements is indicated. $N_1$, $N_2$, $N_3$ nucleotide binding sites on β; S–S disulfide on γ; D and L sulfhydryl groups on γ; ε sulfhydryl group on α; χ L amino groups on α.

measured with both approaches: electron microscopy gives the mass distribution in the enzyme, fluorescence energy transfer measures distance between different sites on the enzyme. The model is based on the following assumptions and considerations:

1. Shape of $CF_0$ and subunit interactions are described already in Fig. 20, *i.e.*, the stalk is mainly formed from subunits I and II.
2. In the top view, $CF_1$ has a subunit arrangement as depicted in Fig. 14; *i.e.*, the α– and β–subunits have pseudohexagonal arrangement, and in the central cavity the α– and ε–subunits are located together on one side, the δ–subunit on the other side.
3. The α– and β–subunits have a spherical shape.
4. From the fluorescence energy transfer measurements it results that the β–subunits are located about 9 nm above the membrane, the α–subunits about 6.4 nm above the membrane. The α– and ε–subunits are located about 4 nm above the membrane between the center and one side of the $CF_1$–part.
5. If the δ–subunit is added to membrane–bound $CF_0$, the conductivity of the membrane is drastically reduced [57–59]. This implies that there is an interaction between δ–sub–unit and subunits I and II. On the other hand, $CF_1$ can be rebound to membrane–bound $CF_0$; however, in this case the function of the enzyme is not

restored [57–59]. This implies that also the other subunits must be in contact with I and II. According to crosslinking experiments, there is contact between subunits I and II and the subunits $\alpha$, $\beta$ and $\gamma$ [60]. This is possibile since subunit I and II are buried about 1.8 nm in $CF_1$ (see Section **6**).

Figure 21 shows the resulting model: the data from electron microscopy are combined with the data from fluorescence energy transfer measurements. The model is drawn to scale.

From all these results and discussions it can be concluded that:

1. The shape of $CF_0F_1$ (and of the other $F_0F_1$–ATP–synthases) is well established now by different independent methods. Also, the subunit arrangement in the $F_1$–part — at least — in the top view seems to be clarified.
2. The subunit interaction in $F_0F_1$ is still a matter of discussion.
3. The structure of the individual subunits of the enzyme and the structure of the holoenzyme is not known and certainly requires an X–ray analysis with high resolution.

**Acknowledgements**

We would like to thank Prof. B. Rumberg for stimulating this work and Prof. E. Zeitler for continuous support. We thank Drs. P.Fromme, M. van Heml and U. Lücken for helpful discussions and M. Glaubitz for her drawings. Our work was supported by the Deutsche Forschungsgemeinschaft (Sfb 312) and the Fonds der Chemischen Industrie.

**References**

[1] M. Nelson, *Curr. Top. Bioenerg.,* **11,** 1 (1981).
[2] L.M. Amzel and P.L. Pedersen, *Annu. Rev. Biochem.,* **52,** 801 (1983).
[3] A.E. Senior and J.G. Wise, *J. Membr. Biol.,* **73,** 105 (1983).
[4] M. Futai and M. Kanazawa, *Microbiol. Rev.,* **47,** 285 (1983).
[5] H. Strotmann and S. Bickel–Sandkötter, *Annu. Rev. Plant Physiol.,* **35,** 97 (1084).
[6] Y. Kagawa, in *Bioenergetics,* L. Ernster (Editor), Elsevier Biomedical Press, Amsterdam (1986), p. 149.
[7] R.H. Filingame, *Curr. Top. Bioenerg.,* **11,** 35 (1984).
[8] J.E. Walker and M. Saraste and M.J. Gay, *Biochim. Biophys. Acta,* **768,** 164 (1984).
[9] P.V. Vignais and M. Sartre, *Mol. Cell Biol.,* **60,** 33 (1984).
[10] Y. Hatefi, *Annu. Rev. Biochem.,* **54,** 1015 (1985).
[11] E. Schneider and K.H. Altendorf, *Microbiol. Rev.,* **51,** 477 (1987).

[12] J.E. Walker, M.J. Runswick and M. Saraste, *FEBS Lett.*, **146,** 393 (1982).
[13] P. Fromme, E. Boekema and P. Gräber, *Z. Naturforsch.*, **42 C,** 1239 (1987).
[14] J. Hennig and R.G. Herrmann, *Mol. Gen. Genet.*, **203,** 117 (1986).
[15] A.L. Cozens, J.E. Walker, A.L. Phillips, A.K. Muttly and J.C. Gray, *EMBO J.*, **5,** 217 (1986).
[16] R.E. McCarty and J.V. Moroney, in *The Enzymes of Biological Membranes,* A.M. Martonosi (Editor), Plenum Press, New York (1985), Vol. **4,** pp. 383
[17] N. Nelson in *Electron Transport and Phosphorylation* J. Barber (Editor), Elsevier Biomedical Press, Amesterdam (1982), p. 81
[18] J.W. Soper, G.L. Decker and P.L. Pedersen, *J. Biol. Chem.*, **254,** 11170 (1979).
[19] J.W. Soper, G.L. Decker and C.J. Arntzen, *Biochim. Biophys. Acta,* **642,** 149 (1981).
[20] E. Mörschel and L.A. Staehelin, *J. Cell Biol.*, **97,** 301 (1983).
[21] E.J. Boekema, G. Schmidt, P. Gräber and J.A. Berden, *Z. Naturforsch.*, **43 C,** 219 (1988).
[22] B. Böttcher, *Diplomarbeit,* Technische Universität Berlin (1988).
[23] E.P. Gogol, U. Lücken, R.A. Capaldi, *FEBS Lett.*, **219,** 274 (1987).
[24] T. Nawroth, *Thesis,* Universität Mainz, 1986.
[25] T. Nawroth, K. Dose and H. Conrad, *Physica B,* **156 + 157,** 489 (1989).
[26] L.M. Amzel and P.L. Pedersen, *J. Biol. Chem.*, **253,** 2067 (1978).
[27] L.M. Amzel, M. McKinney, P. Narayanan and P.L. Pedersen, *Proc. Natl. Acad. Sci. USA,* **79,** 5582 (1982).
[28] E.J. Boekema, M. van Heel and P. Gräber, *Biochim. Biophys. Acta,* **933,** 365 (1988).
[29] E.P. Gogol, U. Lücken, T. Bork and R.A. Capaldi, *Biochemistry,* (1989) in press.
[30] U. Lücken, E.P. Gogol and R.A. Capaldi, *Biochemistry,* **28** (1989), submitted for publication.
[31] M. van Heel and J. Frank, *Ultramicroscopy,* **6,** 187 (1981).
[32] M. van Heel, *Ultramicroscopy,* **13,** 165 (1984).
[33] G. Harauz, E.J. Boekema and M. Van Heel, in Methods in Enzymology, Academic Press, New York (1988), pp. 164, 35.
[34] H. Tiedge, G. Schäfer and F. Mager, *Eur. J. Biochem.*, **132,** 37 (1983).
[35] H. Tiedge, H. Lünsdorf, G. Schäfer and M.V. Schairer, *Proc. Natl. Acad. Sci. USA,* **82,** 7874 (1985)
[36] E.P. Gogol, R. Aggelar, H. Sagermann and R.A. Capaldi, *Biochemistry,* **28,** 1034 (1989).
[37] E.J. Boekema, J.A. Berden and M. van Heel, *Biochim. Biophys. Acta,* **851,** 353 (1986).
[38] C.W. Akey, V. Spitzberg and S.J. Edelstein, *J. Biol. Chem.*, **258,** 3222 (1983).

[39] V.L. Tsuprun, J.V. Mesyanshinova, J.A. Kozlov and E.V. Orlova, *FEBS Lett.*, **167,** 285 (1984).

[40] U. Lücken, T. Bork and S. Remington, *Biochim. Biophys. Acta*, (1989), submitted for publication.

[41] J. Grotjohann and P. Gräber, *FEBS Lett.*, (1989) in press.

[42] P. Fromme, P. Gräber and J. Salnikow, *FEBS Lett.*, **218,** 27 (1987).

[43] J. Grotjohann, *Diplomarbeit,* Technische Universität Berlin (1988).

[44] R.G. Herrmann, *Eur. J. Biochem.*, (1989) submitted.

[45] W. Sebald and E. Wächter, *FEBS Lett.*, **122,** 307 (1980).

[46] J. Alt, P. winter, W. Sebald, J.G. Moser, R. Scheckel, P. Westhoff and R.G. Herrmann, *Curr. Genet.*, **7,** 129 (1983).

[47] J. Kyte and R.F. Doolittle, *J. Mol. Biol.*, **157,** 105 (1982).

[48] P.Y. Chou and G.D. Fasman, *Adv. Enzymol.*, **47,** 45 (1978).

[49] D.M. Engelman, T.A. Steiz and A. Goldman, *Annu. Rev. Biophys. Chem.*, **15,** 321 (1986).

[50] J.K. Rao and P. Argos, *Biochim. Biophys. Acta,* **869,** 197 (1986).

[51] H. Weber, W. Junge, J. Hoppe and W. Sebald, *FEBS Lett.*, **202,** 23 (1986).

[52] W. Sebald, H. Weber and J. Hoppe in *Bioenergetics,* T. Izawa and S. Papa, (Editors), Japan Sci. Soc. Press, Tokyo (1987), pp. 215.

[53] C.A. Kumamoto and R.D. Simoni, *J. Biol. Chem.*, **261,** 10037 (1986).

[54] W. Sebald and J. Hoppe, *Curr. Top. Bioenerg.*, **12,** 1 (1981).

[55] Th. Förster, *Fluoreszenz organischer Verbindungen,* Vandenhoek and Ruprecht, (1951).

[56] R.E. McCarty and G.G. Hammes, *TIBS,* **12,** 234 (1987).

[57] S. Engelbrecht, H. Lill and W Junge, *Eur. J. Biochem.*, **160,** 635 (1986).

[58] S. Engelbrecht and W. JUNGE, *FEBS Lett.*, **219,** 321 (1987).

[59] S. Engelbrecht and W. Junge, *Eur. J. Biochem.*, **172,** 213 (1988).

[60] K.H. Süss, *FEBS Lett.*, **201,** 63 (1986).

[61] L. Tsuprun, E.V. Orlova, J.V. Mesyanzhinova, *FEBS Lett.*, **244,** 279 (1989).

# KINETICS OF PROTON-TRANSPORT COUPLED ATP SYNTHESIS IN CHLOROPLASTS

PETER GRÄBER

*Biologisches Institut, Bioenergetik, Universität Stuttgart*
*Ulmerstrasse 227, 7000 Stuttgart 60 F.R.G.*

Contents



*Bioelectrochemistry III*
Edited by G. Milazzo and M. Blank
Plenum Press, New York, 1990

## Introduction

The coupling of (scalar) chemical reactions with (vectorial) transport of molecules or ions across a membrane occurs at practically all biological membranes, *e.g.*, at the plasma membranes of all cells as well as in cell organelles like mitochondria and chloroplasts. Of special importance is the coupling between proton transport and ATP-synthesis or ATP-hydrolysis. This process is catalyzed by a membrane-bound $F_oF_1$-ATPase (ATP-synthase). The importance of this process is demonstrated by the fact that in man about 50 kg ATP per day are synthesized by this process and that this ATP is then used as energy source for most biochemical reactions in the organism (muscle contraction, molecule- and ion transport, biosynthesis of different molecules).

Also, from a theoretical point of view this reaction is interesting: according to the CURIE-*symmetry principle* [1], in an isotropic space, *i.e.* in a homogeneous solution, the coupling between a scalar and a vectorial process is forbidden. Such a coupling always requires an anisotropic space like a membrane [2].

The structure and the function of $F_oF_1$-ATPases has been investigated during the last years in many laboratories. Although considerable progress has been made, the real *heart of the mechanism* – how the proton flux through the enzyme effects the formation of the phosphate-anhydrate bond between ADP and phosphate – is not yet known. In this chapter a kinetic approach to this problem is described; *i.e.*, the kinetics of this reaction is measured and from the results we can draw conclusions about the mechanism. The basis for our investigation is the chemiosmotic theory [3, 4]. For the photosynthetic membrane in chloroplasts we can formulate the essential of the chemiosmotic theory as follows:

1. Light excitation of the photosynthetic reaction-centers leads to a transmembrane redox reaction. Thereby, an electron is transferred through the membrane from the inside of the vesicle (thylakoids; donor) to the outside. This leads to the generation of a transmembrane electric potential difference, $\Delta\phi$.
2. Subsequent protolytic reactions of the electron donors and acceptors lead to a coupled vectorial transmembrane proton transport and the generation of a transmembrane pH difference, $\Delta$ pH.

chemical reaction

$$ADP + P_i \rightleftarrows ATP$$

transport reaction

$$H^+_{in} \rightleftarrows H^+_{out}$$

driving force

$$\Delta G_p = \Delta G_p^{o'} + RT \ln \frac{[ATP]}{[ADP]\,[P_i]} \quad (1)$$

driving force

$$\Delta\tilde{\mu}_{H^+} = -\,(2.3\ RT\ \Delta pH + F\ \Delta\phi) \quad (2)$$

coupling

coupled reaction

$$ADP + P_i + n\ H^+_{in} \rightleftarrows ATP + n\ H^+_{out}$$

driving force

$$\Delta G = \Delta G_p + n\Delta\tilde{\mu}_{H^+} \quad (3)$$

at equilibrium:

$$n = \frac{\Delta G_p}{|\Delta\tilde{\mu}_{H^+}|} \quad (4)$$

where $\Delta G_p$ is the GIBBS free energy of ATP synthesis, $\Delta G_p^{o'}$ is the standard GIBBS free energy which depends on pH, $Mg^{2+}$ concentration, ionic strength and temperature. At pH 8.2, $[Mg^{2+}] = 10^{-3}\ M$, ionic stregth $= 0.1\ M$ and $T = 25$ °C, it was found $\Delta G_p^{o'} = 33.3$ kJ/mol [10]. $n$ is the number of protons translocated per synthesized ATP, the other parameters are defined in Fig. 1.

## 2. Methods

Our experimental approach to the mechanism of the enzyme requires the measurement of the rate of ATP-synthesis and ATP-hydrolysis as a function of the concentrations of all substrates and products, *i.e.*, [ADP], $[P_i]$, [ATP], $[H^+]_{in}$ and of $\Delta\phi$. The main problem of such experiments is the determination of $[H^+]_{in}$ and $\Delta\phi$. The internal volume of the thylakoids is so small that the internal proton concentration can be measured only indirectly (for review see *e.g.* Refs. 11-13). The methods used are subject to different limitations and produced results that differ by a factor of about 10. Also, the transmembrane electric potential difference is difficult to measure. In this case, too, experimental results obtained with different methods differ widely (for review see *e.g.* Refs. 14-16).

We have, therefore, used the following method, introduced by JAGENDORF and URIBE in 1967 [17]. Thylakoids are incubated in the presence of a permeable buffer at low pH (*e.g.* at pH 5). During the incubation time (about 30 s) an equilibrium of proton-, ion- and buffer concentration is established between internal and external aqueous phases. Then, this acidic suspension is mixed with a strongly buffered basic medium. Proton and ion concentrations in the external aqueous phase are changed within the mixing time; whereas, the internal concentrations correspond to those in the acidic incubation medium. We have thus generated a transmembrane pH difference giving rise to a proton efflux.

The difference in the ion concentrations between the internal and external phases automatically leads to a transmembrane electric potential difference, $\Delta\phi$. In order to get well defined conditions, a $K^+$/valinomycin electric potential difference is generated as follows: thylakoids are incubated in the presence of valinomycin with a low KCl concentration. Valinomycin increases specifically the permeability of the thylakoid membranes for $K^+$. Then, this suspension is mixed with a medium of high KCl concentration. This leads to rapid $K^+$ influx, generating an electric potential difference (positive inside). Using different media, $\Delta$pH and $\Delta\phi$ can be varied, each one independently of the other. The transmembrane $\Delta$pH is then simply measured with a pH electrode. The acidic incubation medium gives the $pH_{in}$, the basic medium, after mixing with the acidic suspension, gives the $pH_{out}$. The transmembrane $\Delta\phi$ is calculated from the GOLDMANN equation [18]:

$$\Delta\phi = \frac{RT}{F} \ln \frac{P_{H^+}\,[H^+]_{out} + P_{K^+}\,[K^+]_{out} + P_{Na^+}\,[Na^+]_{out} + P_{Cl^-}\,[Cl^-]_{in}}{P_{H^+}\,[H^+]_{in} + P_{K^+}[K^+]_{in} + P_{Na^+}\,[Na^+]_{in} + P_{Cl^-}\,[Cl^-]_{out}} \tag{5}$$

where $P$ stands for the permeability coefficient of the ions indicated by the subscript. The other ions in the system are neglected, the permeability coefficients for the ions in equation (5) are known, while $P_{K^+}$ was determined in the presence of valinomycin. The calculation shows that the transmembrane electric potential difference is mainly determined by the internal and external $K^+$ concentration. The condition $\Delta\phi = 0$ is of special importance. This is achieved at $[K^+]_{in} = 0.1\ M$ in the presence of valinomycin. For details see Ref. 18.

From these measurements and considerations the initial $\Delta$pH and $\Delta\phi$ are known. However, this initial energization leads to the phosphorylating proton efflux and to fluxes of other ions so that the initial conditions remain constant only for a relatively short time.

Therefore, if ATP-synthesis/hydrolysis is to be investigated under well known conditions, the reaction time must be so short that within this time neither ΔpH nor Δφ are significantly changed. This can be achieved by using rapid double mixing techniques (*quenched flow*) as described in Fig. 2 [18]. This technique allows variable reaction times between 20 ms and 10 s. Figure 2, center, shows a scheme of the procedure: incubation

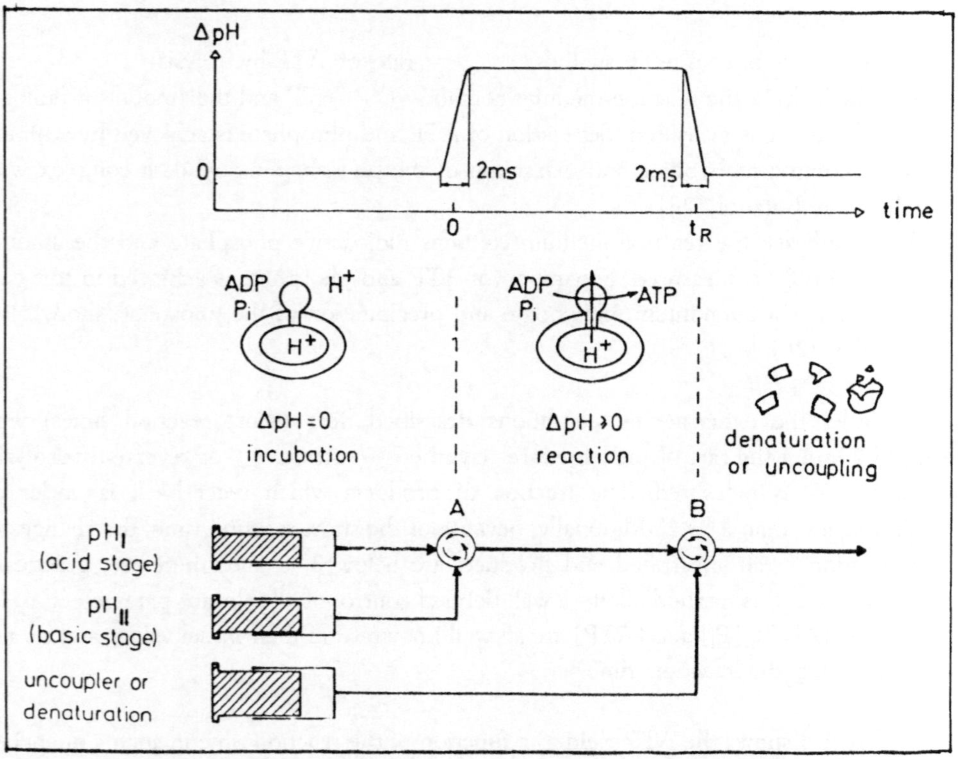


Fig. 2 - Scheme of membrane energization by an acid-base jump with rapid-mixing techniques. Details see text.

for 30 s in a medium with pH 5, mixing with a medium of pH 8.2 and thereby starting proton-transport-coupled ATP-synthesis. Denaturation with trichloroacetic acid, TCA, then stops the reaction. At the top, the transmembrane ΔpH is shown. It is generated within 2 ms during the mixing, then the reaction is allowed to proceed for the time $t_r$ and is finally stopped by denaturation of the thylakoids. At the bottom, a scheme of the apparatus is shown: in syringe I, the thylakoids are incubated in the acidic medium, syringe II contains the basic medium, syringe III the TCA. The syringes are driven forward simultaneously by a pneumatically actuated piston. In mixing jet A the ΔpH is generated and the reaction is initiated, in mixing jet B the reaction is stopped and then the reactions medium is analyzed for ATP.

The amount of ATP synthesized or hydrolyzed has been measured as follows.

1. The ATP-concentration in the reaction-medium was determined by luciferin/luciferase. This method gives the difference between forward and reverse reaction

$$\frac{d[ATP]}{dt} = v_{syn} - | v_{hyd} | \quad (6)$$

where $v_{syn}$ = rate of ATP-synthesis; $v_{hyd}$ = rate of ATP-hydrolysis

2. ATP-hydrolysis: the reaction-medium contains $\gamma(^{32}P)$-ATP and the amount of radioactive phosphate is measured. Separation of ATP and phosphate is achieved by addition of ammonium mobybdate and extraction of the phosphate-mobybdate complex with toluene isobutanol [28].
3. ATP-synthesis: the reaction-medium contains radioactive phosphate and the amount of $\gamma(^{32}P)$-ATP is measured. Separation of ATP and phosphate is achieved in this case by addition of ammonium mobybdate and precipitation of the phosphate-mobybdate complex [28].

Under the experimental conditions described here (short reaction times) with method 2 and 3 the rate of pure forward- (synthesis) reaction, $v_{syn}$ or reverse-(hydrolysis) reaction, $v_{hyd}$, is measured. The fraction of products which react back is under all conditions less than 2 %. Additionally, because of the short reaction time, the changes of concentration in all substrated and products are below 2 % and, therefore, practically constant. Thus, this method allows a well-defined control of all relevant parameters: $pH_{in}$, $pH_{out}$, $\Delta\phi$, [ADP], [$P_i$] and [ATP] are all well-known from their initial values and do not change during the reaction time.

Figure 3 shows the ATP yield as a function of the reaction time in such a quenched flow experiment. Up to about 150 ms a linear increase of the ATP yield with reaction time can be seen. Within this range $\Delta$pH and $\Delta\phi$ remain practically constant and a steady state rate of ATP-synthesis is observed. The slope of the curve directly shows the rate of ATP-synthesis. After this linear range the curve bends and this bending indicates that $\Delta$pH has decreased due to the proton efflux. At long reaction times (compare, *e.g.*, the yield at 500 ms and 10 s) there is no further increase in ATP yield because $\Delta$pH has decayed completely after this time. If the internal buffer capacity is increased at constant $\Delta$pH (by changing the buffer concentration in the acidic medium I) the initial slope, *i.e.*, the rate, is not changed; however, the linear range is extended to longer reaction times. This shows that the rate of ATP-synthesis depends only on $pH_{in}$ and $pH_{out}$, *i.e.* on the driving force, but not on the buffer capacity — in contrast to the ATP yield observed after 10 s reaction time. In the following experiments the rate of ATP-synthesis was

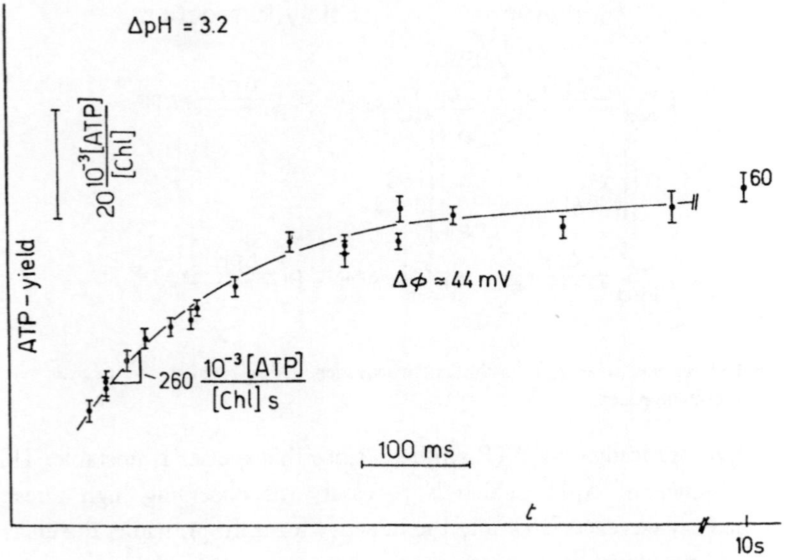


Fig. 3 - ATP yield as a function of reaction time. $pH_{out}$ = 8.2, $pH_{in}$ = 5.0, $\Delta\phi$ = 44 mV. The numbers at the slopes give the reaction rate in units $\{10^{-3}[ATP]/([Chl]s)\}$.

always determined from the slope of ATP yield *versus* reaction time curves in the linear range.

## 3. Regulation of the ATP-synthase

The investigations carried out under these carefully controlled conditions show that it is not only necessary to control substrate- and product concentration and the energization of the membrane but also the exact state of the enzyme must be known. The ATP-synthase catalyzes not only the proton-transport-coupled ATP-synthesis, the enzyme also carries out two other reactions: a redox reaction where a -S-S-group on the γ-subunit becomes reduced or oxidized [19-21], and a protolytic reaction where the ATP-synthase is converted from an inactive into an active state [22-26]. The relations between the different states are depicted in a simplified way in Fig. 4.

The inactive enzyme forms, $E^{ox}_{inact}$ and $E^{red}_{inact}$, can be converted to the active forms by membrane energization. For this process the oxidized form requires a higher energization than the reduced form (see Fig. 9). The inactive and the active enzyme can be converted from the oxidized to the reduced form by reaction with dithiothreitol (*in vitro*) or thioredoxin (*in vivo*). $E^{red}_{act}$ can catalyze high rates of ATP-synthesis and ATP-hydrolysis

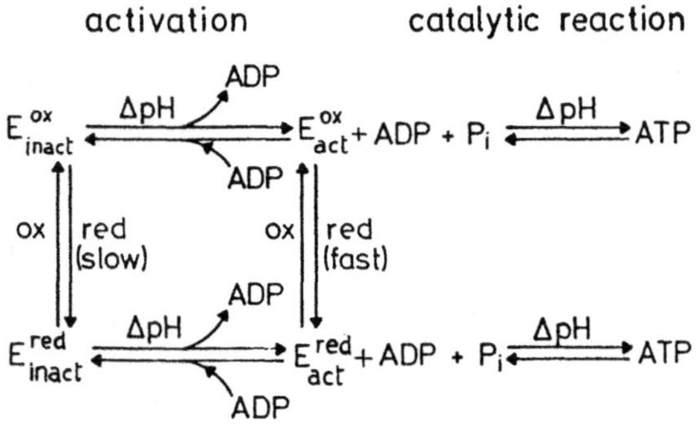


Fig. 4 - Simplified reaction scheme for activation, inactivation, redox reaction and catalytic reaction of the ATP-synthase from chloroplasts.

$E^{ox}_{act}$ can catalyze pratically only ATP-synthesis since this species is unstable. This implies that in the absence of ΔpH (which is necessary for observing high rates of ATP-hydrolysis) the enzyme reacts back into the in active form. In principle, the enzyme shows four different ΔpH-dependent reactions.

## 4. Activation/inactivation of the reduced enzyme

In class II chloroplasts, *i.e.* in thylakoids, the enzyme in usually in the inactive, oxidized state. Illumination in the presence of dithiothreitol converts all enzymes into the active, reduced state. After deenergization of the membrane, *e.g.* in the dark after illumination, the enzyme reacts back to the inactive state. Since only the active enzyme

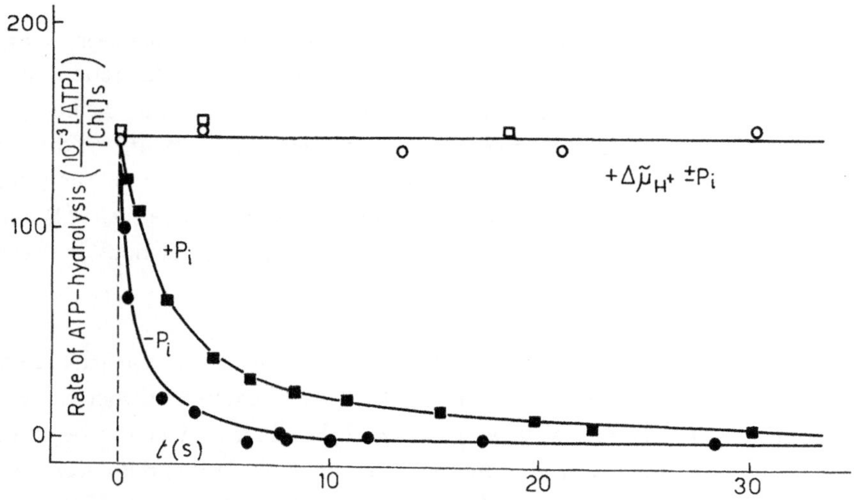


Fig. 5 - Relative rate of ATP hydrolysis as a function of time after illumination (■, $+P_i$ = in the presence of $P_i$; ●, $-P_i$ in the absence of $P_i$. Open symbols (□ and ○, graph $+\Delta\tilde{\mu}_{H+}$, ± $P_i$ stand for activity after reactivation by $\Delta\tilde{\mu}_{H+}$.

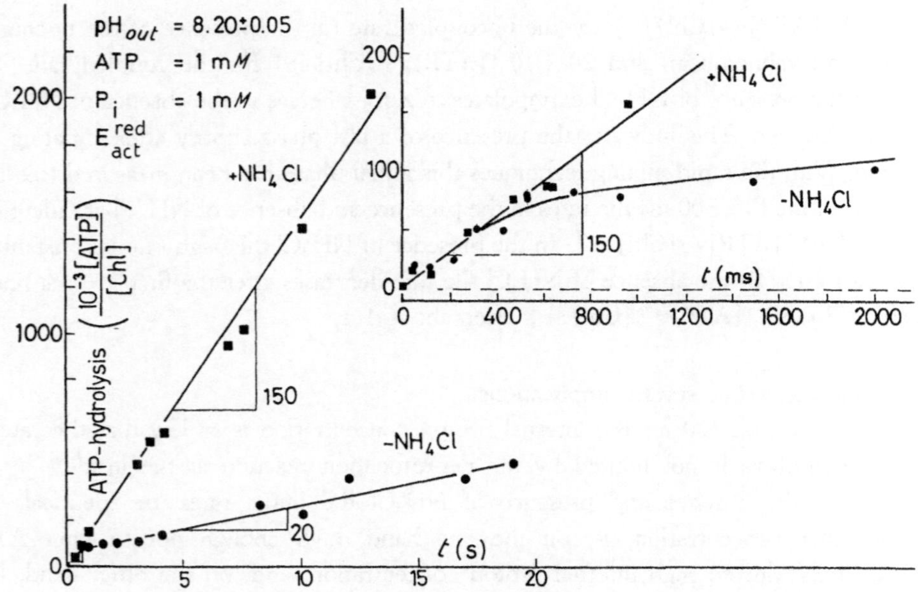


Fig. 6 - Coupled (– $NH_4Cl$) and uncoupled (+$NH_4Cl$) ATP-hydrolysis as a function of reaction time. The numbers at the slope give the rate in units { $10^{-3}$ [ATP]/([Chl]s)}. Inset: ATP-hydrolysis under the same conditions measured with high time resolution with rapid-mixing techniques.

catalyzes ATP-hydrolysis, the inactivation can be investigated by measuring the rate of ATP-hydrolysis [27]. This implies that the concentration of the active, reduced enzyme is proportional to the rate of ATP-hydrolysis [see equation (7)]. Figure 5 shows the rate of ATP-hydrolysis as a function of the time after illumination. The relative rate of ATP-hydrolysis decreases with a half-lifetime of about 30 s in the absence, ($-P_i$), of phosphate and a half-lifetime of about 2 min in its presence ($+ P_i$). If, after the complete decay of the hydrolysis activity, the membrane is energized (*e.g.*, by a high ΔpH) the initial rate is observed again (curve $+ \Delta\tilde{\mu}_{H^+}$, $\pm P_i$). This implies that the inactivation/activation process is completely reversible.

ATP-hydrolysis is coupled with a proton translocation into the internal phase. The maximal rate of hydrolysis is expected when the proton release into the internal phase is not inhibited by a high internal proton concentration. Usually, the maximal hydrolysis rate is measured in the presence of an uncoupler. An uncoupler dissipates the ΔpH by increasing the proton permeability of the membrane. Whether the maximal rate of ATP-hydrolysis is obtained in the presence of an uncoupler, or not, has been checked by measuring ATP-hydrolysis under coupled and uncoupled conditions. This is shown in Fig. 6. The rate of ATP-hydrolysis, *i.e.* the slopes of the curves in Fig. 6 are

150 { $10^{-3}$ [ATP] / ([Chl] s) } for the uncoupled rate (in the presence of the uncoupler $NH_4Cl$ and valinomycin) and 20 { $10^{-3}$ [ATP] / ( [Chl] s) } for the coupled rate. The curve in the presence of $NH_4Cl$ extrapolates to zero, whereas in the absence of $NH_4Cl$ a burst is observed. This indicates the presence of a fast phase shortly after initiating the reaction. With the rapid mixing techniques this initial phase has been measured (see Fig. 6, inset). In the first 300 ms the rates in the presence and absence of $NH_4Cl$ are identical, *i.e.* 150 { $10^{-3}$ [ATP] / ([Chl] s) }. In the presence of $NH_4Cl$ this high rate is constant up to at least 20 s. In the absence of $NH_4Cl$ the rate decreases after the first 300 ms finally reaching 20 {$10^{-3}$ [ATP] / ( [Chl] s) } after about 1 s.

This result has several implications:

1) During the first 300 ms the internal proton concentration is so low that the rate of ATP-hydrolysis is not limited by the deprotonation reaction at the inside.
2) Since in the absence and presence of $NH_4Cl$ the initial rates are identical, the uncoupler concentration, is, on the one hand, high enough not to limit ATP-hydrolysis, due to high internal proton concentration, and, on the other hand, low enough not to inhibit the enzyme.
3) Therefore, the measured rate is the maximal rate of ATP-hydrolysis. With the enzyme concentration the maximal turn-over of the enzyme in hydrolysis direction can be calculated, resulting in 105 $s^{-1}$. This is the highest rate measured up to now with chloroplast ATPase.

On the basis of these results the activation of the reduced enzyme has been measured as follows:

1) Generation of the inactive, reduced enzyme $E^{red}_{inact}$ as shown in Fig. 5.
2) Activation of the enzyme by $\Delta pH$ of variable magnitude.
3) Addition of uncoupler and ATP and measurement of the rate of ATP-hydrolysis.
4) Under these conditions the rate of hydrolysis is proportional to the number of enzymes in the active state, *i.e.*

$$v_{hyd} \propto E^{red}_{act} \; ; \; v_{hyd,\,max} \propto E^{red}_{act,max}$$

The fraction of active enzymes, $\eta^{red}$, is then given by

$$\eta^{red} = \frac{E^{red}_{act}}{E^{red}_{act,\,max}} = \frac{v_{hyd}}{v_{hyd,\,max}} \tag{7}$$

Figure 7 shows ATP-hydrolysis as a function of reaction time after this procedure. ATP-hydrolysis was measured by the release of phosphate from ATP and plotted as

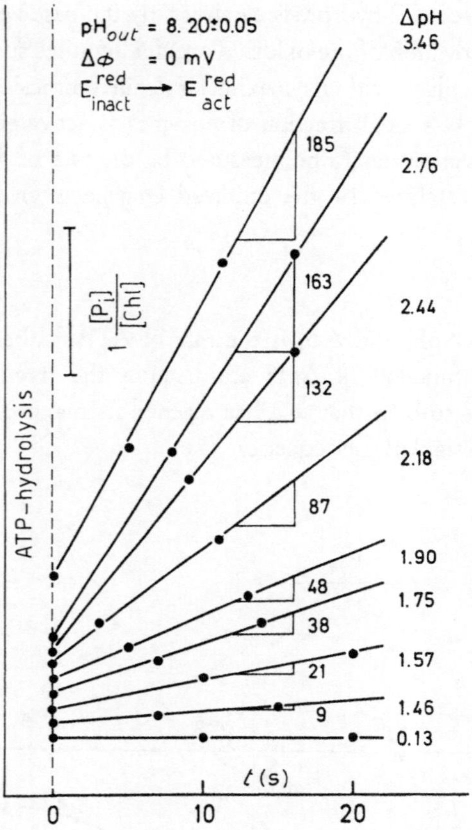


Fig. 7 - ATP-hydrolysis as a function of reaction time. The inactive, reduced state, $E^{red}_{inact}$ of the enzyme was generated; then the membrane was energized by different ΔpH. Subsequently, the uncoupled rate of ATP hydrolysis was measured. The numbers at the slope give the rate of ATP hydrolysis in units $\{10^{-3}[ATP]/([Chl]s)\}$. All curves start at the same zero point and are arbitrarily displaced for clearer presentation.

$[P_i]$ / [Chl]. If the energizing ΔpH is small, no ATP-hydrolysis is observed, *e.g.* ΔpH = 0.13. When ΔpH in the activating step is increased, the rates of ATP-hydrolysis are increased too. From the rates the fraction of activated enzymes has been calculated according to equation (7) and plotted in Fig. 8. A sigmoidal dependence of this fraction on ΔpH is observed with a half maximal value at ΔpH = 2.2.

## 5. Activation/inactivation of the oxidized enzyme

The active state of the oxidized enzyme, $E^{ox}_{act}$, has only a short lifetime after deenergization of the membrane. Therefore, the ΔpH dependence of the activation of the oxidized enzyme cannot be measured by ATP-hydrolysis after a ΔpH jump. It has not been possible to measure ATP-hydrolysis catalyzed by the oxidized enzyme. This can be explained as follows: activation of the oxidized enzyme requires such a high ΔpH that the catalytic reaction takes place with its maximal rate in synthesis direction under these conditions; whereas, only a small fraction of enzymes is activated. Therefore, the ΔpH dependence of the activation has to be measured by the rate of ATP-synthesis[18]. The rate of ATP-synthesis catalyzed by the oxidized enzyme is given by

$$v^{ox}_{syn} = w^{ox}_{syn} \eta^{ox}$$

where $v^{ox}_{syn}$ (function of ΔpH and Δϕ) is the rate of ATP-synthesis per active, oxidized enzyme and $\eta^{ox}$ (also function of ΔpH and Δϕ) is the fraction of active, oxidized enzymes. If the ΔpH is so high that $w^{ox}_{syn}$ has reached its maximal rate, we can calculate $\eta^{ox}$ from equation (8) similarly as equation (7):

$$\eta^{ox} = \frac{E^{ox}_{act}}{E^{ox}_{act,\ max}} \approx \frac{v^{ox}_{syn}}{v^{ox}_{syn,\ max}} \tag{9}$$

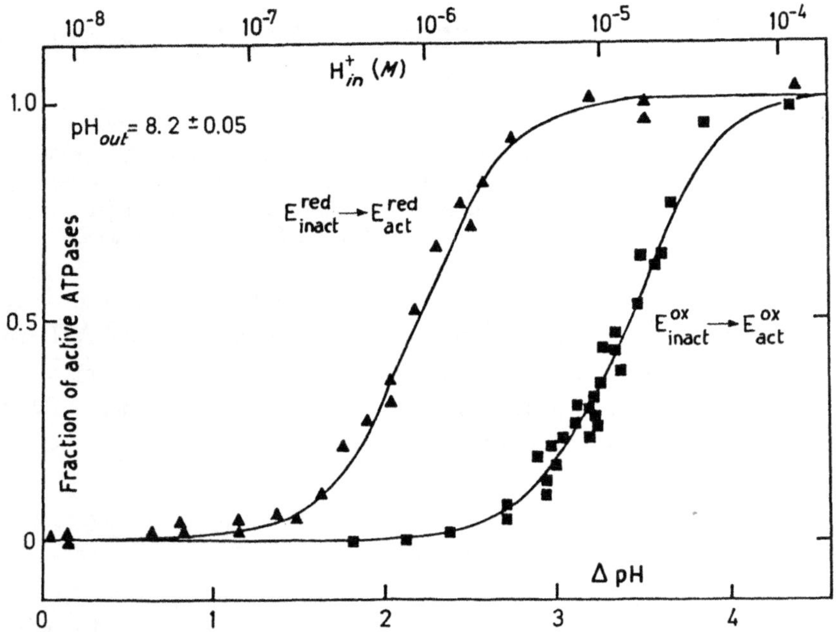


Fig. 8 - Fraction of active ATPases as a function of ΔpH. The fraction of active, reduced ATPase was measured by the relative rate of ATP-hydrolysis (data from Fig. 7). The fraction of active, oxidized ATPases was measured by the relative rate of ATP-synthesis (data from Fig. 9).

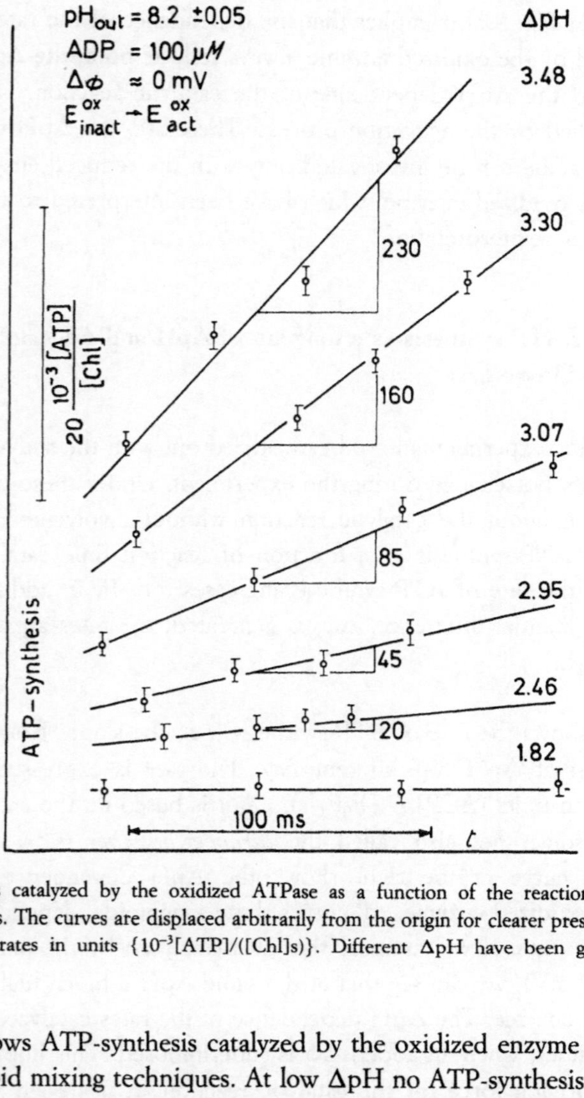


Fig. 9 - ATP-synthesis catalyzed by the oxidized ATPase as a function of the reaction time measured with rapid-mixing techniques. The curves are displaced arbitrarily from the origin for clearer presentation. The numbers at the slope give the rates in units $\{10^{-3}[ATP]/([Chl]s)\}$. Different ΔpH have been generated by acid-base transitions.

Figure 9 shows ATP-synthesis catalyzed by the oxidized enzyme at different ΔpH measured with rapid mixing techniques. At low ΔpH no ATP-synthesis is observed, with increasing ΔpH also the rate increases. From such data the relative rate of ATP-synthesis was calculated, which, according to equation (9), corresponds to the fraction of active enzymes and plotted in Fig. 8 as a function of ΔpH. As in case of the reduced enzyme, a sigmoidal increase of the fraction of active enzymes with ΔpH is observed. However, the half maximal value is now achieved at ΔpH = 3.4. This implies that reduction of the enzyme decreases the ΔpH which is necessary for activation from 3.4 to 2.2. Since the data shown in Fig. 8 are obtained at constant $pH_{out}$ = 8.2, this implies that a higher internal proton concentration is necessary for activation of the oxidized enzyme than for the reduced enzyme. In other words, the reduction of the enzyme has increased the p*K* value of the groups which have to be protonated from the inside for activation.

The result in Fig. 8 also implies that the dependence of the rate of ATP-synthesis on ΔpH catalyzed by the oxidized enzyme always reflects both, the ΔpH dependence of the activation and the ΔpH dependence of the catalytic reaction. Usually, the rate is kinetically controlled by the activation process. Therefore, the ΔpH dependence of the catalytic reaction alone can be investigated only with the reduced enzyme. Older results obtained with the oxidized enzyme, which have been interpreted to reflect the catalytic reaction, require a reinterpretation.

## 6. The rate of ATP-synthesis as a function of ΔpH and Δφ catalyzed by the active, reduced ATP-synthase

The following experiments have been carried out with the active, reduced enzyme, and this state does not change during the experiment. Under these conditions, we now obtain information about the catalytic reaction without involvement of the activation. Figure 10 shows ATP-synthesis as a function of reaction time catalyzed by $E^{red}_{act}$. With increasing ΔpH, the rate of ATP-synthesis increases too. If, in addition to the ΔpH a transmembrane potential difference, Δφ, is generated, the rates are increased considerably (Fig. 10, right).

Figure 11 shows the rate of ATP-synthesis (*i.e.*, the slopes from the curves in Fig. 10) as a function of ΔpH for different Δφ. The rate is expressed here, and in the following figure, in units [ATP] / ($[F_0F_1]$s), *i.e.* it is based on the enzyme concentration $[F_0F_1]$. This is sometimes also called the *turnover number* ($s^{-1}$). Additionally, for a comparison, the curve at the right shows the ΔpH dependence of ATP-synthesis catalyzed by the oxidized enzyme. All curves show a sigmoidal dependence on ΔpH with the same maximal rate. If we compare the rates catalyzed by the oxidized and reduced enzyme (Δφ = 0 mV), we can see that at the same ΔpH a much higher rate is obtained with the reduced enzyme. The ΔpH dependence of the rates catalyzed by $E^{red}_{act}$ are shifted to lower ΔpH values, when an added Δφ is superimposed. This implies that both ΔpH and Δφ act as driving force for the catalytic reaction. It has been shown that for the activation of the oxidized enzyme ΔpH and Δφ act additively [18]. From ΔpH and Δφ we can calculate the transmembrane electrochemical potential difference of the proton, $\Delta\tilde{\mu}_{H^+}$, from equation (2). As usual, we use the *proton-motive force, p,* introduced by MITCHELL [4], *i.e.*, $\Delta p = \Delta\tilde{\mu}_{H^+}/\mathsf{F}$, where $\mathsf{F}$ is the FARADAY constant.

In Fig. 12 the rate of ATP-synthesis is plotted as a function of the proton-motive force (data from Fig. 11) and, in addition, the fraction of active enzymes as a function of *p* (data from Fig. 8 and Ref. 18). Figure 12 shows that the catalytic rate as well as the activation of the oxidized enzyme depend only on $\Delta\tilde{\mu}_{H^+}$, *i.e.*, on the sum of ΔpH and

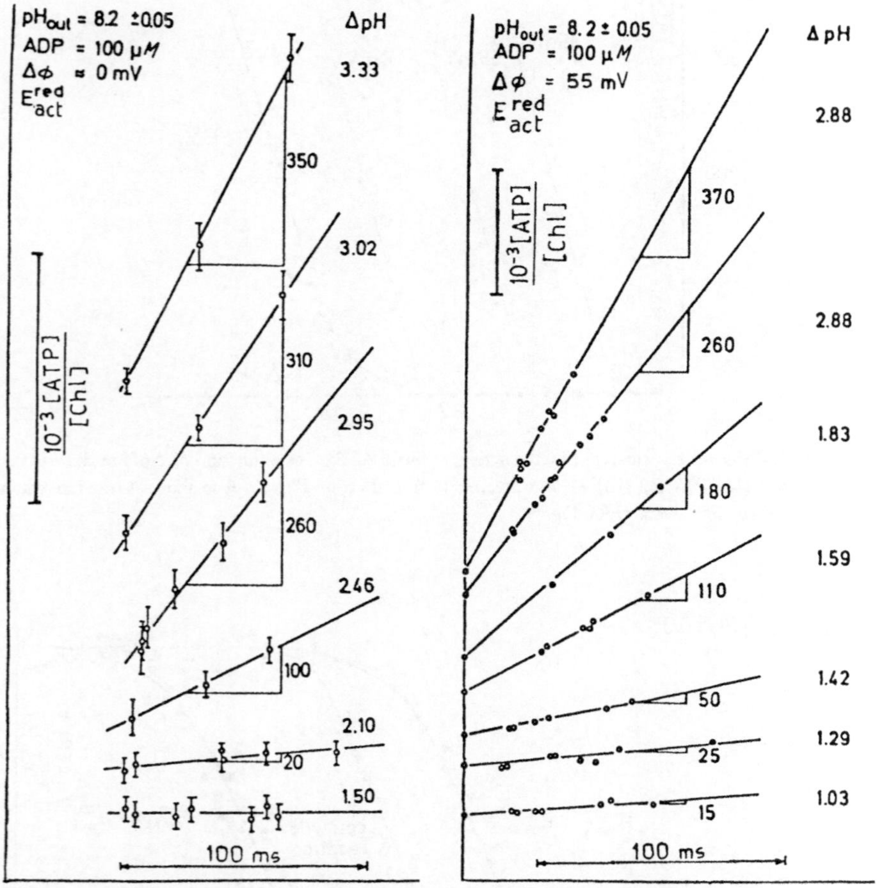


Fig. 10 - ATP-synthesis catalyzed by the active, reduced ATPase as a function of the reaction time measured with rapid-mixing techniques. The curves are displaced arbitrarily from the origin for clearer representation. The numbers at the slopes give the rate in units $\{10^{-3}[ATP]/([Chl]s)\}$. Different ΔpH have been generated by acid-base transition. Left: $\Delta\phi = 0$ mV Right: $\Delta\phi = 55$ mV.

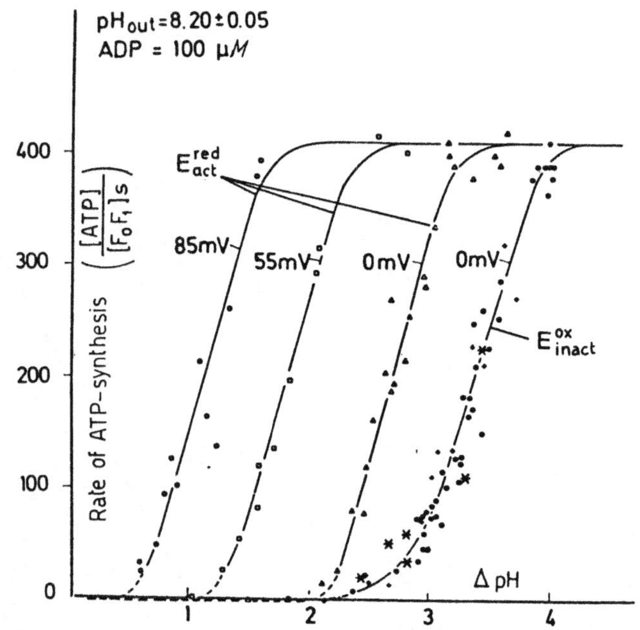


Fig. 11 - Rate of ATP-synthesis catalyzed by the active, reduced ATPase as a function of ΔpH at different Δϕ. The rate is here in units $\{[ATP]/([F_0F_1]s)\}$ *i.e.* in turnovers ($s^{-1}$). Data from Fig. 10. Also shown is the rate catalyzed by the oxidized ATPase, $E^{ox}$. Data from Fig. 9.

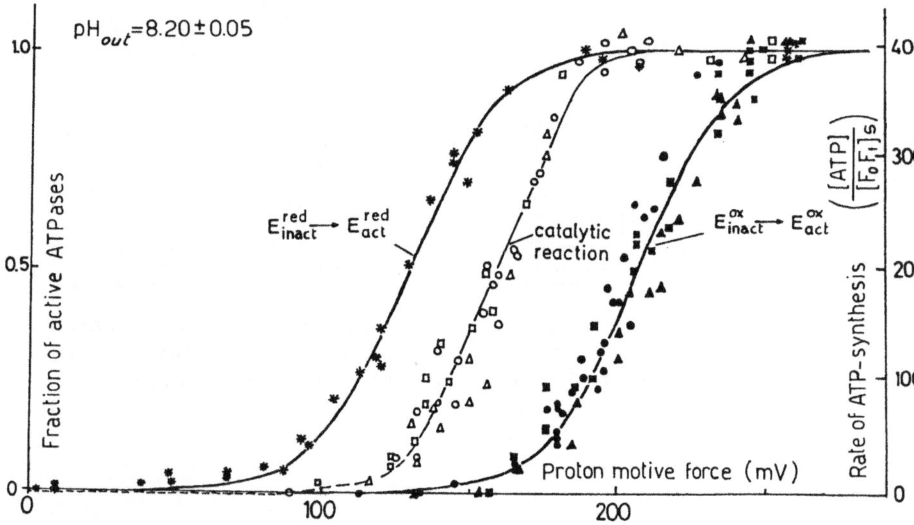


Fig. 12 - Rate of ATP synthesis catalyzed by the active, reduced enzyme (scale on right) and fraction of active ATPases (oxidized and reduced) (scale on left) as a function of the proton-motive force. Data from Figs. 8 and 11 and Ref. 18.

$\Delta\phi$. This implies that $\Delta pH$ and $\Delta\phi$ are kinetically equivalent; *i.e.*, the rate of ATP-synthesis (fraction of active enzymes) is increased by the same factor, when the driving force is increased, *e.g.*, $\Delta pH = 1$ corresponds to $\Delta\phi = 60$ mV.

Figure 12 shows the $\Delta pH$- (or $\Delta p$) dependence of three different reactions of the ATP-synthase. According to the scheme in Fig. 4, we expect four different $\Delta pH$ dependences. However, unfortunately, the catalytic rate catalyzed by the oxidized enzyme cannot be measured because the state $E^{ox}_{act}$ is so unstable that upon deenergization it reacts rapidly to $E^{ox}_{inact}$. Presently, we have no method to measure this rate without involvement of the activation process. The simplest assumption in this respect is that the redox state of the enzyme only influences the activation process but not the catalytic reaction. In this case the curve labeled catalytic reaction in Fig. 12 represents the $\Delta pH$ dependence of the rate for the oxidized and the reduced enzyme.

According to the results shown here, the ATP-synthase from chloroplasts is a strongly regulated enzyme. This regulation is probably necessary to avoid ATP-hydrolysis at the end of an illumination period. *In vivo* the ATP-synthase is reduced by thioredoxin [28, 29]. The redox state of thioredoxin is coupled *via* ferredoxin to the redox state of $NADPH/NADP^+$. The redox state of $NADPH/NADP^+$ is controlled by the rate of its reduction due to the photosynthetic electron transport and the rate of its oxidation in the CALVIN cycle. On the basis of the results of Fig. 12 and the scheme in Fig. 4, the following effects are expected in plants: in dark-adapted leaves the ATPase is usually in the oxidized state. Illumination leads to the generation of $\Delta\tilde{\mu}_{H^+}$ and according to Fig. 12 at 160 mV a part of the ATPases is activated, leading to a rate of ATP-synthesis of about 30 $s^{-1}$. Since, under these conditions, the rate of $CO_2$ fixation is limited by the ATP supply, NADPH will be accumulated and this leads — *via* thioredoxin — to reduction of the ATPase. If all the enzyme is reduced, this would lead at constant $\Delta p$ to an increase of the rate to 300 $s^{-1}$. However, we must take into account that possibly not all ATPases are reduced and that, due to the phosphorylation-coupled proton efflux, the $\Delta p$ will decrease. If *e.g.* the $\Delta p$ decreases by about 0.3 units, the rate will increase only from 30 $s^{-1}$ to 60 $s^{-1}$. Thus, we have the remarkable result that ATP-ynthesis increases, whereas $\Delta p$ decreases. Such a behavior has, in fact, been observed [30, 31]. In our interpretation this is not evidence against the chemiosmotic theory but is due to the fact that the state of the enzyme changes during the experiment [27].

## 7. The rate of ATP-synthesis as a function of the ADP concentration at constant $\Delta pH$

The dependence of the rate of ATP-synthesis has been measured in several laboratories. In most cases MICHAELIS-MENTEN-type kinetics were found. Investigation of the dependence of the $K_M$ values on membrane energization provided different results: the $K_M$ values for ADP decrease when ΔpH is decreased by addition of an electron-transport inhibitor or by decreasing light intensity [32-34]. If ΔpH is decreased by an uncoupler the $K_M$ value increases [33]. These results were interpreted in different ways. However, in a strict sense, these results cannot be interpreted at all. First of all, the state of the enzyme was not controlled during the experiments. Secondly, and more important is the fact that the experiments contain a systematic error: if, at constant light intensity, the ADP concentation is changed from zero to about 100 μ*M*, the ΔpH decreases by about 0.3 units, *e.g.* from ΔpH = 3.0 to 2.7 [35, 36]. The reason for this effect is that parallel to the basal proton efflux a phosphorylating proton efflux occurs which increases with increasing ADP concentration. This leads to a lowering of the steady state level of ΔpH. Therefore, the rate at low ADP concentration is measured at a higher ΔpH than that at high ADP concentrations. A correction of this effect would decrease the rate at

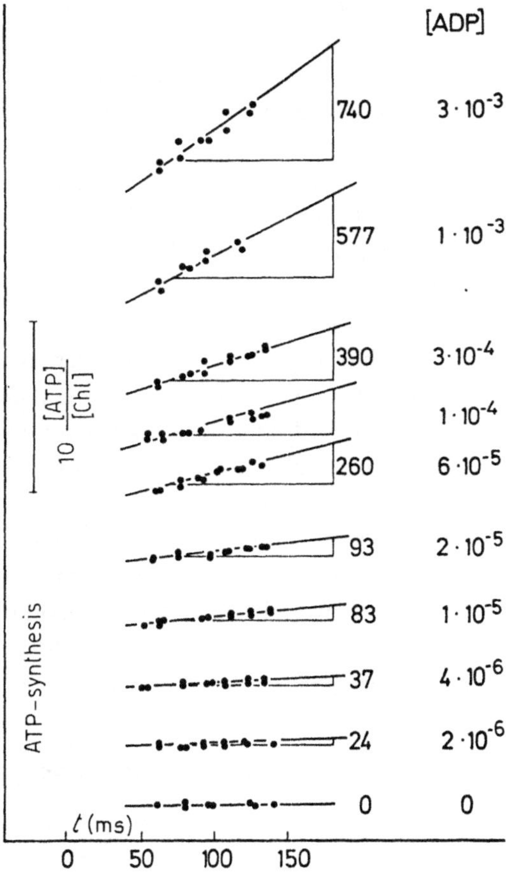


Fig. 13 - ATP-synthesis catalyzed by the active, reduced ATPase as a function of reaction time for different ADP concentrations at ΔpH = 3.2. The numbers at the slope give the rate in units $\{10^{-3}[ATP]/([Chl]s)\}$.

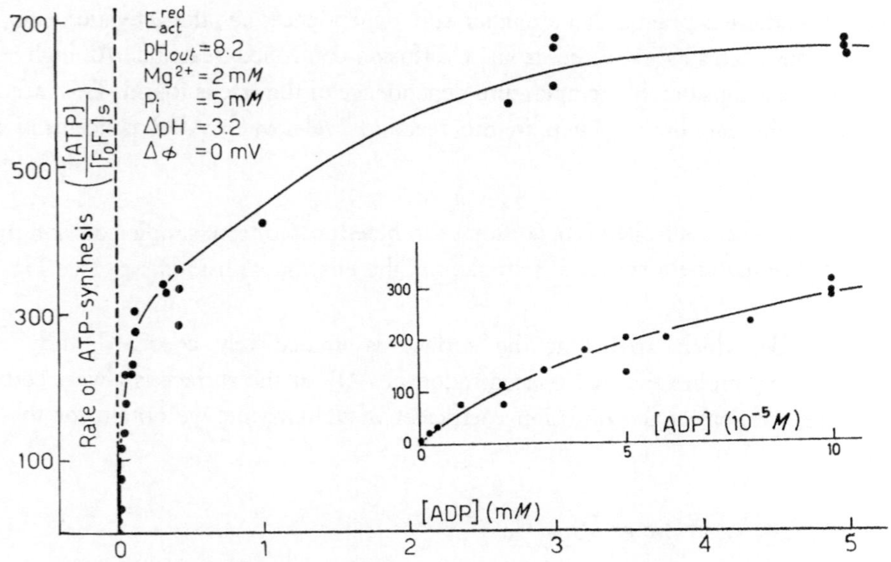


Fig. 14 - Rate of ATP-synthesis as a function of the ADP concentration at constant $\Delta pH = 3.2$. Inset: rate at low ADP concentrations.

low ADP concentration. On the basis of an enzyme-kinetic model it has been tried in Ref. 34 to correct this effect.

The rapid-mixing technique allows measurement of the rate of ATP-synthesis at constant $\Delta pH$ [37]. Figure 13 shows ATP-synthesis as a function of reaction time at $\Delta pH = 3.3$ for different ADP concentrations. Figure 14 shows the rate of ATP-synthesis, *i.e.* the slopes of the curves in Fig. 13 as a function of the ADP concentration. The inset shows the rate at low ADP concentration. In contrast to other experiments the functional dependence under these conditions is not a typical Michaelis-Menten-kinetics. The slope of the curve seems to reflect a negative cooperativity.

The maximum rate observed here is 680 $s^{-1}$. This is the highest rate in the ATP-synthesis direction and represents the turnover number of the enzyme, when all substrates and energization show saturation. It should be noted that such high rates cannot be observed in light-driven phosphorylation because such high proton fluxes cannot be obtained by light energization. The half maximal rate is obtained at 100 $\mu M$, from the slope of the linear increase of the rate at low ADP concentrations we obtain a second order rate constant of $k = 1 \times 10^7 \ M^{-1}s^{-1}$. This high rate constant possibly indicates that the rate of ATP-synthesis under these conditions (*i.e.*, when the enzyme is saturated with all other substrates) is diffusion controlled. Preliminary measurements of the rate of ATP-synthesis at different temperatures have shown that at low ADP concentration there is practically no temperature dependence; *i.e.*, the activation energy is practically zero. This is expected from a diffusion-controlled reaction. At high ADP concentrations a considerable temperature dependence of the rate is found. This indicates a change of the rate-limiting step in the reaction cycle when ADP concentration is changed.

The diffusion-controlled rate constant can be estimated from simple collision theory as follows. We assume a spherical particle, *i.e.* the enzyme with radius $r_E$ (see Fig. 15).

Every molecule which arrives at the surface is immediately adsorbed and reacts chemically. This implies that the concentration of ADP at the surface is always zero. If, furthermore, we neglect the diffusion coefficient of the enzyme, we obtain for the rate constant

$$k' = 4\pi\,(r_E + r_{ADP})\,D_{ADP}\,N_A \tag{10}$$

With $r_E = 5$ nm radius of the enzyme, $r_{ADP} = 0.2$ nm radius of ADP and an estimated diffusion coeffient for ADP $D_{ADP} \approx 5\times10^{-6}$ cm$^2$ s$^{-1}$, $N_A$ Avogadro's number, it results

$$k' \approx 2\times10^{10}\ M^{-1}\ s^{-1}$$

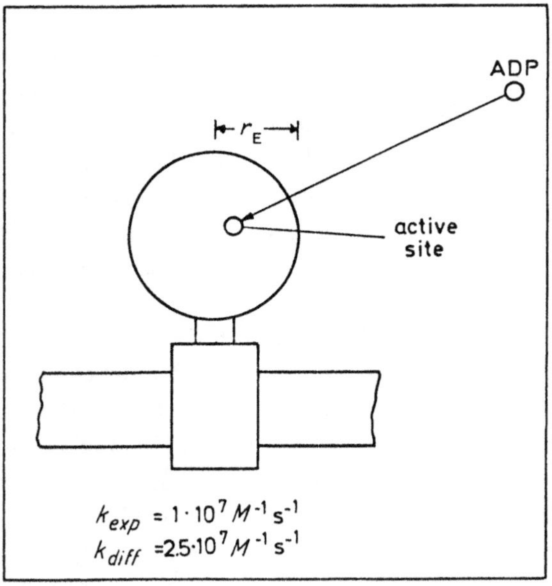


Fig. 15 - Scheme for estimating the rate constant of a diffusion-controlled reaction, indicating the ATPase. Details see text.

The active site for binding of ADP is only a small fraction of the enzyme surface. In the simplest model the area of the active site has the magnitude of the area of ADP. Then, the fraction of active enzyme surface is given by:

$$\delta_E = \frac{\pi r^2_{ADP}}{4\ \pi\ r^2_E} \approx 2.5 \times 10^{-3}$$

Also, not every orientation of ADP leads to a binding at the enzyme. There are at least two different orientations, *e.g.*, the phosphate group or the adenine group, oriented to the binding site. This results at least in a factor $\delta_{ADP} \approx 0.5$. The observed rate constant is then

$$k = k'\ \delta_{ADP}\ \delta_E \approx 2.5 \times 10^7\ M^{-1}\ s^{-1}.$$

This rate constant estimated on the basis of a very crude model is in remarkable agreement with the measured rate constant which is only smaller by a factor two. For a more refined treatment, one must take into account that the ADP molecules not hitting the surface of the enzyme at the binding site do not disappear as assumed in the model, but might be reflected back into the medium and/or might migrate at the enzyme surface to the active site. Also, ADP might diffuse only from one direction to the enzyme, since the other side is shielded by the membrane (see Fig. 15). However, if the rate is indeed diffusion-controlled then MICHAELIS-MENTEN-kinetics cannot be used also at constant $\Delta$pH for describing this process, since it requires an equilibrium between free and bound substrate.

## 8. The function of the reconstituted ATP-synthase

All experiments described above have been carried out with thylakoids; *i.e.*, the enzyme is in its natural environment. It has been argued that this natural membrane is necessary for optimal function of the enzyme. In particular, it has been suggested that proton pathways exist in the membrane so that protons can migrate directly from the electron transport chain to the ATP-ase without equilibration with the internal aqueous phase [5-9]. Therefore, if the natural membrane is replaced, only small rates of ATP-synthesis/hydrolysis are to be expected.

In order to investigate this experimentally, we have isolated and purified the ATP-synthase and then reconstituted it into asolectin liposomes [38, 39]. Figure 16 shows an electron micrograph of these proteoliposomes. The average diameter of the vesicles is about 150 nm; this number depends on the protein to lipid ratio during reconstitution [40]. It is interesting to note that the distribution of the protein is not homogeneous.

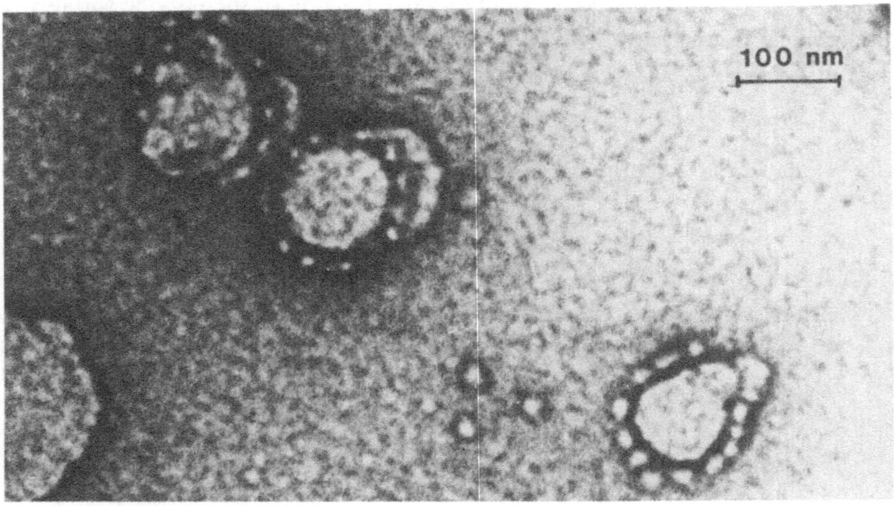


Fig. 16 - Electron micrograph of ATP-synthase reconstituted into asolectin liposomes. Samples were negatively stained with 1 % uranylacetate.

There are a few smaller vesicles containing many ATPases and there are bigger vesicles containing no ATPase at all. With this reconstituted system ATP-synthesis was measured with an acid-base transition using the rapid-mixing techniques. Figure 17 shows ATP-synthesis as a function of reaction time at different ΔpH at constant Δϕ. The slopes of these curves give the rates [ATP]/ ([ATPase]*t*), with time in second, *i.e.* in turnovers ($s^{-1}$). The rate of ATP-synthesis is depicted in Fig. 18 as a function of ΔpH, when different $K^+$/valinomycin transmembrane electric potential differences are generated. Similar to thylakoids, a sigmoidal dependence on ΔpH is found. At constant ΔpH an additional transmembrane electric potential difference (the parameter in Fig. 18 is the ratio between internal and external $K^+$ concentration) increases the rate, unless the maximum has been reached. Since the permeability coefficients of the different ions for asolectin membranes are not known, the transmembrane electric potentials differences cannot be calculated. If the same permeability coefficients are used as for the thylakoid membrane — this is certainly a very poor approximation — transmembrane potentials differences of 0 mV, 47 mV and 68 mV are calculated from equation (5).

The maximal rate observed here was 200 $s^{-1}$. This is almost half of the rate observed in thylakoids (400 $s^{-1}$), under the same conditions. Since the calculation of the rate is based on the amount of protein added for reconstitution, it is possible that not all ATPases are reconstituted with proper orientation. The number given here is a lower limit, since any correction for inactivation would increase the rate. From these results it can be calculated that the natural environment is not necessary for optimal function of the

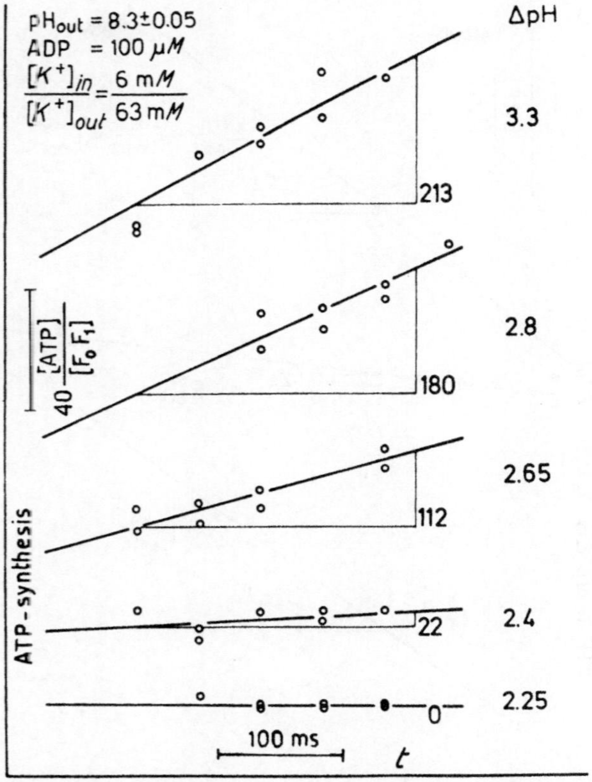


Fig. 17 - ATP-synthesis catalyzed by the reconstituted ATP synthase as a function of the reaction time at different ΔpH. The slopes give the rates in units $\{[ATP]/([F_0F_1]s)\}$ *i.e.* in turnovers. The ratio $K^+_{in}/K^+_{out}$ = 6 m*M*/63 m*M* was used in all experiments.

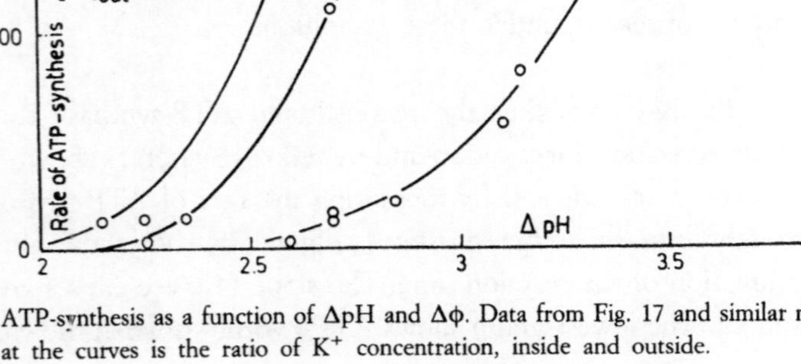


Fig. 18 - Rate of ATP-synthesis as a function of ΔpH and Δϕ. Data from Fig. 17 and similar measurements. The parameter given at the curves is the ratio of $K^+$ concentration, inside and outside.

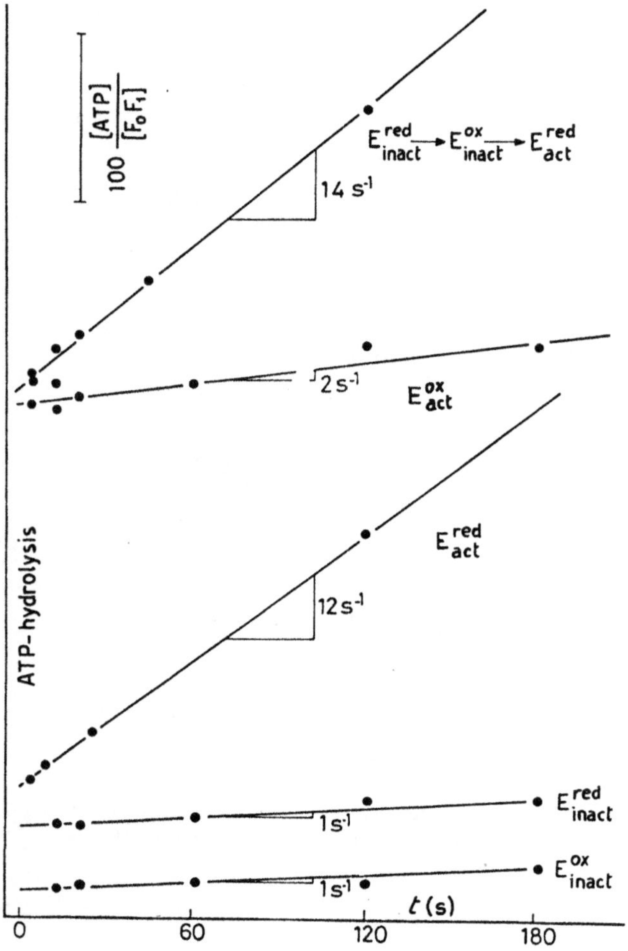


Fig. 19 - ATP-hydrolysis as a function of reaction time catalyzed by different states of the reconstituted ATP synthase. Details see text.

ATPase. Proton transport from the internal aqueous bulk phase to the external aqueous bulk phase obviously can occur with such high rates through the ATPase that the maximal rate can be observed under these conditions.

It may now be asked whether the reconstituted ATP-synthase shows the same strong regulation by activation/inactivation and by redox reaction as shown in the scheme of Fig. 4. This has been investigated by measuring the rate of ATP-hydrolysis after the enzyme was brought into the different states. Figure 19 shows the amount of ATP hydrolyzed as a function of the reaction time. The slope of these curves gives the rate of ATP-hydrolysis in $s^{-1}$. The lowest graph shows ATP-hydrolysis when the enzyme is in the state $E^{ox}_{inact}$. Practically no ATP-hydrolysis is observed. Incubation with dithiothreitol, leads to the state $E^{red}_{inact}$, and again no ATP-hydrolysis is observed. If an acid-base transition, ΔpH, is carried out after reduction of the enzyme, it is transformed into the state $E^{red}_{act}$. Addition of ATP now leads to a high rate of ATP-hydrolysis. If the enzyme is oxidized by iodosobenzoate, and then activated by a ΔpH jump, no ATP-hydrolysis is observed indicating that the state $E^{ox}_{act}$ is very unstable. If the enzyme is first oxidized, then reduced and then activated by ΔpH, ATP-hydrolysis is observed again, indicating the reversibility of the redox reaction.

The results enable us now to carry out investigations of the mechanism under very defined conditions: we have only one purified protein, the ATP-synthase, which is reconstituted into a well characterized lipid. Energization of the membrane is carried out with a well-known ΔpH and the reaction time is so short that also all substrate and product concentrations remain at their initial values or are measured directly.

## 9. Single steps in the reaction cycle of the enzyme

The number of nucleotide binding sites on the ATP-synthase has not been completely clarified up to now. In general, it is assumed that there are six binding sites, three of them having catalytic and three regulatory properties [41]. Under usual conditions in enzyme kinetics there is a big excess of substrates compared to the enzyme concentration. Under such conditions, an *averaged* rate is measured, resulting from all catalytic sites. Mechanistically, there are two extreme possibilities of cooperation between the different sites:

1) all catalytic sites work independently of each other, and
2) the sites work alternatively in a cooperative way [42].

The different mechanisms can be distinguished when the substrate concentration is lower than the enzyme concentration. In such a case only one of the catalytic sites can be occupied (*single site*) and only one turnover of the catalytic cycle can be carried out (*single turnover*). Such *single site, single turnover* conditions have been used to obtain a complete thermodynamic and kinetic description of ATP-hydrolysis catalyzed by the mitochondrial ATPase [43, 44]. Figure 20 shows the minimal reaction scheme for ATP-hydrolysis/synthesis at one catalytic site, where the protolytic steps have been omitted. ATP is bound to the free enzyme, forming the E-ATP-complex; ATP is hydrolyzed on the enzyme leading to the E-ADP-$P_i$-complex. From this species either first ADP and then $P_i$ is released (upper pathway) or first $P_i$ and then ADP (lower pathway), and the free enzyme, E, is regenerated.

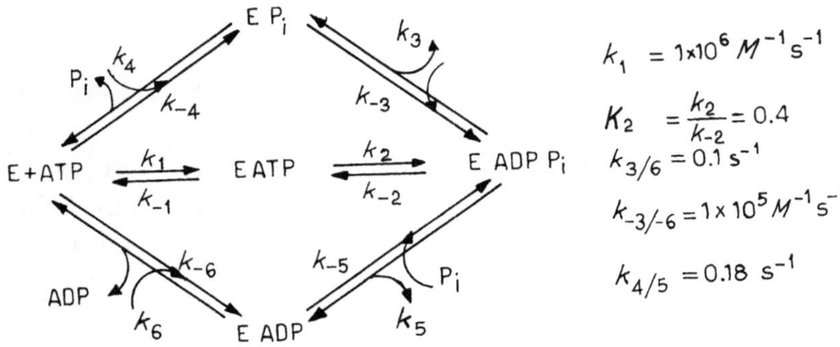


Fig. 20 - Reaction scheme of ATP-synthesis/hydrolysis at one catalytic site of the ATPase. Protonation/ deprotonation steps have been omitted.

Figure 21 shows the principle of the experiments for measuring the free and enzyme-bound concentrations of the different species as a function of time. As discussed above, these investigations must be carried out with the active, reduced enzyme. Therefore, after isolation and reconstitution the enzyme $E^{ox}_{inact}$ is first reduced and then activated by an acid-base transition ($\Delta pH + \Delta\phi$). Then an uncoupler is added in order to prevent any inhibition by protonation-deprotonation reaction; *i.e.*, the experiments are

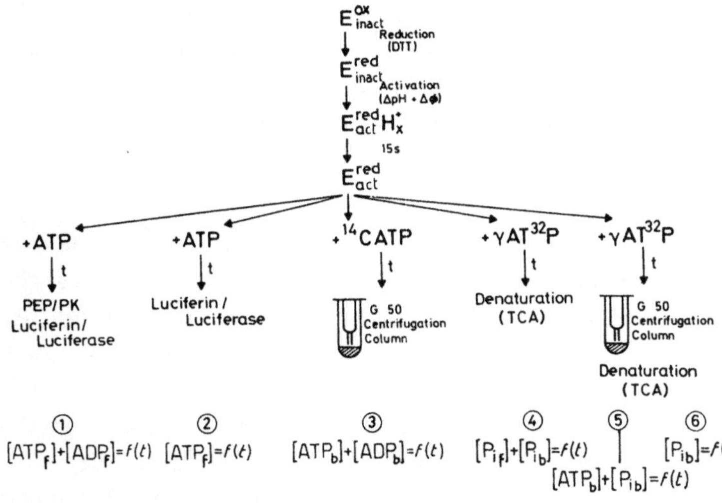


Fig. 21 - Scheme for measurement of the different free (subscript f) and enzyme-bound (subscript b) substrates and products as a function of the reaction time. PEP = phosphoenolpyruvate; PK = pyruvatekinase; TCA = trichloro acetic acid. Details see text.

carried out at $pH_{in} = pH_{out} = 8.2$. Then ATP is added to the enzyme either as cold ATP or $^{14}C$-labeled ATP or $\gamma(^{32}P)$-ATP in a concentration that is similar to the concentration of active enzymes (20 n*M*). Then six different sets of experiments are carried out as described in Fig. 21:

1. Measurement of free ([ADP] + [ATP])
2. Measurement of free [ATP]
3. Measurement of enzyme-bound ($[ATP_b]$ + $[ADP_b]$) nucleotides with ($^{14}C$)-ATP
4. Measurement of free and bound phosphate ($[P_{i,\,f}]$ + $[P_{i,\,b}]$) with $\gamma(^{32}P)$-ATP
5. Measurement of bound [ATP] and bound $[P_i]$, ($[ATP_b]$ + $[P_{i,b}]$) with $\gamma(^{32}P)$-ATP
6. Measurement of bound $[P_i]$, ($[P_{i,b}]$), with $\gamma(^{32}P)$-ATP.

The free substrates and products were separated by using Sephadex-G50 centrifugation columns as introduced by PENEFSKY [45]. This technique permits a very rapid separation of small molecules from the large enzyme molecules or vesicles.

The results of such a set of experiments are summarized in Fig. 22. The numbers refer to the type of measurement from which the data are derived (see Fig. 21). Rate constants for different steps of the reaction scheme of Fig. 20 can be calculated from these data. The second order rate constant $k_1$ for ATP binding, obtained directly from measurement 2, results to be equal to $1\times10^6\ M^{-1}s^{-1}$. The equilibrium constant on the enzyme is found by measuring bound [ATP] and bound $[P_i]$, *i.e.*, $K_2 = 0.4$. In order to determine the rate constant for $P_i$ release, we must assume that the rate constant of $P_i$ release is the same as for species E-ADP-$P_i$ or E-$P_i$. From this we obtain $k_5 = k_4 = 0.2$ $s^{-1}$. Also, we have to assume that the rate constant for ADP release is the same as for species E-ADP-$P_i$ and E-ADP. Therefore $k_3 = k_6 = 0.1\ s^{-1}$. This implies that the two products are released independently of each other. With this assumption the rate constants can be calculated from the rate of release and the concentration of the enzyme-bound products. The results are summarized in Fig. 20. Since an equilibrium constant of about 1 is found between enzyme-bound substrates and products, most of the energy provided by the protonation-deprotonation reactions is necessary for the binding of substrates and release of products as suggested by BOYER [46].

## 10. Coupling between ATP-synthesis/hydrolysis and proton transport

Finally, the results described in sections 6, 8 and 9 must be combined to a complete reaction cycle of proton-transport coupled ATP-synthesis/hydrolysis. Fig. 23 shows a general reaction scheme in which protolytic reactions are coupled with ATP-synthesis/hydrolysis at one catalytic site. The scheme from Fig. 20 is again depicted at the right. At the left, the scheme is coupled with protolytic reactions. The scheme shows a three-step

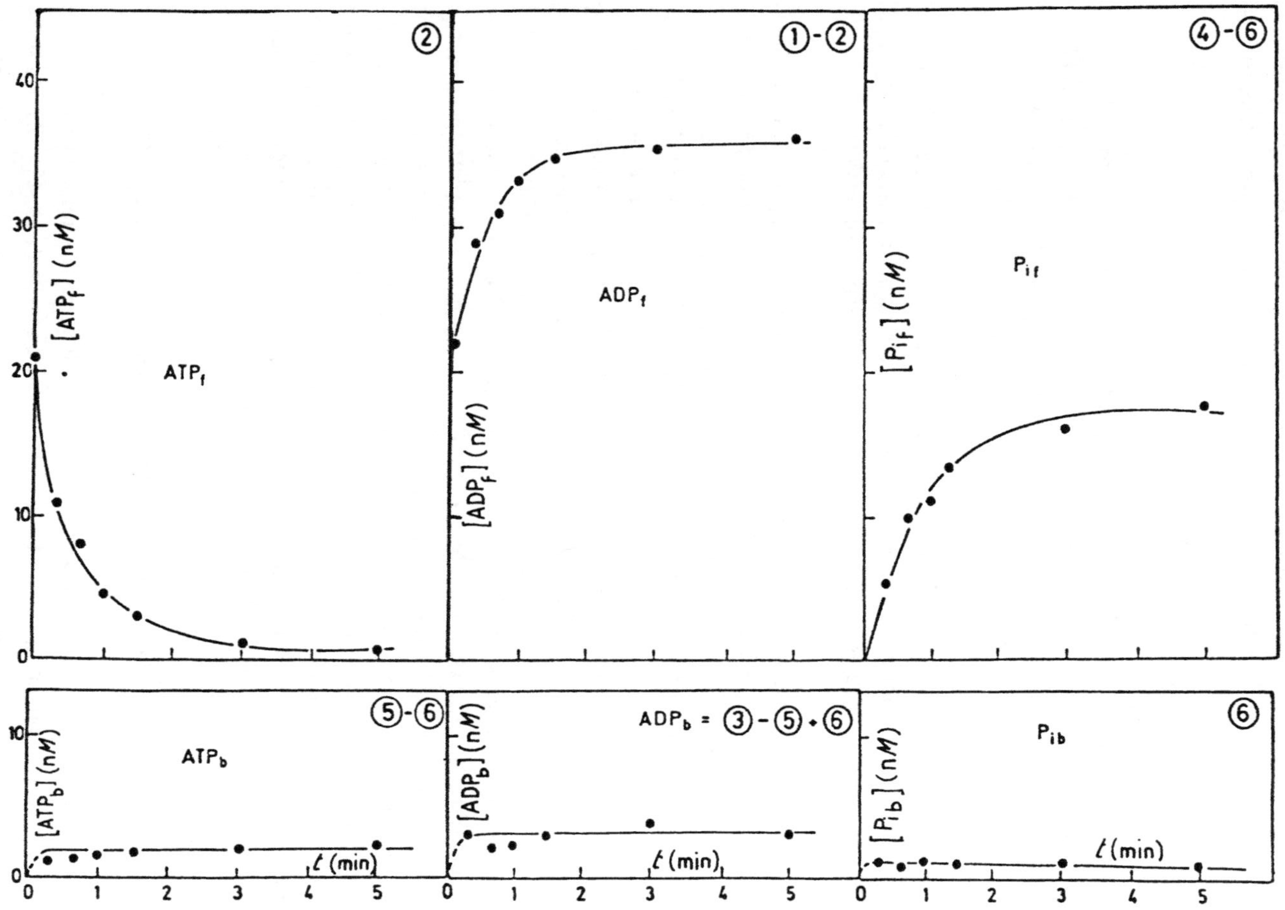


Fig. 22 - Free and enzyme-bound substrates and products as a function of reaction time. The numbers refer to the type of experiment indicated in Fig. 21. The concentration of active reduced enzyme was $2\times10^{-8}$ *M*.

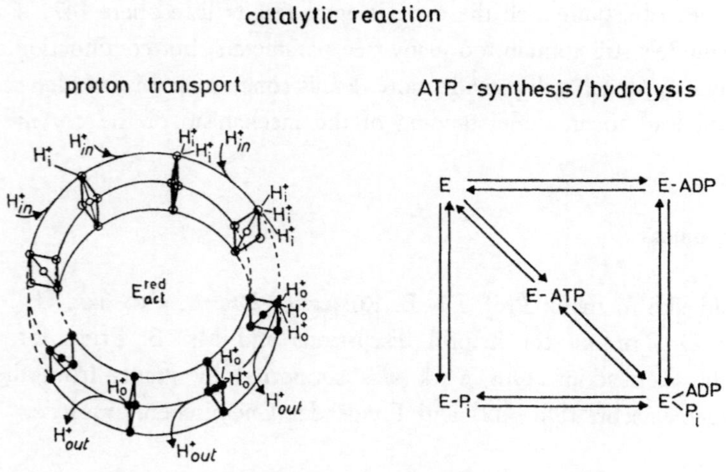


Fig. 23 - Reaction scheme for proton transport and for ATP-synthesis/hydrolysis and their coupling. $[H^+]_{in}$ and $[H^+]_{out}$ are internal and external proton concentrations. The enzyme bound protons are indicated as O-$H^+_i$ when in equilibrium with the internal aqueous phase and as ●-$H^+_o$ when in equilibrium with the external aqueous phase. Details in text.

protonation of the enzyme species E. A conformational change (dotted lines) then gives the protons access to the outside where they are released stepwise. A second conformational change (dotted lines), then returns the enzyme to its original conformation. This sequence corresponds to a proton translocation through the ATPase without coupling to an ATP-synthesis. Therefore, protonation steps and chemical steps must alternate. Based on the results described in this work, the following sequence seems the most plausible at high ΔpH:

1. three-step binding of protons from the inside to E gives the three-fold protonated species $(H^+_i)_3$ E which is in equilibrium with internal proton concentration $[H^+]_{in}$
2. binding of ADP and $P_i$ randomly to the species $(H^+_i)_3E$;
3. equilibrium between enzyme-bound ADP, $P_i$ and ATP;
4. conformational change of the enzyme in state $(H^+_i)_3$E-ATP so that the protonated groups are now directed to the outside $(H^+_{out})_3$E-ATP. This species is in equilibrium with the external proton concentration $[H^+]_{out}$;
5. protons and ATP are then released from the enzyme;
6. a conformational change returns the enzyme to its initial conformation.

The protonation-deprotonation reactions are assumed to be fast in comparison to the other steps of the reaction cycle. We can therefore assume that the protonated enzyme species is in equilibrium with the proton concentration. The corresponding equilibrium constant (p$K$ value) is obtained from the data shown in Fig. 11. Models based on these assumptions describe quite well the experimental results listed here [47]. For the time being, these models still contain too many free parameters, but continuation of this type of experimental approach will provide more details concerning this coupled reaction and, hopefully, will lead to an understanding of the mechanism of the enzyme.

## Acknowledgements

I would like to thank Prof. Dr. B. Rumberg, Drs. P. Fromme, U. Junesch, U. Siggel and G. Thulke for helpful discussions and Ms. B. Pitts for typing and correcting the manuscript. Our work was supported by grants from the Deutsche Forschungs-gemeinschft (Sfb 312) and Fonds der Chemischen Industrie.

## References

[1] I. Prigogine, *Thermodynamics of Irreversible Processes,* Interscience Publisher, Wiley and Sons, New York (1967).

[2] F.A. Sauer, in *Current Topics in Membranes and Transport*, Academic Press, New York (1984), Vol. **19**, p. 21.

[3] P. Mitchell, *Nature (London)*, **191**, 144 (1961).

[4] P. Mitchell, *Chemiosmotic Coupling and Energy Transduction*, Glynn Research, Bodmin, England (1968).

[5] R.J.P. Williams, *J. Theor. Biol.*, **1,** 1 (1961).

[6] H.W. Westerhoff, B.A. Melandri, F.A. Venturoli and D.B. Kell, *Biochim. Biophys. Acta,* **768**, 257 (1984).

[7] Y. de Kouchkovsky, C. Sigalat, F. Maraux and S. Phung Nhu Hung, in *Ion Interactions in Energy Transfer Biomembranes*, G.C. Papageorgiou *et al.,* (Editors), Plenum Press, New York (1985), p. 119.

[8] G.F. Azzone, V. Petronelli, D. Pietrobon, M. Zoratti, in *Achievements and Perspectives of Mitochondrial Research ,* Vol. I., Bioenergetics (E. Quagliariello *et al.* (Editors), Elsevier Science Publishers, Amsterdam (1985), p. 387.

[9] R.A. Dilley, S.M. Theg and W.A. Beard, *Annu. Rev. Plant Physiol.,* **38**, 348 (1987).

[10] J. Rosing and C. Slater, *Biochim. Biophys. Acta*, **267**, 275 (1972).

[11] H. Rottenberg, *J. Bioenerg.,* **7**, 61 (1975).

[12] G.F. Azzone, D. Pietrobon, M. Zoratti, *Curr. Top. Bioenerg.,* **13**, 2 (1984).

[13] H. Rottenberg, this volume.

[14] B. Rumberg, in *Encyclopedia of Plant Physiology, Photosynthesis I, A* (A. Trebst and M. Avron (Editors), Springer Verlag, Berlin (1977), Vol. **5**, p. 405.

[15] J.B. Jackson and D.B. Nicholls, *Methods Enzymol.,* **127**, 41 (1986).

[16] L. Wojtczack, in *Ion Interactions in Energy Transfer Biomembranes*, G.C. Papageorgiou *et al.* (Editors), Plenum Press, New York (1985), p. 7.
[17] A.T. Jagendorf and E. Uribe, *Proc. Natl. Acad. Sci USA,* **55**, 170 (1966).
[18] P. Gräber, U. Junesch and G.H. Schatz, *Ber. Bunsenges. Phys. Chem.,* **88**, 599 (1984).
[19] R.A. Ravizzini, C.S. Andreo, R.H. Vallejos, *Biochim. Biophys. Acta,* **591**, 135 (1980).
[20] C.M. Nalin, R.E. McCarty, *J. Biol. Chem.,* **259**, 7275 (1984).
[21] S.R. Ketcham, J.W. Davenport, K. Warncke, R.E. McCarty, *J. Biol. Chem.,* **259**, 7286 (1984).
[22] W. Junge, *Eur. J. Biochem.,* **14**, 582 (1970).
[23] P. Gräber, E. Schlodder and H.T. Witt, *Biochim. Biophys. Acta,* **461**, 426 (1977).
[24] D.A. Harris and A.R. Crofts, *Biochim. Biophys. Acta,* **502**, 87 (1978).
[25] P. Gräber and E. Schlodder, in *Photosynthesis II*, E. Akoyunoglou (Editor), Balaban International Science Service, Philadelphia PA (1981), p. 867.
[26] B. Rumberg and U. Becher, in *$H^+$-ATPase/synthase; Structure, Function, Regulation*, S. Papa, *et al.* (Editors), Adriatica Editrice, Bari (1984), p. 421.
[27] U. Junesch and P. Gräber, *Biochim. Biophys. Acta,* **893**, 275 (1987).
[28] J.D. Mills, P. Mitchell and P. Schürmann, *FEBS Lett.,* **112**, 173 (1980).
[29] J.D. Mills and P. Mitchell, *Biochim. Biophys. Acta,* **679**, 75 (1982).
[30] J.E. Tillberg, C. Giersch and U. Heber, *Biochim. Biophys. Acta,* **461**, 31 (1977).
[31] C. Giersch, *Biochim. Biophys. Acta,* **725**, 309 (1983).
[32] S. Bickel-Sandkötter and H. Strotmann, *FEBS Lett.,* **125**, 188 (1981).
[33] C. Vinkler, *Biochem. Biophys. Res. Comm.,* **99**, 1095 (1981).
[34] W.P. Quick and J.D. Mills, *Biochim. Biophys. Acta,* **893**, 197 (1987).
[35] B. Rumberg and U. Siggel, *Naturwiss.,* **56**, 130 (1969).
[36] U. Pick, H. Rottenberg and M. Avron, *FEBS Lett.,* **32**, 91 (1973).
[37] G. Thulke, *Thesis,* TU Berlin (1989).
[38] U. Pick and E. Racker, *J. Biol. Chem.,* **245**, 2793 (1979).
[39] G. Schmidt and P. Gräber, *Biochim. Biophys. Acta,* **808**, 46 (1985).
[40] P. Fromme, *Thesis,* TU Berlin (1988).
[41] F.A.S. Kironde and R.L. Cross, *J. Biol. Chem.,* **261**, 12544 (1986).
[42] Z. Xue, J.M. Zhou, T. Melese, R.L. Cross and P.D. Boyer, *Biochemistry,* **13**, 3749 (1987).
[43] G. Grubmeyer, R.L. Cross and H.S. Penefsky, *J. Biol. Chem.,* **257**, 12092 (1982).
[44] H.S. Penefsky, *J. Biol. Chem.,* **260**, 13735 (1985).
[45] H.S. Penefsky, *J. Biol. Chem.,* **252**, 2891 (1977).
[46] P.D. Boyer, in *Membrane Bioenergetics*, C.P. Lee, G. Schatz and L. Ernster (Editors), Addison-Wesley, Reading, MA (1979), p. 461.
[47] U. Junesch, *Thesis,* TU Berlin (1989).

# CHARGE EFFECTS IN ELECTROMAGNETIC STIMULATION OF BIOSYNTHESIS

MARTIN BLANK[1] and REBA GOODMAN[2]

*Department of Physiology and Cellular Biophysics[1] and Department of Pathology[2],*
*Columbia University, College of Physicians & Surgeons*
*630 West 168th Street, New York, NY 10032, USA*

Contents



*Bioelectrochemistry III*
Edited by G. Milazzo and M. Blank
Plenum Press, New York, 1990

## 1. Background

Exposure to low intensity non-ionizing radiation has been linked to a number of significant biological effects, both clinically beneficial (*e.g.,* acceleration of bone fracture healing in cases of non-unions [1]) and potentially harmful (*e.g.,* epidemiological studies showing an increased risk of childhood leukemia [2, 3]). Studies of the stimulation of biosynthesis in cells by electromagnetic fields have demonstrated the presence of signal-specific proteins [4], and have also shown that the intensity thresholds for affecting biosynthesis are different at different frequencies [5]. The effects of electromagnetic stimulation on biosynthesis provide detailed information that is important in unraveling the mechanism of action of electromagnetism on fundamental cellular processes. A description of the mechanism at this level could serve as a basis for understanding the more complex effects of electromagnetism in living systems and eventually defining the effects at different levels of electromagnetic exposure.

Most of the information about the effects of electromagnetism on RNA and protein biosynthesis has been obtained by GOODMAN and HENDERSON on the salivary gland cells of *Sciara* and *Drosophila* using the incorporation of $^{3}$H-uridine to assess RNA synthesis and the incorporation of $^{35}$S-methionine to measure polypeptide synthesis [4]. Unexposed cells were used to establish baselines for biosynthetic activity, and other cells were then exposed to various stimuli that included sine waves, as well as the asymmetric waveforms used for treating a variety of medical problems, *e.g.,* avascular necrosis, osteoporosis and fracture non-unions (Electro-Biology, Inc.). The changes observed in these data were analyzed using an electrochemical model that had been successful in describing the effects of electric fields on potential difference gated channels. This enabled us to identify patterns in the biosynthetic responses to electromagnetic stimulation and to develop the two-pathway hypothesis that considers electromagnetic signals as both:

*a)* a stress akin to the other stresses that stimulate biosynthesis (*e.g.,* thermal, mechanical, chemical, osmotic), and

*b)* a specific perturbing agent that interferes with steps in the biosynthetic processes that are vulnerable to the effects of induced currents.

The properties of electromagnetic stimuli as a physical stress suggest an explanation for the variation of threshold with frequency [5], *i.e.*, the *windows*, or particular ranges of frequency and amplitude in which effects have been reported. Frequency windows suggest links to specific molecular processes, such as the connection between the electrical responses of excitable cells and the time constants of the gating currents of the sodium channels [6]. In more complicated systems, such as the stimulation of ion accumulation in red cells *via* the Na, K-ATPase [7] by exposure to alternating electric fields, we have shown that the reported frequency optimum could relate to the time constants of various charge movements that result in ion concentration changes in an enzyme/channel system [8]. Figure 1 shows the frequency dependence of the sodium ion concentration at the inner surface of a sodium channel when the cell is subjected to

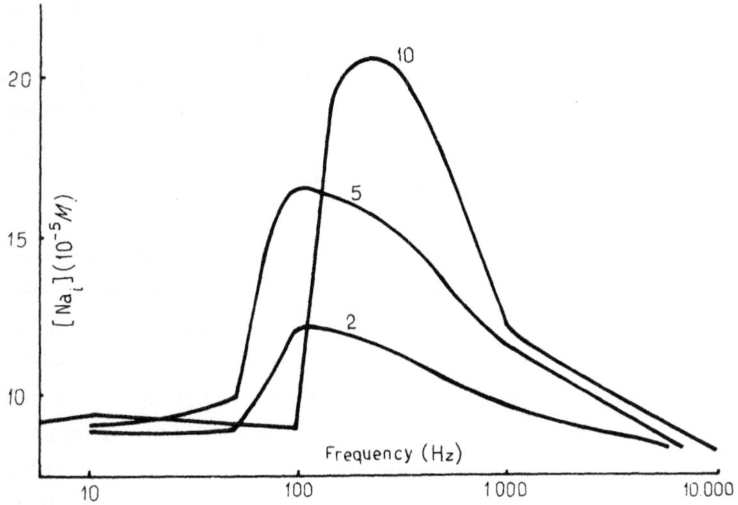


Fig. 1 - A plot of the sodium ion concentration on the inner surface of a sodium channel as a function of the frequency of the imposed alternating electric field. The numbers on the curves correspond to the peak amplitudes (in millivolts) of the imposed fields. Similar effects would be expected for other ions, *e.g.* internal calcium and external potassium.

alternating electric fields of different magnitudes. By focusing on the fundamental electrochemical processes involved in charging the capacitances at membrane surfaces and in ion transport across membranes [9], it has been possible to demonstrate large frequency-dependent ion concentration changes at membrane surfaces in terms of known biophysical mechanisms.

The electric currents that result from electromagnetic stimuli, either directly or as induced from the varying magnetic fields, can lead to ion concentration changes inside the cell that stimulate various processes [10]. They can also directly affect processes that are particularly vulnerable to charging. We feel that the nascent protein chain that grows when the mRNA is attached to the ribosome during translation could be such a site, and the pattern of new proteins found in electromagnetically stimulated cells supports this view. We shall discuss the indirect and direct effects of electromagnetism on biosynthesis below in the context of the electrochemical model.

## 2. Bioelectrochemical analysis

### *2.1. The electrochemical model*

The electrochemical model has been productive in dealing with several aspects of ion transport through channels, including the organization of the channels themselves [11]. The structural similarities between the various membrane protein assemblies, called channels, suggest a common design and mechanism of operation. In the electrochemical model, we have proposed that the relation between surface charge density and protein aggregation provides a basis for understanding membrane channel opening/closing reactions. We have used the surface free energy of protein association to describe the aggregation/disaggregation reactions in the potential difference gated channels of excitable membranes, and have calculated voltage clamp currents that are similar to those observed in the sodium and potassium channels of excitable membranes.

The ionic processes occurring at the channel surfaces were calculated using the Surface Compartment Model as an approximation [9] for the electrical double layer regions at charged surfaces. An important finding of this approach is that the unusual patterns of selectivity seen in ion channels appear to be determined by the gating current. This relation can explain the simultaneous effects of batrachotoxin on gating current and selectivity [12]. Figure 2 shows the calculated effect of gating current upon selectivity. We have also shown that channel inactivation varies with the gating [9], and that both sodium and potassium channels show this property but at different times, as has been observed [13].

In summary, our calculations using the electrochemical model have shown that:

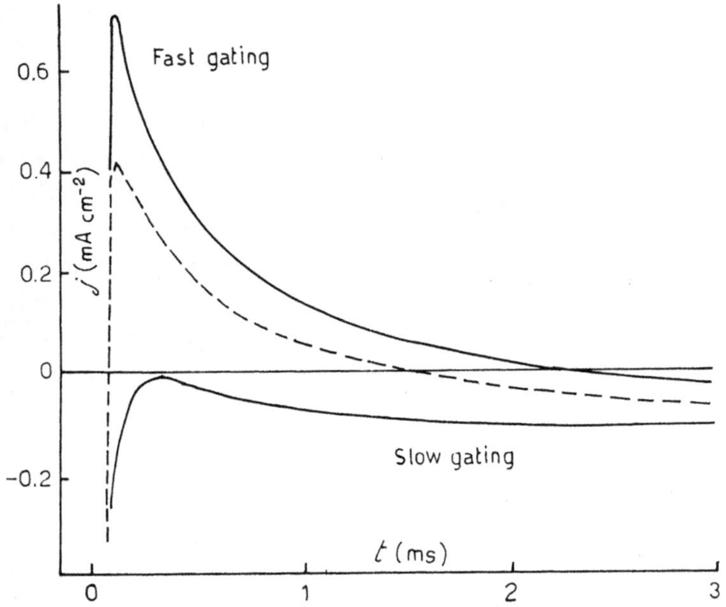


Fig. 2 - A plot of the ionic current density (in mA/cm$^2$), across a potential difference dependent channel as a function of the time, *t* (in ms). The fast gating curve has a gating charge conductance that is 15 times greater than the slow gating curve, so the channel opens more rapidly during a depolarization. The fast gating curve resembles the current in a sodium channel and the slow gating curve resembles the oppositely directed current in a potassium channel, even though the initial steady conditions are the same for the two channels. The dashed line is for an intermediate gating speed that could represent the lowered sodium selectivity seen with batrachotoxin. Reproduced with permission from *Bioelectrochem. Bioenerg.*

1) the ionic currents in excitable membranes can be described by electrodiffusion theory;
2) the selectivity of an ionic channel depends upon the kinetics of channel opening;
3) channel inactivation is also related to the gating current, and
4) the frequency-dependent transient ion concentration changes at membrane surfaces due to oscillating electric fields can affect the concentrations of the ions that stimulate ion pump enzymes and channel proteins.

These contributions to our understanding of ion transport processes at the molecular level have demonstrated the value of the electrochemical model in describing processes in charged macromolecular systems subjected to ionic currents.

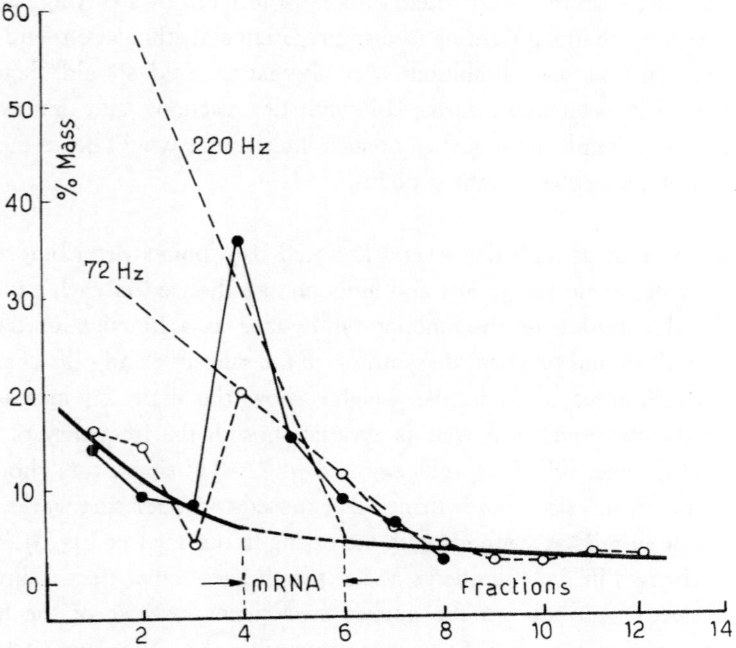


Fig. 3 - The distribution of mass (% mass *versus* fraction) among the different fractions of RNA after 30 minutes of electrical stimulation using 72 Hz (○) and 220 Hz (●) sine waves. The distributions of mRNA in fractions 4, 5 and 6 are indicated by the dashed lines labeled 72 Hz and 220 Hz. (It should be emphasized that *less mRNA is synthesized* with the 220 Hz signal, but the distribution is as shown here). The 220 Hz line starts at a higher % mass but the slope is much greater. The heavy solid line (dashed in the mRNA region) shows a distribution that is similar to the control curve in the absence of electromagnetic stimulation. Reproduced with permission from *Bioelectrochem. Bioenerg.*

## 2.2. *The electrochemical model in biosynthesis*

The electrochemical model should be useful in describing the direct effects of electric fields on biosynthesis, since the same well established ideas that apply to ionic solutions and membranes, and that were effective in analyzing channel processes, should apply to all charged surfaces in solution. The basic ideas relevant to biosynthesis are:

— current flow could lead to the charging of a polyelectrolyte surface;

— charging could lead to the disaggregation of associated polyelectrolytes.

Consider the electrochemical model applied to the interaction of two polyanionic surfaces with adsorbed cations in an alternating electric field. The charged surfaces could be the ribosomes, mRNAs and the nascent polypeptide chains that are being synthesized.

The current, arising from the electric field directly or induced by a varying magnetic field, causes an alternating charging (leading to disaggregation and stimulation) and discharging (leading to aggregation and inhibition). An alternating field should therefore cause frequency-dependent switching during biosynthetic reactions and interference with protein synthesis. A simple consequence of such interference would be an increase in the fraction of lower molecular weight proteins.

To test these ideas with the several hundred data points describing the proteins (molecular weights, isoelectric points and amounts) synthesized in each experiment, we calculated the distribution of the amount synthesized as a function of the molecular weight for both RNA and polypeptide synthesis in the salivary gland cells of the dipteran, *Sciara*. The distributions of molecular weights show the expected increase in lower molecular weight components as well as a variation with the frequency of the applied electric field [14]. The mRNA in cells exposed to 220 Hz sine waves shows a greater augmentation of smaller size classes than cells exposed to 72 Hz sine waves, despite the absolute decrease in mRNA synthesis with increasing frequency (see Fig. 3). This is likely to be due to changes in ionic gradients across membranes rather than a direct effect of the induced electric currents on the nucleic acid chains because of the low energies involved. Data on the synthesis of 53 major polypeptides show the same pattern. There is augmented synthesis of lower (20-50 kDa) molecular weight polypeptides and a reduced synthesis in the 50-90 kDa molecular weight range. Furthermore, the skewing of the distribution increases with the frequency (see Figs. 4 and 5).

These results appear to follow the predictions of the electrochemical model, but similar changes in the patterns of protein biosynthesis occur in the absence of electromagnetic fields with thermal stimuli. This would suggest that physical stimuli may lead to the same general response, probably by a membrane mediated change in internal ion concentration [10]. Electromagnetic stimulation would lead to ion concentration changes that reflect a decrease in the transmembrane gradient and a thermal stimulus would increase the membrane permeability and lead to a similar decrease in a transmembrane gradient.

Consider the experiments of Fig. 6 on *heat shock* (*i.e.*, the stimulation of altered biosynthesis by suddenly raising the temperature 10-15 °C), which show a pattern similar to the electromagnetic experiments of Fig. 4 after subtracting the fraction containing the 70 kDa heat shock proteins. These results suggest that the response of cells to physical stimuli involves two components:

— a general response to stress with a characteristic molecular weight *signature* (*i.e.*, increases and decreases in specific molecular weight ranges);
— modality-related responses (*e.g.*, thermal stimuli will affect heat shock proteins in the

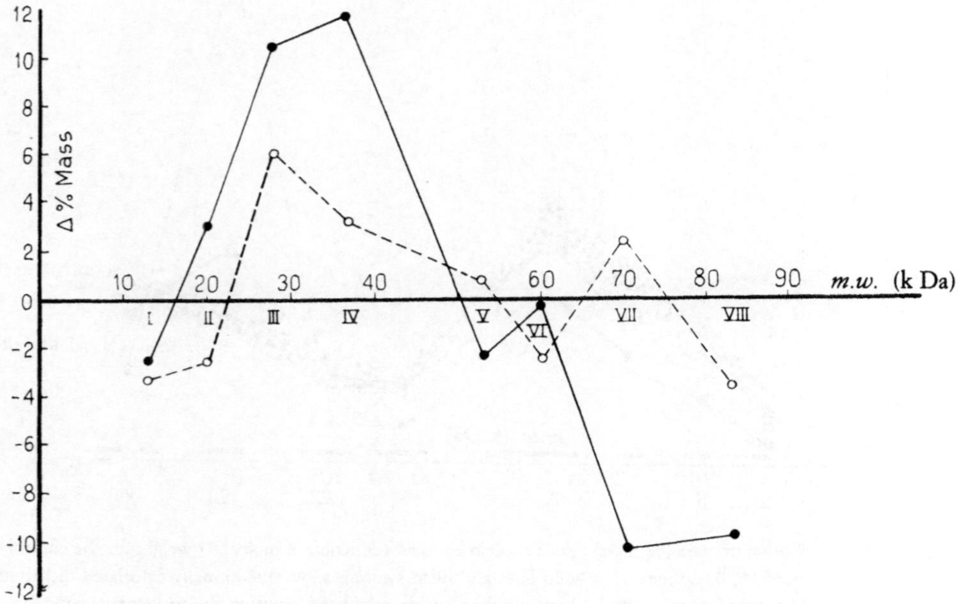


Fig. 4 - Differences in % mass between the electrically stimulated protein synthesis and the control are plotted for the 60 Hz (○) and 72 Hz (●) signals as functions of the molecular weight. Reproduced with permission from *Bioelectrochem. Bioenerg.*

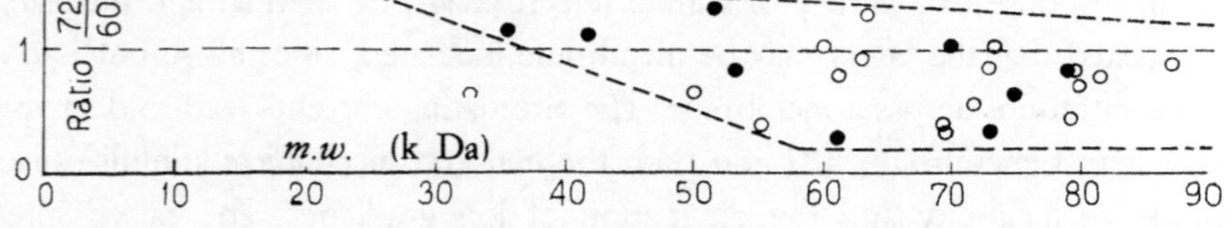


Fig. 5 - The ratio of the amount of a particular protein synthesized at 72 Hz divided by the amount synthesized at 60 Hz, as a function of the molecular weight of that protein. The major proteins (as determined by the number of counts in excess of an arbitrarily chosen level) are full circles (●), while the others are open circles (○). Reproduced with permission from *Bioelectrochem. Bioenerg.*

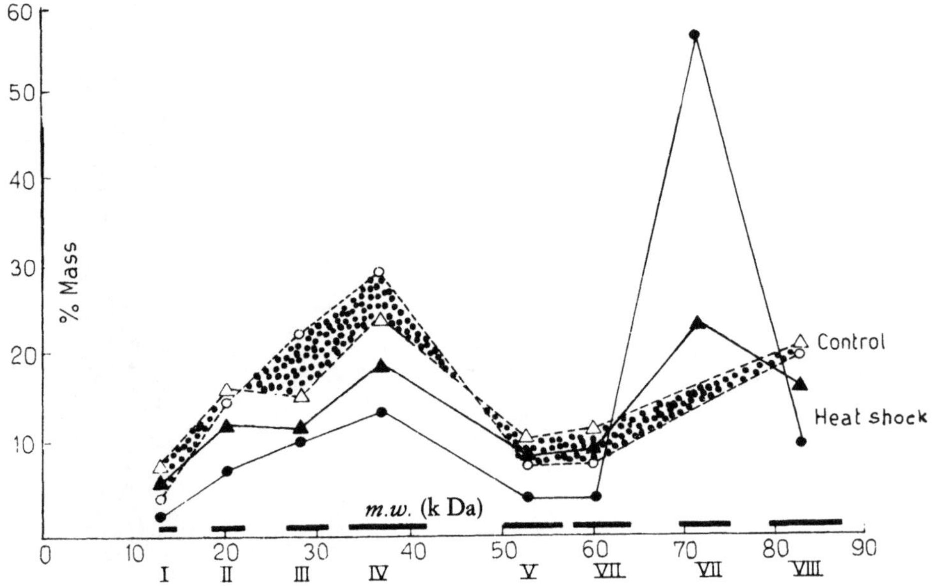


Fig. 6 - The distribution of mass (% mass) *versus* protein band as a function of molecular weight for the control (▲, △) and heat shocked (●, ○) systems. The solid lines and filled symbols show the normally calculated distribution, while the dashed lines and open symbols show the distribution calculated omitting the proteins in band VII. By omitting band VII, the decreases in bands I and II and increases in bands III and IV during heat shock are more easily seen as the stippled areas. Reproduced with permission from *Bioelectrochem. Bioenerg.*

70 kDa range and electromagnetic stimuli will affect the more highly charged proteins as described below).

The general response to stress appears to have the characteristic signature of an augmented synthesis of 20-50 kDa molecular weight polypeptides and a reduced synthesis in the molecular weight ranges below 20 and above 50 kDa. Given that this signature appears in response to a variety of stimuli (electromagnetic thermal and mechanical), we have suggested that the effect may be membrane mediated, since all stimuli can alter the ionic concentrations across a membrane. The alternating currents lead to decreases in ion gradients across membranes [8] and both thermal and mechanical stimuli cause greater membrane permeability and the dissipation of ion gradients. The raised internal ion concentration of $Na^+$ or $Ca^{2+}$ or the decrease in $K^+$ could act as a second messenger to stimulate a common pathway [10].

The common response to stress is composed of the effects of the stimulus on many proteins, each of which may respond in a different way. The cumulative effect is what we have focused on here, but there are unique responses of individual proteins to particular stimuli. When the individual responses are documented, it will be possible to have a better idea of the functions the proteins may fulfill. Because of similarities between the responses described here and the observed shedding of proteins after exposure to microwave radiation [15], the signature of the response to stress may be a mechanism to replace proteins lost by the membranes.

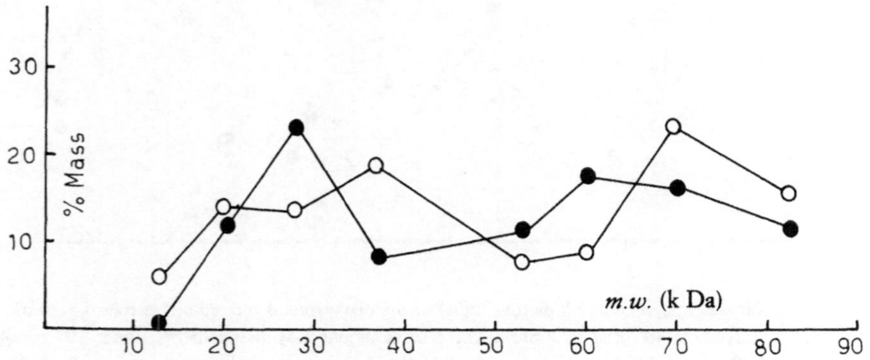


Fig. 7 - The distribution of mass (% mass *versus* molecular weight) is shown for the new proteins (●) and controls (○). Reproduced with permission from *Bioelectrochem. Bioenerg.*

There are also specific modality-related responses in addition to the general response. We have already mentioned that sudden thermal stimuli induce a characteristically large increase in the synthesis of the 70 kDa proteins of *Sciara* salivary gland cells, in addition to the characteristic polypeptide signature. In the case of electromagnetically stimulated cells there are also changes that appear to be linked directly to the electromagnetic stimuli. The new proteins synthesized in response to both electromagnetic and thermal stimuli and absent in the controls, can be regarded as resulting from the interference of the stimuli with biosynthetic processes [16]. Proportionally more new low-molecular-weight proteins are synthesized, and the distribution of their mass is skewed to lower values (see Fig. 7). Whereas control proteins in unexposed samples are largely negative for all molecular weights, the new proteins appear to be distributed more or less symmetrically around $6.0<pI<6.5$ (where pI stands for isoelectric point), which would correspond to zero charge for a negatively charged surface at – 30 to – 60 mV (see Fig. 8). Smaller molecules are more highly charged, both positively and negatively, and the proteins missing under electromagnetic stimulation are more highly charged. The *number* of new proteins divided by the *amount* of new protein, a measure of *diversity* of product, increases with the frequency or repeat rate of the electromagnetic stimulation, while the number of missing proteins decreases (see Fig. 9). According to the electrochemical model, interference should increase with both electromagnetic frequency and

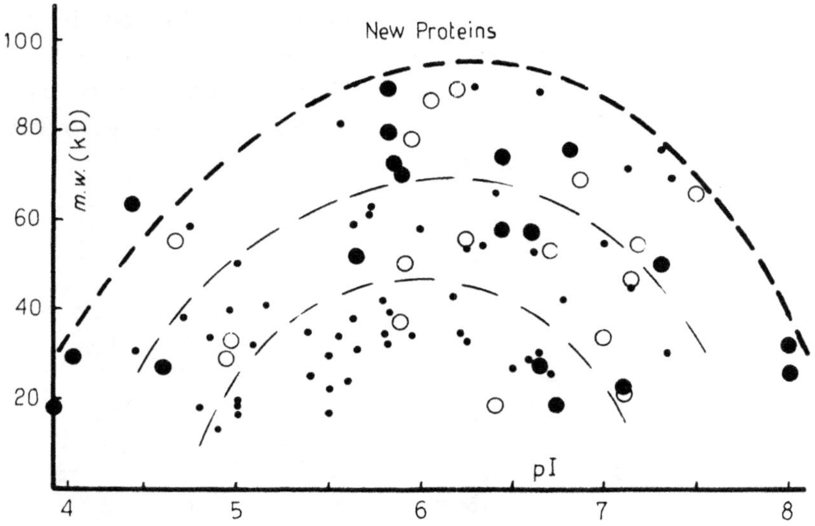


Fig. 8 - The molecular weights, *m.w.* in kilodaltons (kDa) of the new proteins plotted *versus* their isoelectric point, pI's. Three different symbols are used to indicate the relative amounts synthesized: (•) under 100 counts, (○) 100-200 counts and (●) over 200 counts. The envelope of the major points, indicated by the upper dashed line, shows a broad peak around 6.0<pI<6.5. Two other dashed curved lines separate the points into three regions. The upper region contains 57 % of the specific activity, the middle region 28 % and the lower region 15 %. Reproduced with permission from *Bioelectrochem. Bioenerg.*

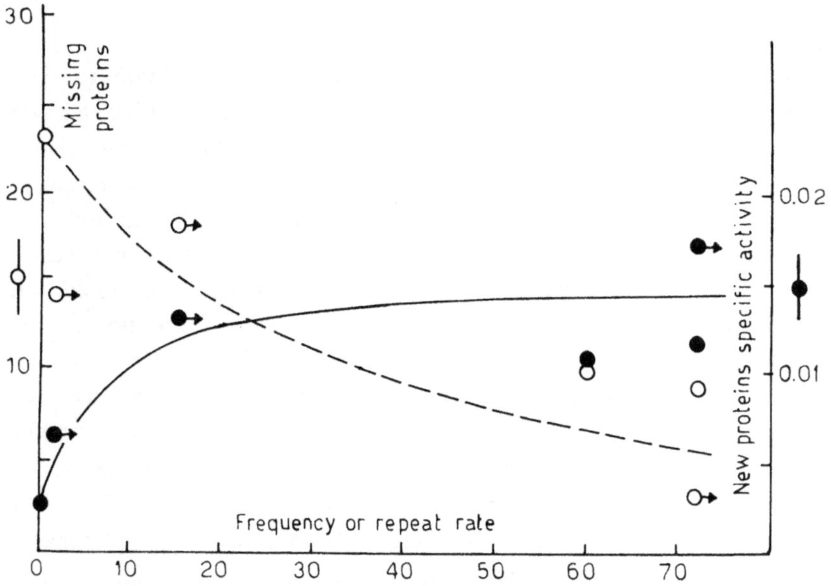


Fig. 9 - The number of new proteins divided by the amount synthesized (•) and the number of missing proteins (○) *versus* the frequency or repeat rate of electromagnetic stimuli. The solid and dashed lines join the points for the sine wave experiments and also the experiments with complex wave forms, which are shown with arrows to indicate the presence of harmonics. The points for the heat shock experiments are shown at zero frequency. Reproduced with permission from *Bioelectrochem. Bioenerg.*

molecular charge, and result in smaller polypeptides than in unexposed control cells. The results from sudden thermal stimuli show the same pattern as with electromagnetic stimuli, suggesting that electrochemical properties play an important role in determining the responses to external physical stimuli.

## 3. Conclusion

From these experimental results it appears that the altered patterns of biosynthesis, seen as a result of the electromagnetic stimulation of cells, can be interpreted in terms of charging phenomena accompanying the induced currents. Charging at membrane surfaces can lead to the stimulation of biosynthesis also seen with other physical stimuli, while charging at the surfaces of biosynthetic structures could lead to changes related to the electrochemical properties of the system. Since both processes occur simultaneously, altered patterns of biosynthesis must be interpreted in terms of both mechanisms. It appears that the charge-mediated disaggregation phenomena found in biopolyelectrolytes (*e.g.*, hemoglobin) or in charged membrane proteins (*e.g.*, ionic channels) may be occurring in biosynthetic structures as well.

## Acknowledgement

We thank the Office of Naval Research and the Electric Power Research Institute for their support of the research described in this chapter.

## References

[1] C.A.L. Bassett, *Clin. Plast. Surg.*, **12,** 259 (1985).
[2] N. Wertheimer and E. Leeper, *Amer. J. Epidem.*, **109,** 273 (1979).
[3] D.A. Savitz, H. Wachtel, F.A. Barnes, E.M. John and J.G. Tvrdik, *Amer. J. Epidem.*, **128,** 21 (1988).
[4] R. Goodman and A.S. Henderson, *Proc. Natl. Acad. Sci. USA*, **85,** 3928 (1988).
[5] K.J. Mc Leod, R.G. Lee and H.P. Ehrlich, *Science*, **236,** 1465 (1987).
[6] Y. Palti and W.J. Adelman, *J. Membr. Biol.*, **1,** 431 (1969).
[7] E.H. Serpersu and T.Y. Tsong, *J. Membr. Biol.*, **74,** 191 (1983).
[8] M. Blank, *J. Electrochem. Soc.*, **134,** 1112 (1987).
[9] M. Blank, *Biochim. Biophys. Acta*, **906,** 277 (1987).
[10] F.T. Marin and F.G. Rothman, *J. Cell Biol.*, **87,** 823 (1980).
[11] M. Blank, *Chemtech.*, **18,** 434 (1988).

[12] B.I. Khodorov and S.V. Revenko, *Neuroscience,* **4,** 1315 (1979).
[13] G. Ehrenstein and D.L. Gilbert, *Biophys. J.,* **6,** 553 (1966).
[14] M. Blank and R. Goodman, *Bioelectrochem. Bioenerg.,* **19,** 569 (1988).
[15] R.P. Liburdy, A.W. Rowe and P.F. Vanek, *Radiation Res.,* **114,** 500 (1988).
[16] M. Blank and R. Goodman, *Bioelectrochem. Bioenerg.,* in press.

# LIST OF PARTICIPANTS

## Directors

BLANK, Martin
Columbia University
College of Physicians and Surgeons
Departement of Physiology
630 West 168th Street
New York, NY 10032, U.S.A.

MELANDRI, Bruno Andrea
Università di Bologna
Dipartimento di Biologia ed Evoluzione Sperimentale
Via Irnerio, 42
I–40126 Bologna, Italy

MILAZZO, Giulio
Honorary President of the Bioelectrochemical Society
Piazza G. Verdi 9
I–00198 Roma, Italy

## Speakers

BAMBERG, Ernst
Max Planck Institut für Biophysik
Heinrich Hoffmann Strasse 7
D–6000 Frankfurt / Main, West Germany

BISSON, Roberto
Università di Padova
Istituto di Patologia Generale

Via Loredan 16
I–35113 Padova, Italy

BLANK, Martin, see Directors

COLOMBINI, Marco
University of Maryland
Department of Zoology
College Park, MD 20742, U.S.A.

FENDLER, Klaus
Max Planck Institut für Biophysik
Heinrich Hoffmann Strasse 7
D–6000 Frankfurt / Main, West Germany

GRÄBER, Peter
Universität Stuttgart
Biologisches Institut
Pfaffenwaldring 57
D-7000 Stuttgart, West Germany

HARTL, Franz Ulrich
Universität München
Institut für Physiologische Chemie
Goethe Strasse 33
D–8000 München, West Germany

HEDRICK, Rainer
Universität Göttingen
Institut für Pflanzenphysiologie
D–3400 Göttingen, West Germany

KESZTHELYI, Lajos
Hungarian Academy of Sciences
Biological Research center
Institute of Biophysics
Odeszai Korut
P.O. Box 521
H–6701 Szeged, Hungary

Malatesta, Francesco
Università "La Sapienza", Roma
Dipartimento di Scienze Biochimiche
Piazzale Aldo Moro 5
I–00185 Roma, Italy

Møller, Jesper
University Aahrus
Institute of Medical Biochemistry
Dk–8000 Aahrus, Denmark

Neumann, Eberhardt
Universität Bielefeld
Physikalische und Biophysikalische Chemie
Postfach 8640
D–4800 Bielefeld, West Germany

Rottenberg, Hagai
Hanemann University
School of Medicine
Department of Pathology, Medicine Laboratory
Philadelphia, PA 19102, U.S.A.

**Attendants**

Aloisi, Giovanni
Università di Firenze
Dipartimento di Chimica
Via G. Capponi 9
I–50121 Firenze, Italy

Benedetti, Luca
Università di Modena
Dipartimento di Chimica
Via Campi 183
I–41100 Modena, Italy

BRUMFELD, Vlad
Weizmann Institute of Science
Department of Membrane Research
76100 Rehovot, Israel

CORAZZA, Sandra
Università di Bologna
Dipartimento di Biologia ed Evoluzione Sperimentale
Via Irnerio 42
I–40126 Bologna, Italy

CRIMI, Massimo
Università di Bologna
Dipartimento di Biologia ed Evoluzione Sperimentale
Via Irnerio 42
I–40126 Bologna, Italy

CZOCHRALSKA, Barbara
University of Warsaw
Department of Biophysics
Zwirkii Wigury 93
Pl–02–089 Warsaw, Poland

DE KROON, Toor
Rijksuniversiteit Utrecht
Center for Biomembrane and Lipid Enzymology
Padulaan 8
3508TB–Utrecht, The Netherlands

DRAIN, Michael
The Rockfeller University
Department of Biophysics,
1230 York Avenue
New York, NY 10021, U.S.A.

FUSI, Paola
Università di Milano
Dipartimento di Biologia

Via Celoria 26
I–20133 Milano, Italy

Galieta, Luis J.V.
Istituto Gaslini
Laboratorio di Genetica Molecolare
Via 5 Maggio 39
I–16148 Genova, Italy

Greenbaum, Nancy
The Rockfeller University
Department of Biophysics
1230 York Avenue
New York, NY 10021, U.S.A.

Guidelli, Rolando
Università di Firenze
Dipartimento di Chimica
Via G. Capponi 9
I–50121 Firenze, Italy

Guirguis Saleh, Osiris W.
Cairo University
Faculty of Sciences
Department of Biophysics
Giza, Egypt

Herzyk, Eugenia
University of Warsaw
Department of Chemistry
Ul. Pasteura 1
Pl–02093 Warsaw, Poland

Kawezynski, Wieslaw
University of Warsaw
Department of Biophysics
Zwirki i Wigury 93
Pl–02093 Warsaw, Poland

KLINGERBERG, Martin
Universität München
Institut für Physikalische Biochemie
Goethe Strasse 33
D–8000 München 7, West Germany

KONAR, Aparna
Dehli University
Maitreyi College
Department of Botany
Chanakyapuri, New Dehli, India

KRYSINSKI, Pawel
University of Warsaw
Department of Chemistry
Laboratory of Electrochemistry
Ul Pasteura 1
Pl–02093 Warsaw, Poland

KWEE, Sianette
University of Aahrus
Department of Medical Biochemistry
Dk–8000 Aahrus, Denmark

MARUTHY, Sai, P.S.
Jawaharlal Nehru University
School of Life Sciences
New Dehli, 110067 India

MASSOTTE, Dominique
Laboratoire Européen de Biologie Moléculaire
Meyerhof Strasse 1
D–6900 Heidelberg, West Germany

MASTRACOLA, Teresa
Istituto Gaslini
Laboratorio di Genetica Molecolare
Via 5 Maggio 39
I–16148 Genova, Italy

McLaughlin, Stuart
State University of New York at Stony Brook
Department of Physiology and Biophysics
Health Science Center
Stony Brook, NY 11794–8601, U.S.A.

Mishra, Kaushmala Prasat
Universität Bielefeld
Physikalische und Biophysikalische Chemie
D–8640 Bielefeld, West Germany

Naumann, Renate
E. Merck Analytisches Zentrallaboratorium
Postfach 4119
D 6100 – Darmstadt, West Germany

Okamura, Melvin
University of California at San Diego
Department of Physics
La Jolla, CA 92093, U.S.A.

Panigrahi, Suryya Sankar
Vidyasagar College
Department of Zoology
39 Sankar Ghosh Lane
Calcutta, 700006 India

Papavero, Federico
Via della Gioventù 40
91025 Marsala TP, Italy

Pistocchi, Rossella
Università di Bologna
Dipartimento di Biologia ed Evoluzione Sperimentale
Via Irnerio 42
I–40126 Bologna, Italy

Szaraz, Sandor
Hungarian Academy of Sciences

Biological Center
Institute of Biophysics
Odeszai Korut 62
P.O.B. 521
H–6701 Szeged, Hungary

Trotta, Massimo
Università di Bari
Dipartimento di Chimica
Via Amendola 173
I–70125 Bari, Italy

Turina, M. Paola
Università di Bologna
Dipartimento di Biologia ed Evoluzione Sperimentale
Via Irnerio 42
I–40126 Bologna, Italy

Virgili, Marco
Università di Bologna
Dipartimento di Biologia ed Evoluzione Sperimentale
Via Belmeloro 8
I–40126 Bologna, Italy

Watala, Cezary
University of Lodz
Chair of Biophysics
Banacha 12/16
Pl–90237 Lodz, Poland

Winterhalter, Mathias
Freie Universität Berlin
Fachbereich Physik
Arnimallee 14
D–1000 Berlin 33, West Germany

Yova–Louka, Dido
National Technical University of Athens
Department of Physics

Heroon Polythecnion 9, Zographou
15773 Athens, Greece

# INDEX

GPSR Compliance

*The European Union's (EU) General Product Safety Regulation (GPSR) is a set of rules that requires consumer products to be safe and our obligations to ensure this.*

*If you have any concerns about our products, you can contact us on ProductSafety@springernature.com*

In case Publisher is established outside the EU, the EU authorized representative is:

Springer Nature Customer Service Center GmbH
Europaplatz 3
69115 Heidelberg, Germany

**Batch number: 10755805**

Printed by Printforce, the Netherlands